W9-CSP-256

Organic and Inorganic Photochemistry

MOLECULAR AND SUPRAMOLECULAR PHOTOCHEMISTRY

Series Editors

V. RAMAMURTHY

Professor
Department of Chemistry
Tulane University
New Orleans, Louisiana

KIRK S. SCHANZE

Professor
Department of Chemistry
University of Florida
Gainesville, Florida

1. Organic Photochemistry, *edited by V. Ramamurthy and Kirk S. Schanze*
2. Organic and Inorganic Photochemistry, *edited by V. Ramamurthy and Kirk S. Schanze*

ADDITIONAL VOLUMES IN PREPARATION

Organic and Inorganic Photochemistry

edited by

V. Ramamurthy
Tulane University
New Orleans, Louisiana

Kirk S. Schanze
University of Florida
Gainesville, Florida

MARCEL DEKKER, INC.   NEW YORK · BASEL · HONG KONG

ISBN: 0-8247-0174-7

This book is printed on acid-free paper.

Headquarters
Marcel Dekker, Inc.
270 Madison Avenue, New York, NY 10016
tel: 212-696-9000; fax: 212-685-4540

Eastern Hemisphere Distribution
Marcel Dekker AG
Hutgasse 4, Postfach 812, CH-4001 Basel, Switzerland
tel: 44-61-261-8482; fax: 44-61-261-8896

World Wide Web
http://www.dekker.com

The publisher offers discounts on this book when ordered in bulk quantities. For more information, write to Special Sales/Professional Marketing at the headquarters address above.

Current printing (last digit):
10 9 8 7 6 5 4 3 2 1

PRINTED IN THE UNITED STATES OF AMERICA

Preface

The science of chemistry has entered an era when techniques of chemical structure analysis are sufficiently sophisticated to render routine the elucidation of amazingly intricate molecular structures. These relatively recent technological advances now allow chemists to carry out detailed investigations of highly complex synthetic and naturally occurring (biological) molecular systems from the molecular perspective. Thus has been born the field of "supramolecular chemistry"—the molecular-level study of complex arrays or assemblies consisting of many basic molecular building blocks held together by weak intermolecular forces. Photochemists have led the charge into this new frontier of chemical science, and therefore it is appropriate that this second volume of the Molecular and Supermolecular Photochemistry series highlights state-of-the-art research of complex naturally occurring and synthetic supramolecular arrays.

During the past decade it has become increasingly clear that the structure and function of polydeoxyribonucleic acid (DNA) can be modified significantly by binding to small molecules. Complexes formed by noncovalent interactions between DNA and small organic and inorganic molecules are examples of supramolecular arrays that provide the basis for drug action and for analytical (sensing) of specific sequences of DNA base pairs. The first two chapters in this volume feature recent work concerning the mechanism by which transition metal complexes bind to DNA and how this binding modifies the photochemical and photophysical properties of the bound complexes and the DNA "scaffold." In particular, both chapters explore the interesting topic of photoinduced electron transfer between intercalated molecules and address the ongoing question as to whether DNA can act as a "molecular wire."

Molecular electron transfer is the basis for many important natural and commercial processes. During the past decade photochemists have relied upon supramolecular arrays of molecules to facilitate their understanding of the chemical and physical basis for this fundamentally important process. It therefore seems appropriate that several chapters in this volume examine thermally and photochemically induced electron transfer in supramolecular assemblies consisting of inorganic molecular building blocks such as covalently linked donor–acceptor "dyads," transition metal clusters, and nanocrystalline semiconductor particles.

Liquid crystalline polymers and molecular organic crystals are examples of supramolecular arrays in which long-range order plays an important role in determining the photophysical and photochemical properties of the system. Two chapters in the present volume explore the properties of liquid crystalline polymers and molecular crystals, with an emphasis on understanding how the structure of the supramolecular organization guides the course of the photochemical reactivity and photophysical properties of the component molecular units.

Progress in supramolecular photochemistry is intimately linked with the progress and understanding of the photochemical behavior of molecular systems. With recent advances in time-resolved techniques it has become possible to study the excited state behavior of reactive intermediates as well as that of higher excited states. A chapter on this topic is also included in this volume.

Although the science of molecular photochemistry remains very active, there is no doubt that in the future photochemists will focus increasingly on supramolecular chemical structures and arrays. This trend is a natural progression as the science of photochemistry meets the demands of new technologies based on advanced materials and as the ability increases to probe complex (photo)biological processes at the molecular level.

V. Ramamurthy
Kirk S. Schanze

Contents

Contributors

David Creed, Ph.D.   Department of Chemistry and Biochemistry, Center for Macromolecular Photochemistry and Photophysics, University of Southern Mississippi, Hattiesburg, Mississippi

Miguel A. Garcia-Garibay, Ph.D.   Department of Chemistry and Biochemistry, University of California–Los Angeles, Los Angeles, California

Prashant V. Kamat, Ph.D.   Radiation Laboratory, University of Notre Dame, Notre Dame, Indiana

Amy E. Keating   Department of Chemistry and Biochemistry, University of California–Los Angeles, Los Angeles, California

W. Grant McGimpsey, Ph.D.   Department of Chemistry and Biochemistry, Worcester Polytechnic Institute, Worcester, Massachusetts

Thomas L. Netzel, Ph.D.   Department of Chemistry, Georgia State University, Atlanta, Georgia

Kirk S. Schanze, Ph.D.   Department of Chemistry, University of Florida, Gainesville, Florida

Eimer Tuite, Ph.D.   Department of Physical Chemistry, Chalmers University of Technology, Göteborg, Sweden

K. Vinodgopal, Ph.D.   Department of Chemistry, Indiana University Northwest, Gary, Indiana

Keith A. Walters   Department of Chemistry, University of Florida, Gainesville, Florida

1

A Comparison of Experimental and Theoretical Studies of Electron Transfer Within DNA Duplexes

Thomas L. Netzel
Georgia State University, Atlanta, Georgia

I. INTRODUCTION

This chapter introduces an active and provocative area of current research. It describes motivations for studying electron transfer (ET) chemistry in general and ET reactions in ($2'$-deoxyribonucleic acid) DNA in particular. Rate expressions for quantum mechanical and semiclassical electron transfer models are given along with a description of appropriate conditions for their use. Formulas describing the distance dependence of electronic coupling (H_{AB}) and solvent reorganization energy are also discussed as a basis for understanding and interpreting recent experiments. Five different long-range ET experiments are presented in detail, and their results are compared with one another and with the results of similar experiments in proteins. Discrepancies and difficulties are noted, and suggestions for new experiments are made. Importantly, the experimental results are analyzed and discussed in light of recent semiempirical CNDO/S level quantum mechanical calculations of H_{AB} in DNA involving super-size sets of valence electrons. The chapter also contains descriptions of how photoinduced ET reactions in RNA are used to understand the operation of ribozymes (catalytic RNA) and of recent studies that demonstate (1) long-range migration of oxidative damage to DNA single strands in DNA–DNA and peptide nucleic acid (PNA)–DNA duplexes, (2) long-range oxidative thymine dimer (T^T) scission in DNA duplexes, and (3) migration of excess photoinjected electrons in DNA duplexes. The chapter concludes with a description of the best measurement todate of the decay of electronic coupling in DNA duplexes with increasing distance of donor/acceptor (D/A) separation, $\beta = 0.64 \pm 0.1$ Å^{-1}. This value of β, however, is for ET transfer from a guanosine nucleotide to the excited state of a doubly tethered stilbene acceptor. Whether β would be as low if DNA bases themselves were not directly involved as electron donors or acceptors is not yet known.

II. WHY ARE ELECTRON TRANSFER (ET) REACTIONS INTERESTING?

From a chemical perspective, ET reactions are the simplest kind of process in which reactants are changed into products. The process, although complex on a microscopic scale, has the important simplifying feature that the nuclei do not move while the chemically significant electron changes its location and turns reactants into products [1,2]. This fact makes detailed theoretical calculations of ET reactions possible, and thus provides chemists with a precise way of summarizing experimental results and of designing new experiments both to refine the understanding of the molecular basis of ET chemistry and to use this chemistry in purposeful ways. From a practical perspective, ET chemistry is responsible for life as we know it on Earth [3,4]. This is true because photosynthesis in plants and bacteria is at its core based on nature's ability to assemble a number of electrochemically active molecules in the right relative locations so that when they absorb light an ET reaction occurs [5]. In fact much research on photoinduced ET reactions in the past three decades has demonstrated that getting the first charge transfer to occur is not hard [6–8]. The key to photosynthesis is that the oxidized donor (D^{+}) and the nearby reduced acceptor (A^{-}) do not immediately back-react to re-form the starting molecules [9]. In fact, in bacterial photosynthesis nearly all of the electrons transferred to the primary acceptor are successfully transferred to secondary acceptors [5,10–14]. The important consequence of this secondary ET chemistry is that plants, upon which we and most other animals depend, grow. However, our dependence on ET chemistry does not end with photosynthesis. When we eat food, we acquire both energy and the molecular building materials needed to sustain our lives. ET reactions in our mitochondria reverse the photosynthetic process by consuming plant-produced carbohydrates and thereby provide us with the energy we need to live [15]. Our respiratory ET chemistry is in fact a highly controlled oxidation (or "burning") of plant fuel.

A. What Is Interesting About ET in DNA?

From a chemical perspective, the double-helix produced by two intertwining strands of oligomeric DNA is a fascinating and unique molecular structure. (See Fig. 1 for a structural model of a 12-base pair duplex of B-form DNA.) In it nucleic acid bases are stacked in pairs one on top of the other with a slight twist reminiscent of a spiral staircase [16]. The unique stacking and overlapping of the n- and π-electrons of DNA bases may provide a preferred path for electron transfer. Similarly, the exceptional closeness of the stacked bases may have important consequences for charge motion in DNA duplexes. Additionally, the

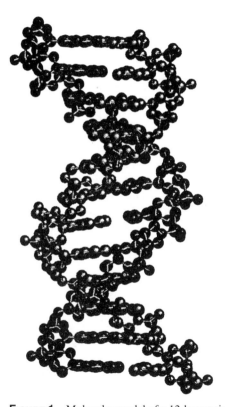

FIGURE 1 Molecular model of a 12-base pair duplex of canonical B-form DNA. The two 12-mer strands that intertwine to form the duplex are colored separately (black and gray). Nucleic acid base pairs are stacked perpendicular to the helical axis at 3.4-Å intervals (center-to-center distance), and the duplex helix repeats its spiral structure every 10 base pairs. (Figure provided by Dr. Carolyn Kanagy using the Sybyl Version 6.3 molecular modeling program from Tripos, Inc. and standard B-form DNA substructures.)

semirigid structure of a short DNA duplex makes it an attractive scaffold for creating an artificial photosynthetic reaction center. Research on both model and natural photosynthetic systems has demonstrated that the precise positioning of redox-active molecules is crucial for minimizing energy-wasting back-reactions and maximizing desired secondary ET reactions [17–24]. Thus, covalently substituting a DNA duplex with redox-active molecules offers a strategy for accurately controlling the position of desired redox-active chromophores.

From a health perspective, both radiation and natural cellular processes damage DNA and create reduced and oxidized ET products [25–34]. Fortunately, much of this damage is repaired shortly after it occurs by DNA repair enzymes. However, in some instances this is not the case, and tumors or cancer result. We would benefit from molecular information on the types of damage radiation produces and on the mechanisms that enzymes use to repair this damage. Radiation damage to DNA involves primary ionization steps as well as migration of charges to trap sites where irreversible chemical reactions occur [27,28,32,33]. Thus charge migration is a key step in both natural photosynthesis and artificial solar energy utilization, as well as in DNA radiation damage. Additional health benefits that are not directly related to ET chemistry also arise from ET research in DNA because the new synthetic methods developed to modify DNA for this research can also be applied to creating covalently modified DNA structures for use as drugs, medical diagnostic agents, or as tools for studying biomolecular processes (see below). For example, DNA-probe based diagnostic tests for bacterial and viral infections and for genetic defects may be improved, since these tests rely on the same DNA modification chemistry as does DNA ET research. Finally, research results on the mechanisms of DNA-mediated ET processes hold promise that entirely new ways of performing DNA diagnostic tests can be developed based on the ET properties of synthetically modified DNA strands.

III. THEORETICAL MODELS OF ET

To understand the factors that are important in controlling the rates of ET reactions, it is best to refer to a specific theoretical model [35]. Choosing a model defines terms and allows us to analyze the results of experiments in precise ways. There are two main types of models: classical and quantum mechanical. One way of smoothly moving from the use of a classical to a quantum mechanical model is provided by semiclassical (Landau–Zener) ET theory [36–38]. At high temperature, quantum mechanical models become equivalent in most respects to semiclassical ones. Thus the appropriate choice of model depends on the type of ET reaction that we are interested in studying. Ones in which the electron donor and acceptor have strong electronic interaction with each other prior to the ET event are well described by a classical model. In these systems an ET reaction always proceeds to products if the reactants reach the top of the reaction barrier (strong-interaction or adiabatic limit). In most biological ET processes (including photo-synthetic ones), there is some chance (perhaps a very large one) that even if the reactants reach the top of the reaction barrier they will not proceed to products, but rather will relax back to starting materials without ET. Not surprisingly this occurs because the electron donor and acceptor have a weak electronic interaction with each other prior to the ET event. These types of ET reactions (weak-interaction or nonadiabatic) must be modeled quantum mechanically or semiclassically. An active area of current ET research involves calculating and measuring the magnitudes of electronic couplings in D/A systems.

A. Quantum Mechanical Models

Brunschwig and Sutin [39] building on the work of De Vault [2] and Jortner [40] give Eq. (1) as a quantum mechanical expression for the rate of electron transfer (k_{ET}) for when two vibrational modes are responsible for bringing the reactant and product configurations into an energy-matching configuration (usually the top of the reaction coordinate). Note that since the electron must transfer from the reactant configuration (D ... A) to that of the product (D$^{\cdot+}$ ... A$^{\cdot-}$) without nuclear rearrangement, the transfer must occur between isoenergetic states for energy is to be conserved.

$$
k_{ET} = \frac{2\pi^{3/2} H_{AB}^2}{h} \left(\frac{1}{\lambda_s kT}\right)^{1/2} \exp\left[-S_f \coth\left(\frac{h\omega_f}{4\pi kT}\right)\right]
$$
$$
\times \sum_{m=-\infty}^{m=+\infty} \exp\left(\frac{mh\omega_f}{4\pi kT}\right) I_{|m|}\left[S_f \operatorname{csch}\left(\frac{h\omega_f}{4\pi kT}\right)\right]
$$
$$
\times \exp\left[-\frac{(\Delta E + mh\beta_f/2\pi + \lambda_s)^2}{4\lambda_s kT}\right] \quad (1)
$$

$I_{|m|}[Z]$ is the modified Bessel function, defined as:

$$
I_{|m|}[Z] = \sum_{k=0}^{k=\infty} \frac{(Z/2)^{m+2k}}{k!(m+k)!} \quad (2)
$$

The following variables are used in Eqs. (1) and (2): H_{AB} is the electronic coupling matrix element that permits ET to occur; h is Planck's constant; k is Boltzmann's constant; T is absolute temperature; λ_s is the reorganization energy of the solvent vibrations associated with ET; ω_f is the angular frequency of the quantized, high-energy molecular vibration associated with ET, such that $\omega_f/2\pi = \nu_f$; m in the summation refers to the number of quanta in the high-frequency mode; S_f is the ratio of the high-frequency mode's reorganization energy (λ_f) to the energy of one quanta of energy of this mode ($h\nu_f$), $S_f = \lambda_f/h\nu_f$; and ΔE is the energy change in an ET reaction on going from reactants to products.

In fact, many vibrations may be involved in the ET reaction, but the above coarse grain theory focuses on two modes of vibrational motion. One mode has low energy (hv_s or $h\omega_s/2\pi$) and is typical of solvent motions. Since quantum modes in this class have energy spacings that are much smaller than the energy of a room temperature bath ($hv_s \ll kT$), they act classically with continuously varying energy configurations. Their contribution to k_{ET} is governed by the λ_s reorganization energy parameter. The other vibrational mode has high energy, hv_f. In some studies its value has been taken to be 1500 cm^{-1} [41,42]. Note that specifying v_s and either λ_f or S_f is sufficient, but it is usually more convenient to assign λ_f because it is conceptually similar to λ_s.

The energy change in an ET reaction is ΔE, and it is negative for exergonic (spontaneous) reactions. Inspection of the Gaussian term in Eq. (1) shows that it is strongly peaked for values of $m = 2\pi$ ($|\Delta E| - \lambda_s$)/$h\omega_f$, which limits the range over which the sum must be calculated. In practice for photoinduced ET reactions, ΔE is obtained from electrochemical and excited state energy data according to [43]

$$\Delta E \approx \Delta G°(ET) = e[E°(D^{\cdot+}/D) \\ - E°(A/A^{\cdot-})] - E_{0,0}(\text{excited state}) + w(r) \tag{3}$$

where e is the charge on an electron; $E°$ is a reduction potential; $E_{0,0}$ is the light-absorbing (D or A) chromophore's reactive excited state energy relative to its ground state; and $w(r)$ is a coulombic interaction term between oxidized donor and reduced acceptor which represents the free energy due to separating the ionic products a distance r relative to each other $w(\infty) = 0$ [44,45]. Generally, in polar media the coulombic term is less than ca. 0.1 eV and is neglected [36,46,47].

H_{AB} is the electronic coupling matrix element that links the initial and final electronic states. Note that k_{ET} depends on the square of this electronic coupling element. To a first approximation the magnitude of H_{AB} depends on the overlap of the donor and acceptor wavefunctions. For long-range electron transfer between donors and acceptors bound to proteins or DNA, H_{AB} can be viewed as depending on the interaction of donor and acceptor orbitals with the intervening bridge orbitals summed over all occupied and unoccupied molecular orbitals of the bridge. Importantly, the interaction with the bridge orbitals is multiplied by the inverse of an energy difference term, $1/(E_N - E_{tun})$, where E_N is the energy of the Nth bridge orbital and E_{tun} is the electronic energy of the activated complex (top of the reaction coordinate). This term has the effect of preferentially weighting low-lying electronic states of the bridge for the case in which E_{tun} is only slightly below or above these bridge states. For cases in which the energy difference between E_{tun} and the nearest bridge orbitals is large, k_{ET} is not expected to depend significantly on the tunneling energy [48,49].

Although the Eq. (1) rate expression appears formidable at first sighting, programs such as *Mathematica* [50] make evaluating it as a function of various molecular parameters straightforward. In the past lack of such tools prompted simplifications of Eq. (1) for ease of use. A strong, but popular simplification involves the use of only one molecular vibration. This single vibration is also usually assumed to be in the high-temperature limit, $h\nu_s \ll kT$. The resulting rate expression is given in Eq. (4) and is fully quantum mechanical yet successfully reproduces a classical model's Arrhenius activation behavior. A less sweeping simplification used by Miller and Closs et al. [41,42,51,52] uses two vibrational motions as in Eq. (1) but assumes that the low-frequency mode is in the high-temperature limit and the high-frequency mode is in the low-temperature limit, $h\nu_s \ll kT \ll h\nu_f$. The latter assumption is one reason for preferring 1500 cm^{-1} over somewhat lower values (800–1300 cm^{-1}) as the energy of the high-frequency mode ($h\nu_f$).

B. A Single-Mode/High-Temperature Limit Quantum Mechanical Model

Equation (4) presents a commonly used ET model that is a special case of Eq. (1) [5,35,36,38,39,53–58]. Here only a single vibrational (nuclear) mode is active, and it is in the high-temperature (or classical) limit. In this model all the reactants that become products pass over the top of the reaction barrier (an activated process), and none of the reaction occurs by nuclear tunneling through the barrier. However, not all of the reactants that reach the top of the barrier actually become products; many simply relax without reacting (nonadiabatic or weak-interaction limit).

$$k_{ET} = \frac{2\pi^{3/2}H_{AB}^2}{h}\left(\frac{1}{\lambda kT}\right)^{1/2}\exp\left[-\frac{(\Delta E + \lambda)^2}{4\lambda kT}\right] \quad (4)$$

Although, ΔE is not generally equal to $\Delta G°$ unless the frequencies of the reactant and product are the same, this assumption is almost universally made. With this assumption classical Marcus ET theory combined with a quantum mechanical (Landau–Zener) treatment of the barrier crossing also yields Eq. (4) [2,36–39]. This derivation of Eq. (4) is called semiclassical ET theory, and therefore in the rest of this paper Eq. (4) will also be referred to as the semiclassical rate expression or the semiclassical model.

Strictly interpreted Eq. (4) cannot distinguish between the nuclear reorganization energy associated with the donor and acceptor molecules themselves, λ_f, and the solvent reorganization energy, λ_s. However, it is also true that if both of the vibrational modes in Eq. (1) are in the high-temperature limit, the resulting rate expression is the same as Eq. (4) with the substitution of $\lambda_s + \lambda_f$ for λ [39]. Equation (4) is frequently used in this latter form, but since both the D/A and solvent motions are treated classically, it is valid only at temperatures at which both of these vibrations are fully excited, $h\nu_s$ and $h\nu_f \ll kT$. In general, D/A vibrations are not likely to be in this limit. Thus, Eq. (1) will usually be required instead of Eq. (4) for cases in which λ_f is not equal to zero.

C. Comparison of Three Quantum Mechanical Models

All three of the above models, Eq. (1), Eq. (4) (semiclassical model), and the Miller-Closs model with two-modes active but with one in the high-temperature limit and the other in the low temperature limit), can equally well fit experimental k_{ET} versus ΔE data provided $|\Delta E|$ is not much larger than the sum of λ_s and λ_f. For

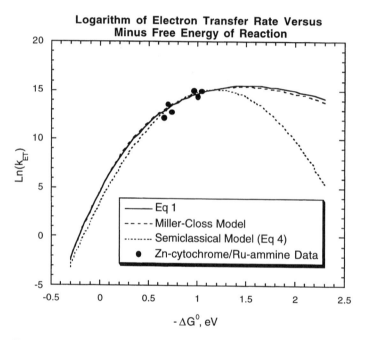

FIGURE 2 Plots of the logarithm of electron transfer rate vs. the negative of the free energy of the reaction for three ET models and six rate measurements. The data are from Refs. 54, 55, 57, 59, 60 for a Zn-substituted *Candida krusei* cytochrome *c* that was also successively substituted at histidine 33 by three $Ru(NH_3)_4L(His\ 33)^{3+}$ derivatives with L = NH_3, pyridine, or isonicotinamide. The shortest direct distance between the porphyrin and imidazole carbon atoms was 13 Å corresponding to the 10-Å edge-to-edge D/A distance. Table 1 presents a summary of the parameters used in the three calculations plotted in this figure. For a β of 1.2 Å^{-1}, Eq. (5) yields H°_{AB} values (± 10 cm^{-1}) of 80 cm^{-1}, 50 cm^{-1}, and 75 cm^{-1}, respectively, for Eq. (1), the semiclassical model [Eq. (4)], and the Miller–Closs model at the above D/A separation distance. The λ_s values were calculated using Eq. (6) with the following parameters: a_D = 10 Å, a_A = 6 Å, and r = 13 Å. The λ_f and H°_{AB} parameters were varied independently to produce the plotted curves.

large driving forces, the "Marcus inverted" region where $-\Delta E > (\lambda_s + \lambda_f)$, Eq. (4) predicts a much more rapid decrease in k_{ET} as driving force increases (increasing reaction exergonicity) than do the other two models. (See Fig. 2 for plots of the logarithm of ET rate vs. the negative of the free energy of reaction for these three models.) While Eq. (1) and the two simplifications of it described above all fit rate vs. driving force data for moderate driving forces very well, the actual parameters required in each case vary. Thus in the absence of other experimental data, it is difficult to know the meaning of particular parameter sets. Table 1 presents the results of using the above three models to fit data by Gray et al. [54,55,57,59,60] for a Zn-substituted *Candida krusei* cytochrome *c* that was also successively substituted at histidine 33 by three $Ru(NH_3)_4L(His\,33)^{3+}$ derivatives with L = NH_3, pyridine, or isonicotinamide. In these experiments photoinduced ET from the Zn–porphyrin (ZnP) triplet state reduced the bound Ru(III) ammine complex, while thermal back-ET within the product complex from Ru(II) ammine to $ZnP^{\cdot+}$ re-formed the starting complex, ZnP–Ru(III) ammine. (See Fig. 3 for a schematic illustration of these electron transfer steps.) The driving force $(-\Delta E)$ in these experiments was moderate, varying from 0.6 to 1.1 eV. (See Fig. 2 for a plot of the Zn–cytochrome/Ru(III) ammine data along with the fits to this data by the three models discussed above.)

Several conclusions are apparent from the results in Table 1 and Fig. 2. The first is that the semiclassical model [Eq. (4)] understates the total nuclear reorganization energy $(\lambda = \lambda_s + \lambda_f)$ relative to the two-mode models. IIf λ in Eq. (4) is identified with $\lambda_s + \lambda_f$ rather than with just λ_s, there is no discrepancy with regard to λ_s. The second conclusion is that the two-mode models do not differ within error in their estimates of H_{AB} for similar values of λ_s and λ_f and for a high-frequency mode as large as 1500 cm^{-1}. However, since there is no need to assume that the high-frequency mode is in the low-temperature limit, Eq. (1) is the preferable of these two models. Whether both high- and low-frequency modes are important cannot be distinguished on the basis of the results in Table 1. However, there are theoretical reasons to believe that the solvent reorganization energy

TABLE 1 Parameters for Three ET Models That Fit ZnP/Ru(III) Ammine ET Rate vs. Free Energy of Reaction Data from Refs. 54, 55, 57, 59, 60

Model	Eq. (1)	Semiclassical model [Eq. (4)]	Miller-Closs Model/ Low-temperature and High-temperature limits
H_{AB} (cm^{-1})	0.20	0.12	0.19
λ_s (eV)	0.75	1.20	0.75
λ_f (eV)	0.85	—	0.80
$h\omega_f/2\pi$ (cm^{-1})	1500	—	1500

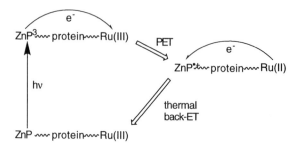

FIGURE 3 Illustration of photoinduced ET (PET) in a Zn-substituted *Candida krusei* cytochrome *c* (ZnP-protein) that was substituted at histidine 33 by a Ru(III) ammine complex. Photoexcitation of the cytochrome's Zn-porphyrin (ZnP) group produced the ZnP triplet state (ZnP³) which reduced the bound Ru(III) complex thorough protein-mediated ET. Subsequently the thermal back-ET within the Ru-substituted protein re-formed the starting ZnP-protein-Ru(III) complex.

should be in the 0.4 to 0.7 eV range for this type of ET reaction [45,57]. The third conclusion is that all three models give reasonably similar electronic coupling matrix elements, but the semiclassical model's value is a little lower than those of the two-mode models.

D. Distance Dependence of D/A ET Rates

Electronic structure calculations are being reported for long-range ET between bound donors and acceptors in both protein and DNA systems. However, comparison of theory with experiment generally requires a coarse grain approach. A useful expression for this is given in Eq. (5), and is based on electron tunneling through a one-dimensional square barrier. It describes the decay of the electronic coupling matrix element with increasing barrier thickness (D/A separation distance).

$$|H_{AB}(r)| \approx |H^{\circ}_{AB}| \exp\left[-\frac{\beta}{2}(r - r^{\circ}) \right] \qquad (5)$$

where H°_{AB} is the electronic coupling when the donor and acceptor are in contact at a center-to-center distance r°; r ($\geqslant r^{\circ}$) is the D/A center-to-center distance of separation; and β is the distance decay parameter of the electronic coupling. However, $H_{AB}(r)$ cannot be measured directly. Instead the distance dependence of k_{ET} for a given D/A pair is measured and Eq. (5) is substituted into a rate expression such as Eq. (1) or (4). If λ_s is assumed to be independent of r (see below), β can be determined from the slope of a plot of ln (k_{ET}) vs. r. β is nonphysical; however, it does provide a concise way of qualitatively comparing electron transfer rates for different types of D/A bridges and intervening media.

Note that electronic coupling decays exponentially with increasing r as $\beta/2$, while k_{ET} decays as β.

Just as Eq. (5) must be substituted into Eq. (1) or (4) to describe the variation of the k_{ET} with D/A separation due to varying electronic coupling, so too the dependence of k_{ET} on λ_s variation due to D/A separation should also be explicitly modeled. In fact it is not proper to assume that λ_s is independent of r (even though this is done many times). Clearly, when both H_{AB} and λ_s are functions of r, a simple plot of $\ln(k_{ET})$ vs. r cannot be used to evaluate β.

Rather full calculations of $k_{ET}(r)$ vs. r for various β values must be compared to the experimental results to determine β. Equation (6) gives a widely used expression for solvent reorganization energy that can be substituted into k_{ET} expressions. It was derived by Marcus over 40 years ago and is both simple and useful [61]. It models the donor and acceptor as two conducting spheres imbedded in a dielectric continuum.

$$\lambda_s = (\Delta e)^2 \left(\frac{1}{2a_D} + \frac{1}{2a_A} - \frac{1}{r} \right) \left(\frac{1}{D_{op}} - \frac{1}{D_s} \right) \qquad (6)$$

where Δe is the charge transferred, a_D and a_A are the radii of the donor and acceptor spherical cavities, D_{op} is the solvent's optical dielectric constant (square of the refractive index), and D_S is the solvent's static dielectric constant.

Two other dielectric-continuum solvent models have also been developed in recent years. The single-sphere model, suitable for analyzing λ_s in the ZnP–Ru(III) ammine experiments discussed above, views the ET as occurring between two redox sites imbedded in a low dielectric spherical cavity (radius ≈ 16 Å for cytochrome c, $\varepsilon \approx 2$) that is itself imbedded in a higher dielectric continuum ($\varepsilon \approx$ 78) [44,45,57,62]. The ellipsoidal model, useful for bridged D/A complexes, views the ET as occurring between two redox sites imbedded in an ellipsoidal cavity itself in a high dielectric continuum [44]. For the applications noted, these latter two models of solvent reorganization energy are more realistic than the Marcus two-sphere model in Eq. (6). Unfortunately the mathematical calculations for the ellipsoidal model are significantly more complex than for the Marcus model. Nevertheless, for bridged D/A systems the two-sphere model can give results that are close to those obtained with the ellipsoidal one [44]. For ET experiments in substituted globular proteins such as cytochrome c, the single-sphere model is appropriate, and accurate closed-form approximations for λ_s are available [62].

IV. EXPERIMENTAL AND THEORETICAL STUDIES OF DNA-MEDIATED ET

Experimental measurements of k_{ET} for D/A groups bound to DNA are sparse. One reason is that the synthetic chemistry involved is more complex than that necessary for studying ET processes between D/A groups bound to proteins. Also, ET studies in substituted proteins began to be reported in 1982, while the first report of ET in a DNA duplex involving bound D/A groups appeared in 1992 by Brun and Harriman [63]. Since then until late 1997 only two more studies were reported, one by Murphy et al. [64] in 1993 and the other by Meade and Kayyem [65] in 1995. There are two ways of fixing the relative location of D/A groups associated with DNA duplexes. The first, used by Brun and Harriman, intercalates them between DNA base pairs. The second, used by Meade and Kayyem, co-valently attaches them to DNA strands with very short tethers. Barton and Turro et al. use a combination of these two techniques and covalently attach the D/A groups to the 5'-ends of complementary DNA strands with 16-atom-long linkers [64,66]. The long linkers themselves do not accurately locate the positions of the D/A groups. However, in this work the D/A groups also intercalate into DNA duplexes. Thus 5'-end tethering combined with intercalation constrains each redox-active chromophore in this work to be within 2 ± 1 base pairs of a du-plex end.

In late 1997 two new kinetics studies of DNA-mediated ET were carried out. In the first one Barton and coworkers [64] again covalently attached inter-calating donor and acceptor groups. In this new system, the photoexcited electron donor was the ethidium cation (EB^+) and the acceptor was a tris-polypyridyl Rh(III) complex [67]. These workers are to be commended for constructing a DNA–D/A system with five different D/A separation distances. However, the interpretation of the emission quenching results obtained so far is not straight-forward. Thus, unambiguous ET rate measurements are not yet available for this new system. Nevertheless, this work is interesting and will be discussed briefly near the end of this chapter. In the second study, Lewis, Letsinger, Wasielewski et al. [68] used 4,4'-stilbenedicarboxamide as a bridge to connect complementary DNA strands of six or seven bases each, which formed duplex hairpins at tempera-tures below 75°C. In these hairpins photoexcited stilbene was reduced by gua-nosine nucleotides, and five hairpins were studied each having zero to four A/T base pairs separating the stilbene-bridge and a single G/C base pair. [See Fig. 4 for molecular structures of five ribonucleosides: 2'-deoxyuridine (dU), 2'-deoxy-thymidine (dT), 2'-deoxycytidine (dC), 2'-deoxyguanosine (dG), and 2'-deoxy-adenosine (dA).] Importantly, this work has produced the best measurement of the distance decay of D/A electronic coupling in a DNA duplex todate. It will be fully discussed at the end of this chapter. However, before doing so the three pre-1997 studies of ET in DNA introduced above will be analyzed and compared with each other.

All three of the pre-1997 DNA studies, in common with all ET studies in proteins, fall into the category of having a large energy difference between E_{tun} and the oxidized and reduced states of the bridge. In terms of the Hartree–Fock approximation, these states are related to the occupied and unoccupied orbitals

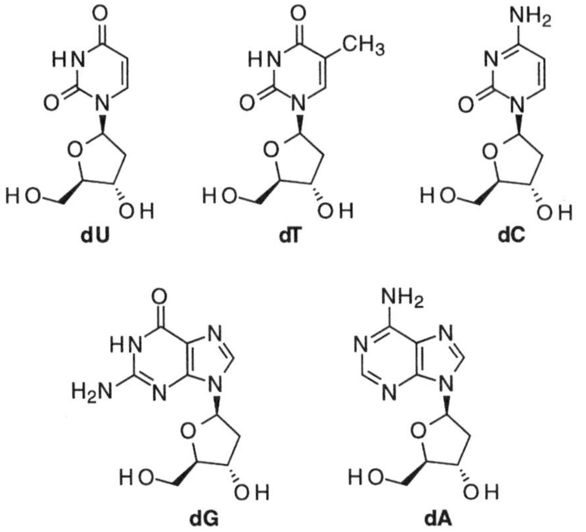

FIGURE 4 Molecular structures of five DNA nucleosides: 2′-deoxyuridine (dU), 2′-deoxythymidine (dT), 2′-deoxycytidine (dC), 2′-deoxyguanosine (dG), and 2′-deoxyadenosine (dA).

of the bridge. It is these virtual (never really occupied) states of the DNA bridge that assist long-range ET [49]. Because H_{AB} is proportional to $1/(E_N - E_{tun})$, DNA-mediated photoinduced ET experiments in which E_{tun} is very close in energy to the oxidized and reduced bridge states may show enhanced electron transfer rates. This latter category of "small-tunneling-gap" ET has yet to be tested significantly with experiments, but is likely to be examinable in systems in which a photo-produced donor is nearly able to reduce bridge bases or a photo-produced acceptor is nearly able to oxidize bridge bases. Recent work by Geacintov et al. [69–71] and by Netzel et al. [72,73] has shown that dG and dU nucleosides can be covalently joined to pyrene with very short linkers, and that upon photoexcitation either the pyrene$^{\cdot-}$/dG$^{\cdot+}$ or pyrene$^{\cdot+}$/dU$^{\cdot-}$ charge-transfer (CT) products are formed in less than 30 ps. (See Figs. 5 and 6 for examples of pyrene-labeled nucleosides.) Thus, substituting pyrene-labeled nucleotides in a DNA duplex offers a way of photoproducing either a donor, dU$^{\cdot-}$, or an acceptor, dG$^{\cdot+}$, for ET studies. Both of these species will have small tunneling gaps for long-range ET in DNA duplexes because dG$^{\cdot+}$ is nearly able to oxidize dA, and dU$^{\cdot-}$ has nearly the same reduction potential as dT and dC. The ≤ 1 nanosecond lifetimes of the pyrene/dG and pyrene/dU CT products studied so far limit their use in long-range ET experiments, but it is likely that pyrene/nucleoside adducts possessing longer-lived CT products can be constructed.

10S (+)-*trans-anti*-[BP]-*N^2*-dG

10S (-)-*cis-anti*-[BP]-*N^2*-dG

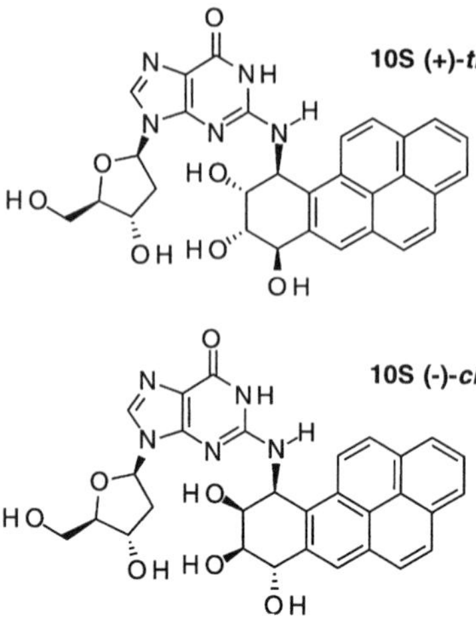

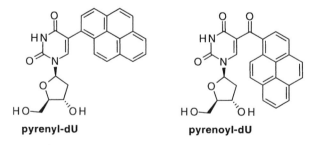

Figure 5 Molecular structures of the (+)-*trans-anti*-[BP]-N^2-dG and (−)-*cis-anti*-[BP]-N^2dG adducts formed from the metabolically activated form of the widespread environmental polutant benzo[*a*]pyrene (BP) and dG. These two adducts have the same S-configuration about the *anti*-[BP]-C10–N^2-dG linkage but different arrangements of the OH groups at the 7, 8, and 9 positions of the *anti*-[BP] residue. Each of these two diastereoisomers also has a corresponding stereoisomer with a 10R configuration at C10 (not shown).

pyrenyl-dU **pyrenoyl-dU**

Figure 6 Molecular structures of two pyrenyl-substituted uridine nucleosides, pyrenyl-dU and pyrenoyl-dU. In polar protic solvents, the lowest-energy electronic excited state of pyrenyl-dU is a pyrene-to-uridine charge-transfer state, $\text{pyrene}^{\cdot+}/\text{dU}^{\cdot-}$.

In late 1996, a new family of DNA-mediated ET experiments began to be reported. In these the ET donor is comprised of one or more DNA nucleotides (either G, GG, GGG, or a covalent thymine dimer) while the other is a covalently attached, photoactivated electron acceptor (either anthraquinone, pyrene, or an tris-polypyridyl M(III) complex, where M = Ru or Rh) [74–80]. These experiments have much in common with the kind of experiments outlined immediately above and with the new hairpin studies recently reported. They will be discussed toward the end of this chapter. They differ from the pre-1997 ET kinetics experiments that will be discussed immediately below in that none of them has reported ET rate measurements. Rather, yields of net photochemistry (either DNA strand cleavage or thymine dimer scission) are presented as evidence that photoinduced ET events occurred. For comparison of experimental results with ET theory it is clear that at a minimum ET rates must be measured and at a maximum several rates should be measured for a given D/A pair at a variety of separation distances.

A burning question in DNA-mediated ET studies is what is the appropriate value for the electronic coupling decay parameter, β, for ET along the helical axis of a DNA duplex. The immediately following sections describe results from each of the three pre-1997 ET experiments on this topic. Of course, it is clear from the above discussion that β cannot be determined quantitatively without addressing the questions of what are the appropriate values of the nuclear reorganization energies, λ_s and λ_f, in these experiments. Thus a large number of additional experiments in this area still need to be done before there will be enough data to know the answers to these questions, but the results to date are both interesting and provocative.

A significant incentive for experimental research in this area is the concurrent development of electronic structure calculations of H_{AB} values for specific DNA bridges [48,49]. The goal of ET experiments in this area is to measure rates for given D/A pairs as a function of the base sequence of the DNA bridge. Once sufficient data are available it is possible that the fiction that Eq. (5) describes the distance dependence of H_{AB} will have to be either abandoned or qualified severely. For example, theoretical calculations (based on an extended-Huckel Hamiltonian) of DNA bridge electronic couplings for the $(dA)_6(dT)_6$ duplex show that while couplings along either strand decay monotonically and approximately exponentially as D/A distance increases [$\beta = +2.1$ and $+1.5$ Å^{-1}, respectively, along the $(dA)_6$- and $(dT)_6$-strands], this is not true for across-strand couplings. In fact for across-strand D/A distances determined by four or fewer base pairs for 3′-3′ substitutions, the bridge electronic coupling *increases* a factor of 220 as the D/A separation distance increases from 11.2 to 13.4 Å, $\beta = -4.9$ Å^{-1}, and then drops precipitously (2×10^3-fold) on going from 13.4 to 14 Å; the coupling does not increase with increasing D/A distance for 5′-5′ substitutions. Additionally, for 5′-5′ across-strand substitutions, the bridge coupling falls a factor of 1.1×10^5 as separations increase from 14.4 to 17 Å, $\beta = +9.0$ Å^{-1} [81]. Solvent molecules are not included in these calculations, and whether or not they provide efficient electronic coupling pathways in the duplex grooves is not known. Thus while these theoretical results suggest that D/A ET rates will be highly sensitive to DNA duplex structure, experimental confirmation is needed to assess their correctness.

A. Ru-to-Ru ET Along an Eight Base Pair Duplex

Meade and Kayyem recently published a communication that describes the successful construction of an eight base pair duplex with ruthenium metal complexes coordinately bound to the 2′-ribose positions at the 5′-end of each strand [65]. To do this the normal 2′-OH group in uridine was converted into a primary amine and protected with trifluoroacetyl. This nucleoside was then converted into a phosphoramidite reagent for use on an automated DNA synthesizer. After synthesizing complementary 8-mer oligonucleotides with terminal primary amines, it was necessary to attach the ruthenium complexes. This step was potentially problematic because the ruthenium complexes used, $Ru(bpy)_2CO_3$ (followed by addition of imidazole) and $Ru(NH_3)_4(py)(H_2O)^{2+}$, where bpy = 2,2′-bipyridine and py = pyridine, were capable of reacting not only with the terminal 2′-amines but also with the heterocyclic nitrogen atoms on the bases. To solve this dilemma, the reactive sites on the bases were first protected by complexation to a complementary DNA strand that did not have a 2′-amino group. Under these conditions each of the above ruthenium complexes added readily to terminal 2′-amino positions while unwanted reactions with amines on the bases were either eliminated or substantially lessened.

The exact base sequences in this experiment are 5′-Ru(II)(bpy)$_2$(imidazole)-(2′-NH$_2$-U)GCATCGA-3′ and its complement 5′-Ru(II)(NH$_3$)$_4$(pyridine)(2′-NH$_2$-U)CGATGCA-3′. (See Fig. 7 for a structural model of this doubly labeled duplex.) The metal-to-metal distance between the two ruthenium sites in the 8-mer duplex is 20.5 Å. However, it is not reasonable to view the ET event as a purely metal-to-metal process in these coordinate covalent complexes. Thus the relevant edge-to-edge distance for the ET process is on the order of 12.5 Å. There are several ways that ET processes can be measured in this system, and the authors provide no experimental details in their communication other than to say that both direct photoinduced and flash-quench techniques were used. The flash-quench technique was developed earlier for intramolecular ET studies in Ru-substituted proteins by Chang, Gray, and Winkler [56]. (See Fig. 8 for a schematic llustration of the flash-quench technique for measuring intramolecular ET rates in Ru-substituted cytochrome c.) In their experiment, photoexcited Ru(bpy)$_2$(imidazole)-(amine)$^{2+}$ was quenched during its 80-ns excited state lifetime by added Ru(NH$_3$)$_6^{3+}$. The Ru(bpy)$_2$(imidazole)(amine)$^{3+}$ product formed in the quenching reaction was subsequently reduced by the ferroheme in cytochrome c. Importantly, the kinetics

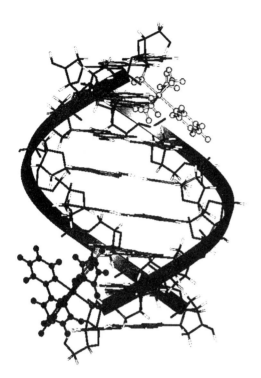

FIGURE 7 A molecular model of an octanucleotide duplex comprised of 5'-Ru(III)(bpy)$_2$-(imidazole)(2'-NH$_2$-U)GCATCGA-3' and its complement 5'-Ru(II)(NH$_3$)$_4$(pyridine)(2'-NH$_2$-U)CGATCGA-3'. The metal-to-metal distance between the two ruthenium sites is 20.5 Å. The Ru(II) ammine donor is at the top, and the Ru(III) polypyridine acceptor is at the bottom. (Figure provided by Drs. David Beratan and Satyam Priyadarshy reprinted with permission from *J. Phys. Chem.* 1996, *100*, 17678. Copyright 1996 Amer. Chem. Soc.)

of the Fe(II)-to-Ru(III) ET process were followed by transient absorbance spectroscopy, thus providing direct evidence for ET product formation. The time window for observing intramolecular ET was set by the time of recombination of bimolecularly formed Ru(NH$_3$)$_6^{2+}$ and Ru(bpy)$_2$(imidazole)(amine)$^{3+}$. Direct photoexcitation of Ru(bpy)$_2$(imidazole)(amine)$^{2+}$ in Fe(III) cytochrome *c* produced the same charge-separated product as the flash-quench experiment, Ru(bpy)$_2$(imidazole)(amine)$^{3+}$/Fe(II). Meade and Kayyem report the ET rate between the acceptor Ru(bpy)$_2$(imidazole)(amine)$^{3+}$ and the donor Ru(NH$_3$)$_4$(pyridine)(amine)$^{2+}$ to be $1.6(\pm0.4) \times 10^6$ s^{-1}.

Meade and Kayyem [65] also point out that use of semiclassical ET model [Eq. (4)] with $\Delta G° = -0.7$ eV, $\lambda = (\lambda_s + \lambda_f) = 0.9$ eV, and their measured ET rate allows calculation of the maximum rate (k_{ET}^{max}) for this reaction, 2.5×10^6 s^{-1}

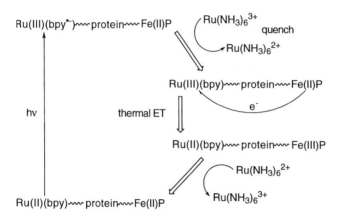

FIGURE 8 Illustration of the flash-quench technique for measuring intramolecular ET rates. Photoexcitation of $Ru(II)(bpy)_2(imidazole)(amine)^{2+}$, Ru(II)(bpy), bound to ferrocytochrome c, Fe(II)P, produced an 80-ns lived metal-to-ligand charge transfer (MLCT) excited state, $Ru(III)(bpy^{\cdot-})$, which was oxidatively quenched by bimolecular reaction with $Ru(NH_3)_6^{3+}$. The resulting Ru(III)-complex was then reduced by the Fe(II)P through thermal, protein-mediated ET. Finally bimolecular reaction of the Ru(II)/Fe(III)P product with $Ru(NH_3)_6^{2+}$ re-formed the starting Ru(II)-protein-Fe(II)P complex.

(obtained when $-\Delta G° = \lambda$). Interestingly, this maximum rate is comparable to those in cytochrome c with histidines 33 or 39 modified by ruthenium substitution. There the maximum ET rates and edge-to-edge distances for histidines 33 and 39 (with $\lambda = 0.8$ eV) are, respectively, 3.3×10^6 s^{-1} at 12.3 Å and 2.7×10^6 s^{-1} at 11.1 Å [82]. Thus k_{ET}^{max} along the length of a DNA duplex even with its unique π-stacking of bases is much the same as that found at comparable distances in protein systems, albeit ones with efficient ET processes. This conclusion accords with reports of protein studies that the presence of aromatic groups between donors and acceptors does not of necessity produce faster ET processes than would an entirely aliphatic intervening medium [82,83].

As was done above, experimentalists many times use the semiclassical ET model [Eq. (4)] when discussing k_{ET}^{max} and H_{AB}. This is done for several reasons. (1) This simple model requires fewer parameters than more accurate multimode quantum mechanical models and is extremely easy to use. (2) While λ_s values are moderately uncertain, λ_f and $h\omega_f/2\pi$ values are very uncertain; thus it is tempting to avoid specifying them. (3) This model fits rate k_{ET} vs. driving force data very well for small to moderate driving forces. Therefore, so long as the reaction driving force is moderate ($|\Delta E| \leq \lambda$) and the D/A vibrational mode is inactive or in the high-temperature limit, the semiclassical model should give good results.

When these conditions are not true, Eq. (1) should be used. So far little work has been done to establish the frequencies of D/A vibrational modes in ET reactions; thus it is not known whether they are sufficiently low to justify using the semi-classical model.

B. ET Between Intercalated Donors and Acceptors

Brun and Harriman made a careful study using steady-state fluorescence, time-resolved fluorescence, and transient absorbance spectroscopies of the rates of excited state ET between two sets of donors and acceptors intercalated into calf-thymus DNA (CT–DNA) [63]. Importantly, the formation of the ET products was directly monitored. While there was good agreement between the kinetics of formation of the ET products as seen by transient absorbance spectroscopy and the excited state emission decay kinetics, the latter data had a much higher signal-to-noise ratio and were therefore used to determine the ET rates. One D/A set used ethidium bromide (EB$^+$) as the photoexcited electron donor and N,N'-dimethyl-2,7-diazapyrenium dichloride (DAP^{2+}) as the electron acceptor [$\Delta G°(ET) = -0.26$ eV], and the other D/A set used protonated acridinium orange (AOH$^+$) as the donor and DAP^{2+} as the acceptor [$\Delta G°(ET) = -0.67$ eV]. (See Fig. 9 for structural models of these three intercalators.) Studies with both D/A sets used high CT–DNA concentrations (700 μM and 250 μM base pairs, respectively) such that ca. 99% of the donors and acceptors were intercalated into the CT–DNA duplex. (See Fig. 10 for a molecular model of EB$^+$ and DAP^{2+} intercalated into a short DNA duplex.)

 The weakest parts of this study are that unknown base sequences intervene between D/A pairs at any given separation distance and an indirect argument must be made to establish the D/A distances. Dye binding studies show that at saturation EB$^+$ and AOH$^+$ have binding ratios of 0.5 dye/base pair, while DAP^{2+} has a

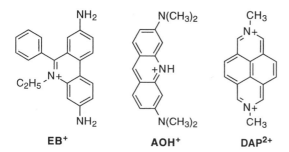

FIGURE 9 Molecular structures of ethidium bromide (EB$^+$), protonated acridinium orange (AOH$^+$), and N,N'-dimethyl-2,7-diazapyrenium dichloride (DAP^{2+}). Each of these molecules is known to intercalate into duplex DNA.

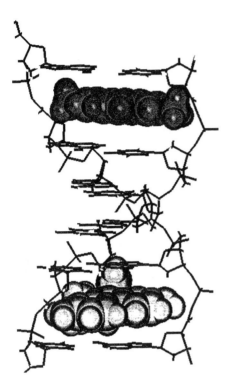

FIGURE 10   A molecular model of a hexamer duplex $(dA)_6(dT)_6$ with the intercalated donor EB^+ in shown as a space filling model at the bottom and the intercalated acceptor DAP^{2+} shown as a space filling model at the top. The donor and acceptor are separated by four base pairs, 13.6 Å. (Figure provided by Drs. David Beratan and Satyam Priyadarshy and reprinted with permission from *J. Phys. Chem.* 1996, *100*, 17678. Copyright 1996 Amer. Chem. Soc.)

ratio of 0.25 dye/base pair. Thus EB^+ and AOH^+ each exclude from binding one additional base pair on either side of their binding site, while DAP^{2+} excludes an additional two on either side of its binding site. The combination of DAP^{2+} with either EB^+ or AOH^+ bound to CT–DNA therefore excludes at maximum loading three additional pairs between each D/A dye pair. Of course, at less than maximum dye loading on the duplex there are more than three base pairs separating some of the D/A pairs. Assuming that there are no cooperative binding interactions between the D/A dyes, this argument establishes that the minimum D/A edge-to-edge separation distances correspond to 3, 4, and 5 base pairs or 10.2, 13.6, and 17.0 Å. These distance estimates assume that the duplex remains fully coiled and does not elongate appreciably upon D/A intercalation. The fact that random base sequences intervene between donors and acceptors means that the distance dependent ET rates are averages over many intervening base sequences. To the extent there is preferential sequence selectivity involved in the D/A binding process, the ET rates would not be averages but would be due to specific but unknown intervening base sequences. Table 2 lists the observed photoinduced ET rates found for the above two D/A pairs at three separation distances. Reverse ET to reform the starting D...A configuration was also measured, but the kinetics were complex and depended on the concentration of DAP^{2+}. Thus bimolecular processes appeared to be involved, and these data are not presented here.

By substituting Eq. (5) into the semiclassical ET rate expression [Eq. (4)] and neglecting the distance dependence of the solvent reorganization energy, one can show that

$$k_{ET} = A \exp\left[-\beta(r - r^\circ)\right] \tag{7}$$

where A is a proportionality constant. Brun and Harriman used Eq. (7) and obtained a β value of 0.9 Å^{-1} by plotting $\ln[k_{ET}]$ vs. $(r - r^\circ)$ for data for both D/A pairs. This value of β is the same as that found in studies of ET in ruthenated sperm whale myoglobin [54]. It is also close to that found in ET studies of ruthenium-modified derivatives of the blue copper protein *Pseudomonas aeruginosa* azurin. There $Ru(bpy)_2(imidazole)(histidine)^{2+}$ complexes bound to azurin were studied by the flash quench technique as described above, and the observed ET rates were analyzed according to Eq. (7). The β value obtained in the azurin protein studies is 1.1 Å^{-1} through this β-strand protein [84]. Theoretical studies of ET in proteins predict that the distance dependence of ET rates in β-strand and α-helix proteins should fall off exponentially with distance with β values, respectively, of 1.00 and 1.26 Å^{-1} [84]. These differences in β between the two types of proteins are understandable because β-strands lengthen nearly linearly while α-helices lengthen with coils. Thus for a given D/A separation fewer bonds will connect donors and acceptors on a β-strand than on an α-helix, and the coupling through a β-strand should be larger than through an α-helix.

TABLE 2 Rate Constants for Photoinduced ET Between Intercalated Donors and Acceptors in CT–DNA

$(r - r^\circ)$, Å	EB^+-to-DAP^{2+}, s^{-1}	AOH^+-to-DAP^{2+}, s^{-1}
10.2	1.3×10^9	4.3×10^9
13.6	7.2×10^7	2.3×10^8
17.0	2.5×10^6	1.3×10^7

Source: Data from Ref. 63.

In view of the above protein studies, the Brun and Harriman result for β appears reasonable. However, the absolute ET rates in Table 2 are a little high. For example, the driving forces and D/A edge-to-edge separation distances are nearly the same in the Meade and Kayyem experiment as in the AOH^+/DAP^{2+} experiment, respectively, $-\Delta G° = 0.7$ eV and 12.5 Å vs. $-\Delta G° = 0.67$ eV and 13.6 Å, yet k_{ET} in the latter experiment is at least 140-fold greater. Following are three possible explanations. (1) In the latter experiment, the π-electrons of the intercalated donors and acceptors overlap extensively with the n- and π-electrons of the DNA bases. Thus their couplings to the bridge may be significantly greater than those of the ruthenium complexes where each metal complex is joined to the bridge through a single sp^3 hybridized $C2'$-ribose atom. (2) The nuclear reorganization terms (λ_s and λ_f) may be smaller in the latter experiment than in the former. (3) Perhaps the D/A distances are not as large as originally assumed. This would be the case if the minimum D/A separation distance was in fact less than three base pairs. Clearly more ET experiments in this area are needed, but comparison of these two results is intriguing. If future experiments confirm the value of β in DNA as equal to 0.9 $Å^{-1}$, it will mean that DNA is not substantially different from proteins in its ability to facilitate long-range ET for the case in which the tunneling energy (E_{tun}) is well separated from the oxidized and reduced states of the bridge. A related study [85,86] of ET quenching of emission from intercalating Ru(phen)$_2$-(dppz)$^{2+}$ enatiomers, where phen = 1,10-phenanthroline and dppz = dipyridophenazine, by intercalating and nonintercalating viologen acceptors finds that (1) externally bound viologen quenches better than intercalated viologen and (2) ET might take place between intercalated donors and acceptors separated by as much as 3 to 6 base pairs as in the Brun and Harriman experiment, but not when they are more than 20 Å apart (edge-to-edge). [See Fig. 11 for structural models of the Λ and Δ enatiomers of the Ru(phen)$_2$(dppz)$^{2+}$ complex.] Using linear dichroism spectroscopy, this work also shows that at low binding ratios ([viologen]/[DNA base] = [Ru]/[DNA base] = 0.04) the D/A groups bind independently and that

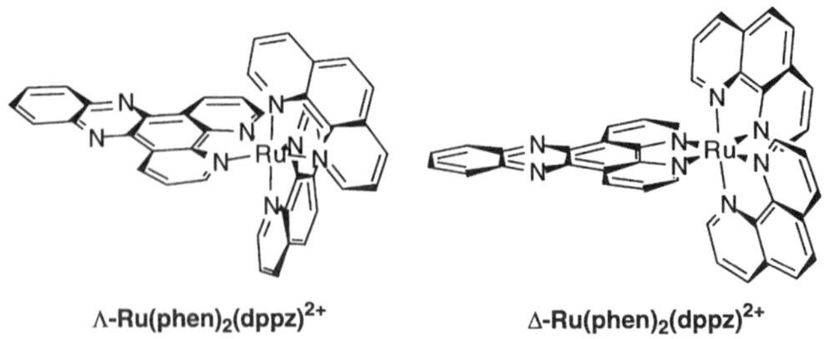

Λ-Ru(phen)$_2$(dppz)$^{2+}$ Δ-Ru(phen)$_2$(dppz)$^{2+}$

FIGURE 11 Molecular structures of the Λ and Δ enantiomers of the intercalating electron donor Ru(phen)$_2$(dppz)$^{2+}$.

Ru*-quenching is not likely to be caused by displacement of ruthenium from the duplex (see below) [87]. Thus Tuite, Norden, Lincoln, and Becker conclude that "... the stacked basepairs represent a relatively poor medium for electron transfer and that long-range electron transfer between intercalated reactants separated by $\geqslant 20$ Å is unlikely...." [87]. A fuller description of this work can be found in Chapter 2 of this volume written by Tuite [88].

C. Ru-to-Rh ET for a Tethered and Intercalated D/A Pair

Barton and Turro et al. have reported an extremely interesting measurement of long-range ET through a DNA duplex [64]. In this work complementary 15-mer strands of DNA were synthesized and modified at their 5′-ends to yield strands each having a 5′-NH_2-$(CH_2)_6$-OPO_3-terminus (L). The 5′-amino ends of each strand were reacted separately with either Ru(phen′)$_2$(dppz)$^{2+}$, where phen′ = 5-(amido-glutaric acid)-1,10-phenanthroline and dppz = dipyridophenazine, or Rh(phi)$_2$(phen′)$^{3+}$, where phi = phenanthrenequinone diimine. (See Fig. 12 for structural drawings of these three ligands as well as of the common metal chelator 2,2′-bipyridine, bpy.) Both of these metal complexes were chiral having Λ and Δ forms, but in this study racemic mixtures were used.

The Ru(phen)$_2$(dppz)$^{2+}$-chromophore is the electron donor in this study. It has absorbance maxima in aqueous solution at 372 and 439 nm, respectively, $\varepsilon =$ 2.48 and 2.23×10^4 M^{-1} cm^{-1}. The absorbance at 439 nm arises from a metal to ligand charge transfer (MLCT) transition that is common to tris(bipyridyl)-ruthenium(II) complexes, while the absorbance at 372 arises from an intraligand

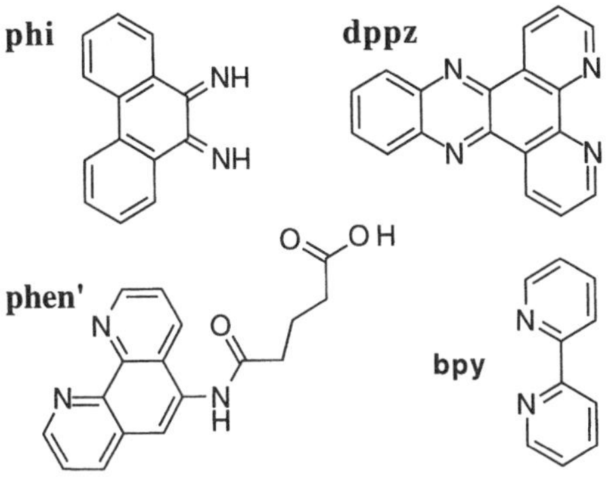

FIGURE 12 Molecular structures of four bischelating ligands commonly used to form octahedral Ru- and Rh-complexes for attachment to DNA: phenanthrenequinone diimine (phi), dipyridophenazine (dppz), 5-(amido-glutaric acid)-1,10-phenanthroline (phen′), and 2,2′-bipyridine (bpy).

(IL) transition localized on dppz [89]. By analogy to the electronic structure of Ru(bpy)$_2$(dppz)$^{2+}$ [90,91] excitation of the bipyridyl-MLCT absorbance at 439 nm is thought to rapidly populate a lower-energy dppz-MLCT state with Ru(III)/dppz$\cdot^-$ character [89]. This latter MLCT state at 2.3 eV is the lowest energy excited state of Ru(phen)$_2$(dppz)$^{2+}$ and has unusual photophysical properties. In water its emission lifetime is 250 ps, and in D$_2$O this lifetime increases to 550 ps [92]. Thus proton transfer from solvent appears to provide a major route for excited state relaxation. Additionally, the Λ and Δ forms of Ru(phen)$_2$(dppz)$^{2+}$ have strikingly longer emission lifetimes and higher quantum yields when intercalated into DNA duplexes than when free in aqueous solution [93].

Each Ru(phen)$_2$(dppz)$^{2+}$ enatiomer in DNA has two lifetimes with the amplitude of the long lifetime components increasing as the metal/DNA base pair ratio (R) increases. Thus the long components appear to arise from sequences of closely spaced metal complexes and the short ones from relatively isolated complexes. Linear dichroism (LD) studies show that the orientation of the average transition moment remains constant for each enatiomer up to $R = 0.4$. Both enantiomers have similar binding constants of $K \approx 10^8$ M^{-1}, limiting binding ratios of $R = 0.5$, and binding geometries with the molecular plane of the intercalated dppz ligand parallel to the plane of the DNA bases and close to perpendicular to the DNA helix. The roll angle between the normal to the plane of dppz and the helix axis is also the same within experimental error ($\pm 5°$) for the two enantiomers. Significant differences in their LD spectra, however, do exist and are a consequence of the spectroscopic diastereoisomerism arising from the different orientations of their phenanthroline ligands relative to the DNA helix [85]. Lifetimes in CT-DNA for the Δ-enantiomer vary from 135 to 56 ns for the short component and from 733 to 516 ns for the long component as R increases from 0.04 to 2.0, while lifetimes for the Λ-enantiomer vary from 35 to 67 ns for the short component and from 192 to 391 ns for the long one over the same R range. Thus for racemic mixtures of Ru(phen)$_2$(dppz)$^{2+}$ in DNA, most of the emission comes from the Δ-enatiomer.

The electron acceptor in the Baron and Turro experiment is Rh(phi)$_2$-(phen$'$)$^{3+}$. Related tris(bpy) and tris(phen) Rh(III) complexes have strong ligand centered (LC) π,π^* absorbance bands in the UV, but are nearly colorless in the visible [94]. Both weak metal-centered d,d* emission at 588 nm (2.1 eV) and $^3\pi,\pi^*$ emission at 455 nm (2.7 eV) have been reported for Rh(phen)$_3$$^{3+}$ at room temperature [95]. Population of low-energy d,d* states is frequently responsible for photoinduced ligand dissociation in coordination complexes. Thus photodetachment of the tether joining DNA and Rh(phi)$_2$(phen$'$)$^{3+}$ must be guarded against in this experiment. Rh(phi)$_2$(phen$'$)$^{3+}$ absorbs strongly at 360 nm, but also significantly in the visible tailing off ca. 550 nm, ε(532 nm) = 1230 M^{-1} cm^{-1} [92]. This visible absorbance band is characteristic of the phi chromophore [96].

Thus visible light can excite the electron acceptor as well as the electron donor of this Ru/Rh-DNA experiment. Three excited states have been directly observed in Rh(phi)$_2$(phen')$^{3+}$, a $^3\pi,\pi$(phi) state at 3.0 eV, a $^3\pi,\pi*$(phen) state at 2.8 eV, and an intraligand charge-transfer state (ILCT) at 2.0 (± 0.2 eV). The latter state is assigned to ET from one of the phi ligand's nitrogens to its phenanthrene ring [96]. Emission from the $^3\pi,\pi*$ states is observable at 77 K but not at room temperature. The ILCT state does not emit but has a characteristic absorbance band at 460 nm and decays at room temperature in water with a lifetime of 175 ns. In addition a metal-centered $d,d*$ state similar to that seen in the homoleptic Rh(III)-complexes is probably close in energy to the ILCT state.

The free energy ($\Delta G°$) for ET from Ru(phen)$_2$(dppz)$^{2+*}$ to Rh(phi)$_2$-(phen')$^{3+}$ based on Eq. (3) is -0.75 eV. On the other hand the free energy for ET from Ru(phen)$_2$(dppz)$^{2+}$ to Rh(phi)$_2$(phen')$^{3+*}$ is only ≈ -0.45 eV, because the Rh(III)-complex's ILCT excited state has an energy of only 2.0 eV versus 2.3 eV for the MLCT state of the Ru(II)-complex.

The DNA sequences used in this ET study were 5'-Ru(phen')$_2$(dppz)$^{2+}$-(L)-AGTGCCAAGCTTGCA-3' and its complement 5'-Rh(phi)$_2$(phen')$^{3+}$-(L)-TGCAAGCTTGGCACT-3'. (See Fig. 13 for a molecular model of this doubly labeled duplex.) Indirect arguments based on DNA strand-cleavage results and molecular modeling of the Ru/Rh-duplex concluded that closest edge-to-edge separation between the intercalated ligands of the two metal complexes, dppz and phi, was 37.4 Å, corresponding to 11 intervening base pairs. If ET is viewed as occurring exclusively between the intercalated portions of the metal complexes, this is the distance of ET in this duplex. A more general view is that the relevant ET distance is the overall edge-to-edge separation of the D/A complexes. From this perspective the minimum ET distance between the Ru- and Rh-complexes in this duplex is 26 Å. The remarkable result of this experiment was that no emission was detected when this Ru/Rh-labeled duplex was excited in the visible (420–480 nm). Additionally, attempts to measure the donor's emission lifetime showed only that its emission could not be seen using a kinetics system with 300 ps time resolution. A control experiment in which the Ru(II)-donor was bound to the same 15-mer duplex but without the Rh(III)-acceptor, showed normal Ru(phen')$_2$-(dppz)$^{2+}$ emission. This strikingly rapid and extensive emission quenching in the Ru/Rh-labeled duplex was interpreted as evidence of extremely fast ET from the intercalated Ru(II)-donor to the intercalated Rh(III)-acceptor ($k_{ET} \geqslant 3 \times 10^9$ s^{-1}).

Although Barton and Turro did not provide direct proof that ET was the mechanism of Ru(II)* emission quenching in this tethered Ru/Rh-DNA system, recent work involving bimolecular ET quenching [97] leaves little doubt that closely related racemic-Ru(DMP)$_2$(dppz)$^{2+}$, where DMP = 4,7-dimethylphen-anthroline, and Δ-Rh(phi)$_2$(bpy)$^{3+}$ intercalate into duplex DNA, exhibit static Ru(II)* emission quenching ($\tau < 10$ ns), and form a Ru(III) photoinduced ET

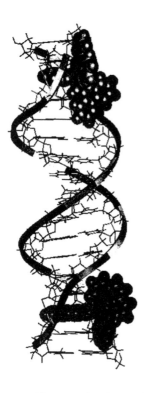

FIGURE 13 A molecular model of a 15-mer duplex comprised of 5'-Ru(phen')$_2$-(dppz)$^{2+}$-(L)-AGTGCCAAGCTTGCA-3' and its complement 5'-Rh(phi)$_2$(phen')$^{3+}$-(L)-TGCAAGCTTGGCACT-3' with the Ru(II)-donor shown on the bottom and the Rh(III)-acceptor shown at the top. The two 16-atom tethers that join a phen-ligand in each metal-complex to the 5'-end of a DNA strand are not shown. The intercalated ligands of the donor and acceptor are separated by 11 base pairs, 37.4 Å, and the overall closest edge-to-edge distance between the ruthenium and rhodium complexes is 26 Å. (Figure provided by Drs. David Beratan and Satyam Priyadarshy and reprinted with permission from *J. Phys. Chem.* 1996, *100*, 17678. Copyright 1996 Amer. Chem. Soc.)

product. However, transient absorbance data in this experiment were not recorded beyond 460 nm, so the ΔA increase at 540 nm that is characteristic of a Rh(II) ET product was not observed [97].

It is tempting to use static Ru(II)*-emission quenching in bimolecular studies involving donors and acceptors intercalated into DNA to learn about the distance dependence of the ET quenching reaction. However, these studies are always open to two interpretations. One interpretation is that the donors and acceptors intercalate independently of each other (i.e., randomly), and the other is that their binding is cooperative (i.e., once one metal complex is intercalated it is more likely for the next one to bind closer rather than farther away from the first complex). The above bimolecular study is a good example of this. For a Ru/Rh ratio of 1:2, 40% of the Ru(II)*-emission is quenched in less than 10 ns. Yet assuming random binding, only 25% of the D/A separations are within 20 Å. Another bimolecular study used $Ru(phen)_2(dppz)^{2+}$, $Rh(phi)_2(phen)^{3+}$, and 28-mer duplex DNA [98]. It reports that with 10 μM Ru, 40 μM Rh, and 500 μM DNA base pairs more than 80% of the Ru(II)*-emission was quenched on the nanosecond (and very likely picosecond) time scale. Yet, the average distance between D/A metal centers is calculated to be $\geqslant 35$ Å.

Recently nearly the same bimolecular system, consisting of $Ru(phen)_2$-$(dppz)^{2+}$ and $Rh(phi)_2(bpy)^{3+}$, was reinvestigated by Tuite, Lincoln, and Norden with the important difference that emission quenching for a wider range of Ru/DNA base pair ratios was examined each vs. the concentration of Rh(III) [86,99]. This latter study reports similar emission-quenching vs. Rh-concentration behavior as the former one [98] for similar Ru/DNA base pair ratios (ca. 0.02). However, as the Ru/DNA base pair ratio is increased, the slope of emission quenching as a function of Rh concentration decreases substantially. Thus for low Ru/DNA base pair ratios, the data support the conclusion that the effective ET transfer distance is as large as 70 Å, *if the donor and acceptors complexes bind randomly*. However at larger Ru/DNA base pair ratios, Rh-acceptors quench Ru(II)*-complexes only over very short distances. All the data spanning a wide range of Ru/DNA base pair ratios can be accounted for by assuming that the Rh-complexes preferentially bind next to Ru-complexes. An extended McGhee–von Hippel [100,101] cooperative-binding model demonstrates that a Rh-complex is 55 times more likely to bind adjacent to a Ru-site than away from it, and spectral perturbations due to adjacent Rh/Ru-binding are also seen in CD spectra for this system [99]. Thus Lincoln, Tuite, and Norden conclude that their biomolecular ET quenching study provides "… strong spectroscopic evidence that Δ-$Ru(phen)_2(dppz)^{2+}$ and Δ-$Rh(phi)_2(bpy)^{3+}$ bind cooperatively to a $[polydA-dT]_2$ duplex and that the remarkable quenching properties of Δ-$Rh(phi)_2(bpy)^{3+}$ can be quantitatively described by a cooperative binding model where ET occurs only from donors bound adjacent to acceptors" [99]. A fuller discussion of these experiments can be found in this volume in Chapter 2 written by Tuite [88]. [See Fig. 14 for structural models of two frequently used Rh(III) electron acceptors, Δ-$Rh(phi)_2(bpy)^{3+}$ and Δ-$Rh(phi)_2(phen)^{3+}$.]

Before leaving the subject of bimolecular studies of DNA-mediated ET, it maybe helpful to note that there are also two more reported studies of photo-induced ET between Δ-$Ru(phen)_2(dppz)^{2+}$ and Δ-$Rh(phi)_2(bpy)^{3+}$ bound to DNA [102,103]. Most of the experiments in these two studies used calf thymus DNA. The first study by Barton, Barbara et al. presented the experimental data and

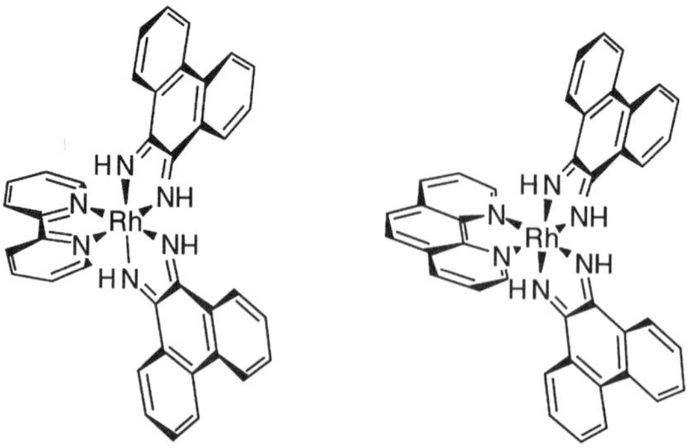

Δ-Rh(phi)₂(bpy)³⁺   Δ-Rh(phi)₂(phen)³⁺

FIGURE 14 Molecular structures of the intercalating electron acceptors Δ-Rh(phi)$_2$-(bpy)$^{3+}$ and Δ-Rh(phi)$_2$(phen)$^{3+}$.

concluded that, "The cooperative binding of donor and acceptor is considered unlikely on the basis of structural models and DNA photocleavage studies of binding. These data show that the DNA double helix differs significantly from proteins as a bridge for electron transfer" [102]. The second study by Olson, Hu, Hormann, and Barabara presented a quantative model of the experimental results of the first study [103]. In this work the intercalation of D/A molecules was simulated by Monte Carlo methods and the DNA duplex was modeled as a one-dimensional lattice of intercalation sites for the Ru/Rh-complexes. Several conclusions were reached. (1) The kinetics results could not be accounted for using a protein-like distance decay parameter, β, of 1.0 Å^{-1} unless the metal complexes bound cooperatively to DNA. (2) Long-range ET models using β values < 0.7 Å^{-1} agreed poorly with all of the picosecond and nanosecond kinetics data and with the steady-state emission results. (3) In contrast, an ET model using a protein-like $\beta = 1.0$ ($\pm$ 0.3) Å^{-1} and cooperative D/A binding, donor binding next to acceptors 13-fold more likely than at other sites, agreed well with the experiments. For this model calculations of both the fraction of fast charge-recombination reactions and the shape of the charge-recombination kinetics agreed with experiment. In contrast, simulations using either a long-range ET model or a noncooperative D/A binding model failed to agree with experiments. Thus this latter work by Barbara et al. [103] came to essentially the same conclusion as did the study by Lincoln, Tuite, and Norden [99] for the same Ru/Rh bimolecular ET system.

In light of the above bimolecular studies of static and dynamic Ru(II)*-emission quenching by donors and acceptors intercalated in DNA, it is even more important to focus on the results of the tethered Ru/Rh–DNA duplex described above. The basic result is that Ru(II)*-emission quenching is interpreted as signaling that the photoinduced rate of ET from Ru(II) to Rh(III) is $\geq 3 \times 10^9$ s^{-1} over a minimum edge-to-edge D/A distance of 26 Å with a driving force ($-\Delta G°$) equal to 0.75 eV [64,104]. This result can be compared to that of Mead and Kayyem for a tethered Ru/Ru–DNA duplex. In this latter study the driving force is similar (0.7 eV), but the D/A distance is much shorter, only 12.5 Å. Despite the appreciably smaller D/A separation, the observed Ru/Ru ET rate is more than 1900-fold slower, 1.6×10^6 s^{-1}. This is especially striking when two facts are recalled. First, Eq. (5) implies that H_{AB} decreases exponentially and not linearly with D/A separation, while Eq. (1) shows that the rate of ET depends on the square of H_{AB}. Thus a 26-Å D/A separation should have an enormously reduced ET rate compared to one with an 11-Å separation, all other things being equal. The second fact to be recalled is that k_{ET} for the Meade and Kayyem Ru/Ru–DNA duplex is very similar to those observed in efficient protein systems with similar driving forces and D/A separations.

The ET rate for the tethered Ru/Rh–DNA duplex can also be compared to rates found by Brun and Harriman for intercalated donors and acceptors. The driving force for the AOH$^+$/DAP^{2+} system is similar, 0.67 eV, and a comparable ET rate, 4.3×10^9 s^{-1}, is found but at a D/A separation distance of 10.2 Å. For these two experiments to make sense, H_{AB} would have to remain constant as the D/A separation in a duplex increases from 10.2 to 26 Å. This violates the expectation of Eq. (5) based on electron tunneling through a one-dimensional square barrier that H_{AB} should decay exponentially with distance. It was earlier pointed out that in the Brun and Harriman experiments β for intercalated donors and acceptors in a DNA duplex had a value of 0.9 Å^{-1}. According to this value of β, k_{ET} in the tethered Ru/Rh–DNA duplex (with 15.8 Å greater D/A separation) should be 1.5×10^6 fold smaller than is inferred from the rate of Ru(II)*-emission quenching.

The semiclassical ET model [Eq. (4)] permits another way of thinking about k_{ET} in the Ru/Rh–DNA duplex if one can assign a maximum value for k_{ET} when the donor and acceptor are in contact (k_{ET}^{max}). It is also helpful but not necessary to assume that this is a nearly activationless process in which $-\Delta G° \approx \lambda$ for all D/A separations. Once k_{ET}^{max} at $r°$ is known, a value of β for the implied decay of H_{AB} with D/A separation distance can be calculated using Eq. (5). Barton and Turro suggest k_{ET}^{max} should be ca. 10^{13} s^{-1} [64]. This value may be as low as ca. 9×10^{12} s^{-1} if the important solvent vibration is on the order of 30 cm^{-1}. Using $k_{ET}^{max} = 10^{13}$ s^{-1} and $k_{ET} \geq 3 \times 10^9$ s^{-1} at 26 Å, yields a β value of ≤ 0.31 Å^{-1}. Clearly this value of β disagrees fundamentally with the one measured by Brun and Harriman. Also, it is dramatically smaller than ones discussed above as typical for β in proteins, generally 0.9–1.4 Å^{-1} [5,53,82–84,105].

More experiments need to be done on the tethered Ru/Rh–DNA system in order to understand it [85]. Some experiments that would be useful include both transient absorbance and emission kinetics for a number of D/A separation distances and intervening bases sequences. It would also be of interest to examine ET in duplexes with mismatched intervening bases. These same comments also apply to the tethered Ru/Ru–DNA system. The fact that the former experiment involves intercalated donors and acceptors and the latter does not remains an important difference. However, the tethered Ru/Rh–DNA system is not as simple as it appears at first sighting. For example, both the donor and the acceptor have Λ and Δ enantiomeric forms, thus the experiments should be done with selected stereoisomers and not with racemates. It is also true that visible wavelengths excite both the donor and the acceptor. Thus there are competing reaction paths, one having a driving force $(-\Delta G°)$ of 0.75 eV and the other having a driving force of 0.45 eV. Changing from $Ru(phen)_2dppz^{2+}$ to $Os(phen)^2dppz^{2+}$ as the electron donor may permit selective excitation of the donor in the presence of a Rh(III)–phi acceptor [104]. In experiments such as the one described above employing racemic donors and acceptors, eight separate photoinduced ET processes can occur at one time. That different enantiomers cannot be ignored is underscored by a recent bimolecular study involving $Ru(phen)_2(dppz)^{2+}$, $Rh(phi)_2(bpy)^{3+}$, and CT–DNA. In this work the amount of Ru(II)*-emission quenching for the same concentrations of metals and DNA varied from 20 to 75% for different enantiomeric D/A combinations [92].

D. Semiempirical Hartree–Fock Calculations of H_{AB}

Given that the experimental results have significantly different implications for the ability of a DNA duplex to facilitate long-range ET, it is not reasonable to expect a theoretical model to resolve their differences. However, theoretical calculations of electronic coupling pathways have been successful in interpreting a large number of ET experiments in proteins. Recently Beratan et al. have reported electronic coupling results for DNA-mediated ET based on a self-consistent field Hartree–Fock calculation at the semiempirical level of quantum theory [48,49]. These are very large-scale calculations including all valence electrons (as many as 3300) in the DNA-duplex systems described above. Ab initio computations on this scale are not possible at present, but the semiempirical method based on complete neglect of differential overlap with a spectroscopic parameterization for core electron integrals (CNDO/S) has been used successfully by others in small molecule and protein fragment studies of long-range ET. The molecular orbital energies and wavefunction coefficients needed to evaluate H_{AB} in a two-level (reactants and products) approximation are produced by the CNDO/S calculation. Importantly, these computations give reliable ionization energies and electron affinities for individual DNA bases compared to both experimental and ab initio results. Because the energies of the bridge virtual orbitals, E_N (oxidized and reduced states), appear in the H_{AB} calculation as a denominator, $1/(E_N - E_{tun})$, it is critical that they be properly estimated. The CNDO/S method does this better than the above described extended Hückel method [81].

A few other comments about these very large-scale CNDO/S semiempirical calculations are worth making. Since all valence electrons and molecular orbitals of the donor, acceptor, and DNA bridge are explicitly treated, the final H_{AB} result does not depend on a tunneling pathway approximation, nor does it depend on how the D/A separation distance is measured. In fact the computation includes contributions from all possible ET paths that might couple the donor and the acceptor. Thus these calculations are ideal tools for asking questions concerning the relative importance of base stacking vs. backbone bonding for facilitating ET and whether or not DNA contact with nonintercalated ligands on a metal complex make important contributions to H_{AB}.

Good agreement between the CNDO/S semiempirical H_{AB} calculation and the experimental k_{ET} for the Ru/Ru–DNA duplex is found. Of course, this comparison requires use of Eq. (4) and a specified value of λ (0.9 eV) in addition to the measured driving force of 0.7 eV. Combining these data yields a calculated $k_{ET} = 7.1 \times 10^6$ s^{-1} compared to the experimental $k_{ET} = 1.6 \times 10^6$ s^{-1}. Extensive use of the same ruthenium complexes as D/A groups in protein studies means that there is not much uncertainty in λ (ca. ±0.2 eV).

With this reassuring result in mind, the semiempirical calculations can be used to explore additional aspects of DNA-mediated long-range ET in this 8-mer duplex with D/A attachments sites on opposite strands. One is that for C2′-ribose distances > 17 Å (corresponding to the 5′-amino-group attachment sites of the Ru-complexes), the average electronic coupling decay β is 1.2 Å^{-1}. This value is in good agreement with the β-range noted above for protein ET studies, 0.9–1.4 Å^{-1}. If H_{AB} is calculated for the same Ru/Ru-labeled duplex without the presence of the bases (a phosphate–ribose backbone only duplex), two observations are made: (1) The magnitude of H_{AB} is nearly two orders of magnitude smaller, and (2) the coupling decays more rapidly with increasing D/A separation than when the bases are included. Thus, the bases dominate electronic coupling over the backbone for opposite strand donors and acceptors, and the bases themselves are not exceptionally strong ET mediators. In other words, from the perspective of the CNDO/S semiempirical calculation, DNA is no more a molecular wire than is a protein. The electronic coupling is larger for donors and acceptors substituted on the same strand than on the opposite strands of the above 8-mer duplex. For example, at 20-Å separation between the two C2′-ribose carbons, the same-strand coupling is *ca.* 150-fold greater than is the across-strand coupling. Additionally, the same-strand coupling is reduced 7000-fold when the bases are excluded from the calculation (backbone only duplex) vs. when they are included in it [48]. One final result is that in order to get the full value of H_{AB}, calculations must include interactions between D/A orbitals and DNA bases 4 to 5 base pairs away from the tethering C2'-ribose position. Thus all portions of the metal complexes contribute to the electronic coupling.

Comparison of the semiempirical calculations with the results of the Brun and Harriman experiment involving EB^+ and DAP^{2+} is neither clearcut nor completely satisfying. One difficulty is that the sequence of the intervening CT–DNA bases is unknown. However, calculations of H_{AB} for different base sequences suggest that there are not dramatic differences among them. With this in mind we note that the β value based on oligo(dT)·oligo(dA) duplexes with EB^+ and DAP^{2+} intercalated at 3, 4, and 5 base pair separations is 1.6 $Å^{-1}$. This is not good agreement with the experimentally determined value of 0.9 $Å^{-1}$. However, uncertainty about the D/A separations in the experimental system limits the usefulness of detailed comments on this discrepancy. If λ = 0.9 eV is used along with the experimentally determined free energy of reaction in Eq. (4), the experimental and calculated electron transfer rates at 17-Å D/A separation are in good agreement. However, due to their different electronic coupling decay constants (β), the agreement at smaller separations is not good. Additionally, Brun and Harriman suggest that λ for their intercalated donors and acceptors is as small as 0.2 eV [106]. Thus there is also considerable uncertainty concerning the size of the nuclear reorganization energy for this system.

Comparison of H_{AB} from semiempirical calculations with k_{ET} found in the Ru/Rh–DNA duplex experiment depends in this case only to a minor extent on the choice of λ (see below). Barton and Turro et al. suggest 0.4 eV should be used [64]. Using this value of λ, the experimental free energy (−0.75 eV), and the calculated H_{AB} value in Eq. (4) yields $k_{ET}^{max} = 2.6 \times 10^2$ s^{-1} vs. an experimental rate of $\geq 3 \times 10^9$ s^{-1}. This discrepancy of 10^7 in ET rate demonstrates that important aspects of this D/A–DNA system are not understood. In particular the separation of the intercalated Ru- and Rh-complexes needs to be directly verified. In short, today's highest level of quantum theory and the ET rate results from two closely related experiments vastly disagree with the ET result from the Ru/Rh–DNA experiment. As found in CNDO/S calculations on the Ru/Ru-system, calculations on the Ru/Rh–duplex also show that interactions of nonintercalated metal-ligands with DNA bases 3 to 4 base pairs away from the intercalated ligands make significant contributions to H_{AB}. This result too disagrees with the idea that intercalated redox chromophores uniquely couple to the DNA bases and achieve "wire-like" ET efficiencies.

Why may ET along the helix of a DNA duplex be not much better than that found in efficient proteins? Part of the answer may be that the above experiments have large energy gaps between the oxidized and reduced bridge states and the tunneling energies of their D/A pairs. This same situation occurs in ET studies of proteins. Another part of the answer may be that the nearly end-to-end (sigma) interaction of the p-orbitals of the DNA bases in van der Waals contact along the helix is only about one-third as large as the π-interaction between *p*-orbitals on neighboring atoms within the bases [49]. Thus the electronic coupling of p-wavefunctions between stacked DNA bases is not expected to be as large as that found for coupling through covalent bonds. Although there are 14 p-orbitals per base pair they do not all overlap both maximally and constructively with the p-orbitals of flanking base pairs. Thus on balance over a 15-Å distance, D/A electronic coupling along a full duplex with donors and acceptors located on opposite strands is calculated to be only as good as that along a DNA single strand lacking bases (ribose–phosphate backbone only) [49].

V. AN APPLICATION OF PHOTOINDUCED ET FOR RIBOZYME STRUCTURE AND FUNCTION DETERMINATION

While a number of intriguing questions remain open concerning DNA-mediated long-range ET reactions, new research is likely to answer them soon. In parallel with these studies, significant progress is being made in studying the dynamics and operation of the ribozyme (catalytic RNA) L-21 Sca I from *T. thermophila* based on modulation of ET quenching of a fluorescent pyrene probe [107–111]. Model studies of pyrene labeled nucleosides, oligonucleotides, and duplexes show that pyrene* emission can be rapidly quenched by ET to and from nucleic acid bases [69–73,112,113]. Additionally, these studies show that the relative quenching rate varies among nucleosides as dC $\approx$ dT > dG >> dA, and that in general single strands quench covalently attached pyrene* labels more effectively than do duplexes [114]. This latter observation is consistent with more restricted access by the pyrene* label to reactive bases in duplex structures than in freely moving single strands. In ribozyme studies fluorescence from pyrene labels attached to short oligomeric substrates (such as 5'-pyreneCCUCU-3') increases as much as 20-fold upon binding to the ribozyme [111].

Fluorescent probes whose lifetimes and quantum yields are sensitive to their microenvironment are extremely useful for studies of RNA dynamics because they can be used at low concentrations, permit detection of molecular motion in the absence of ribozyme chemical reaction, and allow collection of numerous time points after ribozyme and substrate mixing [107]. The fact that ET quenching between pyrene* and RNA bases occurs at various rates for different bases and at different D/A distances makes substrate-linked pyrene* emission exquisitely sensitive to the structure of the host ribozyme. To date these studies have found evidence of both cooperative and anticooperative substrate binding [110], that docking of substrate is not diffusion controlled but is driven by a favorable change in entropy [109], and that cofactor guanosine 5'-monophosphate (pG) facilitates docking of the substrate's 5'-cleavage site into a high free-energy binding configuration that results in substrate destabilization [108]. Ongoing studies show that the 2'-OH group of pG is crucial for correctly locating the substrate in the ribozyme's catalytic site and are directed at producing a model structure for the first step of RNA splicing [107].

VI. CHEMICAL EVIDENCE OF LONG-RANGE CHARGE MIGRATION IN DNA DUPLEXES

A number of interesting DNA-assisted charge migration experiments have recently been carried out based on radiolabeling of DNA-strand ends and gel electrophoretic separation of various lengths of photocleaved strands. These experiments are appealing because it is not hard to ^{32}P-label the 5'-ends of DNA strands, and photographic detection of the radioacive decay of ^{32}P labels is extremely sensitive. Since the photoinduced base modifications that lead to cleavage are irreversible the yield of cleaved strands can be increased arbitrarily by extending the irradiation time. As a result ^{32}P-labeled experiments that detect DNA strand cleavage frequently employ irradiation times of an hour or more and can detect photochemical events that occur with quantum yields as low as 10^{-7}. This ultrahigh sensitivity to net photochemical reaction has both good and bad aspects. The obvious good aspect is that even low-yield photoinduced ET processes can be detected. A frequently overlooked bad aspect is that the observation of a very-low-yield reaction product does not give insight into the mechanism by which the reaction occurred. If the quantum yield for DNA strand cleavage is very small, it is possible that some very low probability dynamical event could be responsible for the observed ET-induced photochemistry. To the extent the quantum yield of the observed strand cleavage (or other irreversible photochemical reaction such as thymine dimer scission) is large, the less likely is a rare dynamical fluctuation of the DNA–D/A system to be the cause of the primary ET event [115]. Two other bad aspects of these experiments are: (1) no kinetics measurements, whose results could be compared to theoretical calculations, are made and (2) no reaction intermediates, such as primary ET products which would allow the mechanism of the reaction to be determined, are observed. Keeping the above caveats in mind, it is worth describing briefly some recent "net photochemistry" experiments on DNA-assisted long-range charge migration.

A. Long-Range Photooxidation of 5'-GG-3' Bases

Guanine is the DNA base that is easiest to oxidize followed by adenine [114,116–120]. In keeping with this, G sites in DNA are prime targets for oxidative attack [78]. The initial oxidative damage to G sites can be further developed by treatment with hot piperidine, which causes release of the damaged base and strand cleavage. However, some G sites are easier than others to oxidize depending upon their 3'-neighboring base [121,122]. Ab initio quantum mechanical calculations of short stacks of DNA bases in B-form geometry find that 5'-G sites increase in ease of oxidation in the series $GT \approx GC \ll GA < GG < GGG$ [123].

Barton et al. [124] tethered a Rh-complex, Δ-Rh(phi)$_2$(bpy')$^{3+}$ where bpy' is 4-butyric acid 4'-methyl 2,2'-bipyridine that was linked via the acid and a nonamethylene alkyl amine linker to a terminal nucleoside's 5'-O, to a DNA strand to serve as an intercalated photooxidant. The ^{32}P label (*) was placed on the 5'-end of a complementary unlabeled strand with the following sequence *5'-ACG^{3}GCATG^{8}GCT^{11}T^{12}C^{13}GT-3'. The metal complex was assayed by direct photocleavage experiments using 313-nm excitation as being intercalated via a phi ligand mostly between the T^{12} and C^{13} bases. Lower-energy excitation of the metal complex at 365 nm produced long-range oxidative damage principally at three sites that increased in cleavage yield in the series T^{11} < G^8 < G^3. However, some cleavage was seen at all sites from N^3 to N^{12}. The quantum yield for overall guanine damage for one hour of irradiation was 5×10^{-7}. In this work the major oxidation product found following enzymatic digestion of an irradiated DNA strand was the nucleoside 8-oxo-7,8-dihydro-2'-deoxyguanosine (8-oxo-dG). Consistent with this finding was the observation that no base damage or strand cleavage occurred when the irradiation was performed under an argon atmosphere. Apparently in the dominant cleavage mechanism, dioxygen somehow trapped a guanine cation and converted it into the nucleotide 8-oxo-G, which was later reacted with piperidine to produce strand cleavage [122].

The ET mechanism that produces 8-oxo-G chemistry at G^3 and G^8 sites is unclear. To the extent that the duplex is intact and the Rh-complex is intercalated when this chemistry occurs, this work demonstrates long-range oxidative chemistry at sites distant from the location of the primary photooxidant. Assuming that this is long-range photochemistry, it is still a question whether it occurs via long-range ET in which the metal complex is reduced in concert with guanine oxidation or whether it occurs as result of sequential base oxidations (hole hopping). However the oxidations of distant guanines occur, it is surprising that the most distant guanine of two GG sequences, G^3, is damaged more than is the proximal one, G^8. Clearly, many chemical steps intervene between the initial photoexcitation event in this experiment and the observation via gel electrophoresis of piperidine-induced strand cleavage.

Additional experiments using the Δ-Rh(phi)$_2$(bpy')$^{3+}$ tethered intercalator have recently been reported by Hall and Barton [80]. The ^{32}P-labeled complementary family of strands in this study had the following base sequence: *5'-ACGG^{dist}CACXTACGG^{prox}CTCGT-3', where X = null (no bulge), A, AA, AAA (A$_{1-3}$), T, TT, TTT (T$_{1-3}$), or ATA. The base sequence of the Rh-labeled strand was always kept complementary to the above sequence with X = null. Thus when X was equal to A$_{1-3}$ or T$_{1-3}$ a bulge was present in the duplex resulting in a globally bent structure. The goal of this work was to examine the relative yields of GG oxidative damage at sites distal (GG^{dist}) and proximal (GG^{prox}) to the intercalated Rh-complex. Photoexcitation of the Rh-complex at 313 nm again located the major region of intercalation as between the third and fourth bases from the 3'-end of the ^{32}P-labeled strand. Note that an NMR study of the structure of a duplex with an ATA bulge showed that (1) the bases in the bulge maintained continuous stacking, (2) the helix bent 50–60° around the bulge, and (3) the bases that flanked the bulge on the complementary strand sheared apart [125].

Table 3 presents Hall and Barton's data [80] and shows that the yield of cleavage at the distal GG pair relative to the proximal GG pair appears to decrease monotonically as the bulge increases from one T to three T's. However, within error all of the T-bulges (T_{1-3}) show the same decrease in distal to proximal cleavage yield relative to the no bulge case. Similarly, the decrease in relative distal GG cleavage is the same for bulges comprised of one to three A-bulges (A_{1-3}). Finally, the largest reduction in relative distal GG cleavage was found for the ATA bulge. However, within error all three of the three base bulges (T_3, ATA, and A_3) exhibited the same two-thirds loss of distal relative to proximal GG cleavage yield. Thus while the single T-bulge has the most distal to proximal GG cleavage and the ATA bulge the least, all of the other five bulges have the same amount of distal to proximal GG cleavage. Clearly, introducing a one to three base-bulge decreases the distal to proximal GG cleavage yield ratio; however, this cleavage ratio (within the above error limits) depends only weakly on the number and kind of bulge bases used. Since the overall guanine cleavage quantum yield is only 5×10^{-7} and a number of chemical reactions intervene between photoexcitation and separation of cleavage products, it is difficult learn about the photo-

TABLE 3   Distal to Proximal GG Cleavage Ratio as a Function of Bulge Composition[a]

Bulge Composition (X)	Distal/Proximal GG Cleavage Ratio
Null (no bulge)	1.2 (± .1)
T	0.8 (± .2)
TT	0.7 (± .2)
TTT	0.5 (± .2)
ATA	0.3 (± .1)
A	0.5 (± .1)
AA	0.5 (± .1)
AAA	0.4 (± .1)

[a]T = 2'-deoxyadenine (dT) and A = 2'-deoxythymine (dA).
Source: Data from ref. 80.

induced ET processes that give rise to the observed cleavage products. However, it is consistent with naive expectations that disrupting the duplex structure by introducing a bulge would decrease the distal to proximal GG cleavage ratio if either duplex-mediated long-range ET transfer or hole (base oxidation) migration were an important step leading to strand cleavage.

To improve the quantum yield of GG oxidation in duplexes with tethered and intercalated photooxidants, Arkin, Stemp, Pulver, and Barton [79] recently reported results based on experiments using a Ru(II)-intercalator, Δ-Ru(phen)-(bpy')(Me$_2$dppz)$^{2+}$ where Me$_2$dppz = 9,10-dimethyldipyridophenazine. These were flash-quench experiments as discussed above in which the Ru(II) excited state was quenched with either Ru(NH$_3$)$_6^{3+}$ or methylviologen (MV^{2+}). The resulting Ru(III) complex was then able to oxidize G or GG sites in the duplex in which it was intercalated. A significant result was that the quantum yield of GG strand breaks induced by piperidine treatment was 1×10^{-5} and 2×10^{-4}, respectively, for the Ru(NH$_3$)$_6^{3+}$ and MV^{2+} quenchers. When these yields were normalized to account for the fact that only quenched Ru(II)* which produced Ru(III) could serve as an oxidant, the quantum yields of GG strand breaks per Ru(III) were, respectively, 1.3×10^{-5} and 2×10^{-3}. This latter quantum yield is 4000-fold larger than that found in experiments using the Rh(III)-tethered intercalator.

The goal of this work was to learn about hole migration in DNA duplexes [79]. To do this two kinds of DNA sequences were examined. In the first the non-ruthenated strand had the following composition: *5'-TGATCGNTGCGTCT-GAGACT-3', where N = G or C. The ruthenated strand always had a complementary base sequence. The location of the Ru(II)-complex was assayed by Ru(II)* production of diffusible $^1O_2$ which reacted primarily with the G nearest the 3'-end of the ^{32}P-labeled strand. Thus the Ru-complex was intercalated most likely between the fourth and third or the third and second bases at the 3'-end of the ^{32}P-labeled strand. The cleavage pattern for the flash-quench experiments with N = G showed that 50% of all G cleavage occured at the 5'G of the GG site with all other G sites damaged equally. When N = C, all G sites were cleaved equally, and the total yield of cleavage chemistry was comparable to that found when N = G. These results suggested that somehow Ru(III) was able to oxidize all G sites along a duplex roughly equally if they had similar redox potentials and a given site preferentially if it was easier to oxidize than all others. One possible mechanism is that the site of base oxidation migrates or hops along the duplex from site to site faster than it is trapped at a given base (usually guanine) by irreversible chemistry (most likely reaction with O$_2$ and H$_2$O).

The second kind of DNA duplex examined had the following base sequence for the nonruthenated strand: *5'-ACGACGGTGACG^{12}CTGAGACT-3' [79]. In these experiments G^{12} on the ^{32}P-labeled strand was either matched with C (G/C) or mismatched with A (G/A) or T (G/T) on the complementary ruthenated strand.

In the case of G/C base-pair matching, the same results as described above for a single GG pair on the ^{32}P-labeled strand were obtained: 50% of all G site cleavage was found on the 5'-G of the GG pair. Interestingly, for the G/A and G/T base-pair mismatches, enhanced cleavage at the mismatched G^{12} site was found in addition to the usual large amount of cleavage at the 5'-G of the GG pair more remote from the Ru-complex. In addition, the cleavage yield for the G/A mismatch was much larger (3–4×) than for the G/T mismatch with G/A mismatch cleavage 50–60% as great as the dominant 5'-G cleavage on GG site of the ^{32}P-labeled strand. These latter results show that base-oxidation migration and trapping at G sites is sensitive G base-pairing. Whether this effect is due to a change in the oxidation potential of G, accessibility to the trapping reactant, or some other factor is not clear. However, it is still possible that this cleavage sensitivity to G base-pairing may be useful for mismatch detection in DNA duplexes [79].

The above studies of DNA cleavage in duplexes compile a significant amount of evidence to support the following conclusions: (1) Base damage and piperidine-induced strand cleavage can occur at sites as far away as ca. 11 base pairs from an initially created photooxidant; (2) this long-range damage is mediated by a duplex; (3) the amount of base damage is equally distributed along a duplex with a number of equally oxidizable damage sites; (4) the amount of damage is varied by intervening bulges; (5) the amount of damage increases with the ease of oxidation of a G site, and (6) G-site damage is sensitive to the nature of the pairing base on the opposite strand. Because these experiments do not provide direct observations of primary ET products or kinetics information, they do not tell us about the distance dependence of concerted long-range ET in DNA duplexes or the mechanism of base-damage migration (although base photooxidation has been directly observed) [78,126], including the times for base-oxidation (or hole) hops and the mechanism of trapping [78,122] oxidative base damage at the cleavage sites.

B. 5'-GG-3' and 8-oxo-G Long-Range Photooxidation in PNA–DNA Duplexes

Armitage and Schuster et al. [74] recently reported the first experiments involving a covalently attached compound that is intercalated into an internal position in a Peptide Nucleic Acid (PNA)–DNA duplex. PNA oligomers are analogs of DNA and RNA oligomers in which the naturally occuring sugar phosphate backbone that links the bases is replaced by a peptide backbone [127–132]. (See Fig. 15 for a structural drawing of a PNA oligomer and of an anthraquinone subunit.) The work by Armitage and Schuster et al. [74] provides the first example in which a photoactivated PNA strand is used to recognize and cleave a DNA target strand. In this work two kinds of experiments were performed. In the first experiment the correspondence of preferential 5'-GG-3' photooxidative damage between DNA/

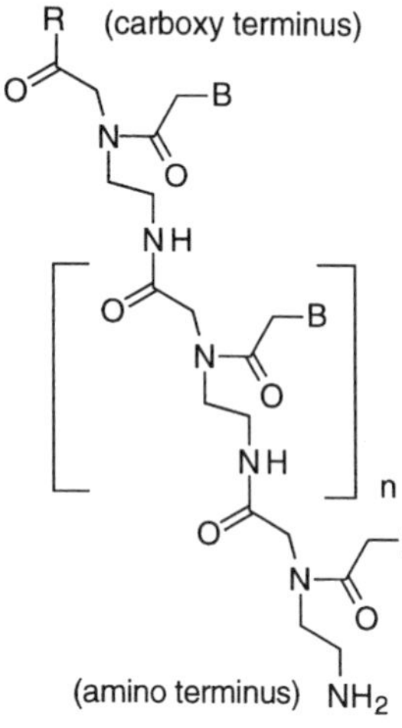

FIGURE 15 Molecular structures of a PNA oligomeric (n) strand showing the carboxy and amino termini and of an anthraquinone (AQ) subunit which can be inserted into a PNA strand in place of a nucleic acid base (B).

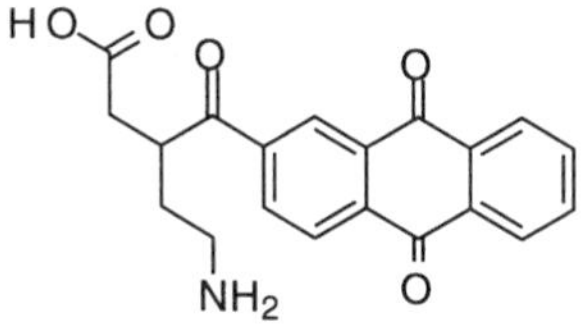

DNA and PNA/DNA duplexes was examined. The ^{32}P-labeled DNA target strand was *5'-TCGCTGG^{C}AAZAAGG^{B}TAGG^{A}A-3' where Z = an abasic site and three different 5'-GG-3' sites were present (GG$^{A–C}$). The PNA strand's base sequence was complementary to the DNA target strand's sequence with the exception that anthraquinone (AQ) was bound to the PNA strand in place of a normal DNA base opposite the abasic site (Z) in the DNA target strand.

Irradiation of the above PNA–DNA duplex showed preferential strand cleavage at the 5'G of the GG^A and GG^B sites, which was much the same result as occurred in DNA/DNA duplexes under similar conditions. Somewhat in contrast to this result, DNA strand cleavage occured equally at each G of the GG^C site. The 5'-G of GG^A was 6 bases away from the abasic site (ca. 20.4 Å), where AQ was intercalated, yet as much damage occured there as at the 5'-G of GG^B and each G of GG^C. Both of these latter two sites were only two bases away from the AQ site; thus distance from the site of photoexcitation was not an imporatant factor in determining strand cleavage.

Even though guanosines are easier to oxidize that adenosines, photoexcited AQ in its lowest triplet state (^{3}AQ) has about 2.76 eV of energy [74] and is capable of oxidizing either nucleoside [$\Delta G°$(ET) = -0.67 eV to form AQ$^{\cdot -}$/dG($-$H)$^{\cdot}$ and -0.54 eV to form AQ$^{\cdot -}$/dA($-$H)$^{\cdot}$ at pH 7] [133]. In this situation an intially produced deprotonated dA($-$H)$^{\cdot}$ neutral radical could back-react with AQ$^{\cdot -}$ or further react by oxidizing a neighboring adenosine or guanosine nucleoside. If sequential base-oxidation migration (a one-dimensional random walk) is the mechanism for damage migration in PNA–DNA or DNA/DNA systems, placing a very easy to oxidize trap in the duplex at the 5′-G location of the **GG**^{B} site should have a marked influence on the DNA cleavage pattern. In particular, damage at the more distant **GG**^{A} site should be greatly attentuated, while damage at the **GG**^{C} site should be unchanged. Fortunately the 8-oxo-dG nucleoside is 0.4–0.5 V easier to oxidize than dG [134] and can be substituted for dG in a duplex with little deformation expected since the modifications to dG are remote from its base-pairing sites.

With these thoughts in mind, Armitage and Schuster et al. [74] performed a second kind of experiment in which they substituted the 8-oxo-G nucleotide for the 5-G′ of the **GG**^{B} site. Not surprisingly, an enormous amount of cleavage was found at the 8-oxo-G site. However, cleavage at the more distant **GG**^{A} site was now almost entirely eliminated, while cleavage at the **GG**^{C} site was unchanged. The clear directional effect (i.e., damage migration in both 5′- and 3′-directions) and the supertrapping of damage at a proximal GG site at the expense of damage trapping a distal site relative to the site of AQ intercalation, strongly support a discrete base-oxidation hopping mechanism for damage migration in PNA/DNA duplexes. Since PNA–DNA duplexes have significantly different base stacking geometries than DNA–DNA duplexes [135], it is not yet known what the results of a similar experiment in a DNA–DNA duplex would be.

C. Long-Range Photooxidative Thymine Dimer Scission

Ultraviolet light can damage DNA by initiating a [2 + 2] photocycloaddition between adjacent thymine dimers on the same strand [136]. The product is a cyclobutyl thymine dimer (T^T). The cyclobutyl ring can be cleaved and the dimerization damage repaired by either a one electron reduction or oxidation [137]. (See Fig. 16 for an illustration of bimolecular oxidative cleavage of a thymine dimer model compound.) Danliker, Holmlin, and Barton [77] tethered the Rh-intercalator described above, Δ-Rh(phi)$_2$(bpy′)$^{3+}$, to the 5′-ends of several DNA duplexes and used it as a long-range photooxidant to repair thymine dimer damage. The decrease in the concentration of DNA strands containing T^T and the corresponding increase in those with repaired thymine bases (TT) was followed by HPLC C$_{18}$ chromatography over a 6-hour time period. In the best case all of the T^T damage was repaired, and the quantum yield for the reaction was 2×10^{-6}

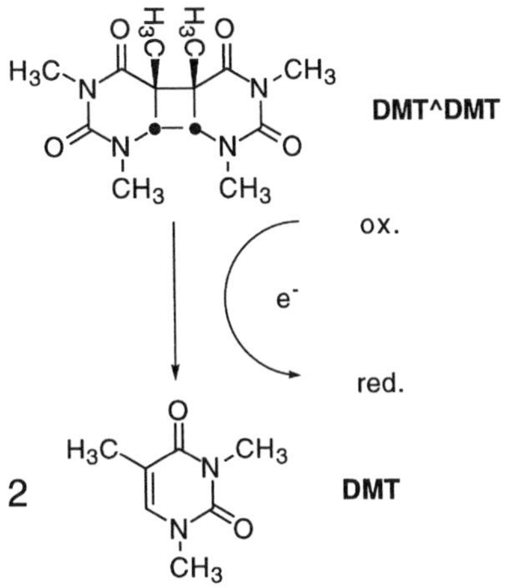

FIGURE 16 Molecular structures of a thymine dimer (T^T) model compound and of its oxidative cleavage products. Here bimolecular oxidation of the *cis-syn* 1,3-dimethyl-thymine dimer (DMT^DMT) yields two 1,3-dimethylthymine monomers (DMT).

using 400-nm light. Interestingly the quantum yield for same reaction with a stoichiometric amount of noncovalently bound $Rh(phi)_2(DMB)^{3+}$, where DMB = 4,4'-dimethyl 2,2'-bipyridine, was 1×10^{-4}. Apparently the detailed stacking arrangements for both T^T and the Rh-intercalator, which could easily have differed in these two experiments, were important in determining the photochemical yields. This appeared to be the case because increasing the number of base pairs separating the Rh-complex and T^T from 4 to 6 increased rather than decreased the quantum yield for dimer repair for the Rh-tethered duplexes (see below). Thus proximity of the Rh-complex and T^T in the tethered duplexes was not an advantage.

Direct photocleavage of the Rh-tethered duplexes using 313-nm light as discussed above showed that the most probable site of intercalation of the Rh-complex was two bases from the tethered end, but significant cleavage also was found at the fifth base. Thus the site of the photooxidizer (●) was best described as 3 (±) bases from the end of the tethered strand. Table 4 lists the sequences of the T^T-containing strands in the three fully complementary duplexes studied and their corresponding percent dimer repair after six hours of irradiation. The data in this table show there is a small increase in the precent of dimers repaired as the number of bases separating the Rh-complex and T^T is increased. This may be due to increasing overlap of the DNA bases in this region as the length of the duplex increases [77]. The surprising result is that the repair efficiency did not decrease as the number of bases separating the electron donor and acceptor

TABLE 4   Percent Thymine Dimer Repair in DNA Duplexes With a Tethered Rhodium Complex at the 5′-End of the Complementary Strand

T^T Strand Base Sequence[1]	D/A Separation Distance[2] ($\pm$ 3.4 Å)	% Dimer Repair[3] ($\pm$ 5%)
5′-ACGTGAGT^TGAGA•CGT-3′	13.6	79
5′-ACGTGATGT^TGTAGA•CGT-3′	17.0	91
5′-ACGTGCATGT^TGTACGA•CGT-3′	20.4	100

[1]The most probable sites of Rh-intercalation are indicated with a "•", and the base sequences for the strands with the Rh-tethers at their 5′-ends are complementary to the ones shown.
[2]The separation distance is measured from the edge of T^T to the edge of the intercalated bpy ligand on the Rh-complex.
[3]After 6 hours of irradiation with 400-nm light.
Source: Data from ref. 77.

increased. This result is similar to those discussed above for DNA/DNA and PNA/DNA duplexes where oxidative damage as multiple G or GG sites in a duplex was roughly equally distributed and the distance between the damage site and the photooxidant did not determine the amount of damage at a given site.

To test the hypothesis that base stacking in duplexes was important in determining the amount of T^T repair, Danliker, Holmlin, and Barton [77] created single-base bulges in the region separating T^T and the Rh-complex. To do this they held the base sequence of the bottom strand in Table 4 constant and first buldged the C's flanking T^T by pairing it with a Rh-labeled strand that was complementary to all of the other bases but had no bases opposite the C's. Second they buldged both the T's and C's flanking T^T in a similar manner. The single C buldge between T^T and the Rh-complex decreased the dimer repair to 80% compared to 100% for the unbuldged case. The double buldge (both T and C) further decreased the dimer repair to 47%. These results are similar to those discussed above for GG oxidative damage by the same tethered Rh(III)-photooxidant. There introducing bulges between the electron donor (a distal 5′-GG-3′) and the acceptor (Rh^{3+*}) also decreased oxidation of the donor.

D. Migration and Trapping of Photoinjected Electrons in DNA Duplexes

The above studies of charge migration in DNA duplexes focused on the translocation of oxidative damage (possibly involving a hole hopping mechanism). Complementing these studies Shafirovich and Geacintov et al. [75] recently designed an experiment to learn about the migration of an excess electron in DNA. To do this they reacted highly polymerized CT–DNA with the racemic *trans-anti*-benzo-[*a*]pyrene derivative 7r,8t-dihydroxy-t9,10-epoxy-7,8,9,10-tetrahydrobenzo[*a*]-

pyrene (BPDE). Benzo[a]pyrene epoxides react at their C10 positions predominantly with the exocyclic amino group (N^2) on dG residues and to a lesser degree with the exocyclic amino group (N^6) on dA residues. (See Fig. 5 for structures of two of the four possible *anti*-BP-dG-isomers.) Thus the exact pyrene substitution pattern on the CT–DNA duplex was not known and random sequences of bases intervened between substitution sites. The resulting ratio of bound *anti*-benzo[a]-pyrene (BP) molecules per DNA base pair (1:60) was determined spectroscopically using characteristic BP and DNA absorbance bands at 348 and 258 nm, respectively.

Control studies with degassed solutions showed that at low excitation levels the mean fluorescence lifetime of the pyrenyl residues was ca. 3 ns. This short mean emission lifetime was due to ET quenching of pyenyl emission by nearby DNA bases as discussed above. Thus added MV^{2+} at concentrations < 0.4 mM could quench pyrenyl emission only if the MV^{2+} was bound to the DNA duplex, most likely at its surface. (See Fig. 17 for a structural model of MV^{2+}.) Interestingly at low excitation levels no ET products were observed at times > 10 ns because the back reactions of products arising from pyrenyl emission quenching, such as reduced $MV^{·+}$, oxidized or reduced DNA bases, and oxidized or reduced pyrenyl residues, were very fast. However, at higher excitation levels (≤ 10 mJ/cm^2 per laser pulse) there was sufficient photon flux to ionize the pyrenyl residues directly via two photons and produce pyrene$^{·+}$ and hydrated electrons. In the presence of added MV^{2+} at < 0.4 mM concentration, there was an immediate (< 7 ns) jump in absorbance due to $MV^{·+}$ followed by a much slower $MV^{·+}$ absorbance increase. The latter slow process coincided with the disappearance of hydrated electrons and was not of much significance. Importantly, the immediate

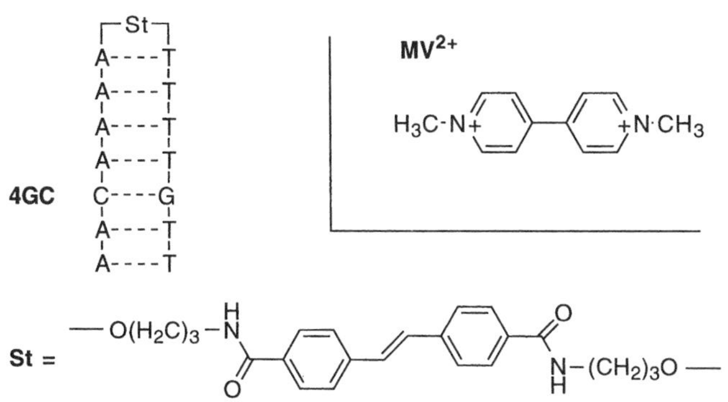

FIGURE 17   Molecular structures of the electron acceptor methylviologen (MV^{2+}) and of a DNA hairpin duplex (4GC) containing a 4,4'-stilbenedicarboxamide bridge (St).

jump in $MV^{\cdot+}$ formation occured only for duplex DNA and not for single strands. Consistent with this the immediate jump in $MV^{\cdot+}$ formation also did not occur when the DNA-bound MV^{2+} molecules were displaced from the duplex by addition of Mg^{2+} ions.

A reasonable interpretation of the prompt ($\leqslant 30$ ns) reduction of MV^{2+} in BP-subsituted duplexes was that excess electrons, arising from photoionization of the pyrenyl residues, migrated from their site of production to MV^{2+} trapping sites bound to DNA. No such migration was seen for BPDE-subsituted single strands with added MV^{2+}. To learn about the distance of excess electron migration in duplexes, Shafirovich and Geacintov et al. [75] studied the yield of prompt $MV^{\cdot+}$ products at 30 ns, $[MV^{\cdot+}]/([MV^{\cdot+}] + [\text{hydrated electron}]$, as a function of the concentration of added MV^{2+} ($> 95\%$ bound to DNA in the absence of Mg^{2+} under the conditions used). A plot of the yield of prompt $MV^{\cdot+}$ vs. the average number of DNA base pairs separating MV^{2+} molecules, $N_{bp} = [DNA_{bp}]/[MV^{2+}]$, was well described by the function $\exp(-N_{bp}/N_{\beta})$, where N_{β} is a base-pair yield decay parameter. For CT–DNA labeled with an average of one pyrenyl residue per 60 base pairs, N_{β} was equal to 24 base pairs. Assuming that the surface-bound MV^{2+} electron traps were randomly distributed and efficient, this result meant that the average distance between a photoexcited pyrenyl residue and an MV^{2+} trap was 12 base pairs or ca. 40 Å.

Although 24-ps duration 355-nm excitation pulses were used in this work, the digital oscilloscope and R928 photomultiplier detection system had only 7-ns resolution. Further the $MV^{\cdot+}$ yields were measured at 30 ns. Thus any $MV^{\cdot+}/$ pyrene$^{\cdot+}$ products that recombined in less than ca. 3 ns were not seen. This had the advantage that ground-state complexes, if present, between pyrenyl residues and MV^{2+} molecules were not observed, since their ET products likely recombined in less than 20 ps [138,139]. Similarly, this experiment favored observation of longer lived ET products and underrepresented shorter lived ones. This may have skewed the observed yield distribution as function of added MV^{2+} somewhat, but it does not affect the striking conclusion that a photoinjected excess electron in a CT– DNA duplex can travel several tens of angstroms to reach a surface-bound cationic trap. How much of this migration is due to concerted long-range ET and how much to base hopping is not known. Additionally, the critical pathways for either process are also unknown (e.g., through base stacks, through the DNA back bone, or through waters of hydration). It will be very interesting to learn of the results of this novel experiment in covalently modified DNA duplexes with site specifically located pyrenyl and MV^{2+} groups.

VII. TWO NEW LONG-RANGE ET EXPERIMENTS

The above described recent experiments examining the yields of net oxidative damage, T^T repair, and $MV^{\cdot+}$ formation show that with quantum yields in the 10^{-4} to 10^{-7} range both holes (oxidized DNA sites) and excess electrons can migrate in DNA duplexes away from their initial sites of generation to more distant trapping sites. These experiments raise numerous questions concerning the detailed mechanisms of these hole and electron migrations including the extent to which site hopping is involved, the times of such hops, and the mechanisms of final trapping. However, none of them provide any insight into the problem of the mechanism of long-range DNA-mediated ET. To do this two kinds of experiments must be performed. First, the same pair of electron donors and acceptors must be covalently attached to a series of DNA duplexes that differ only in the number of bases that separate the D/A groups. Second, direct transient absorbance measurements must be made of the formation and decay of the ET products. So far none of the above discussed experiments have met these two conditions. In fact, only the photoinduced ET experiments in DNA hairpins to be discussed last in this chapter have met both of these conditions.

A. A Tethered EB$^{+\prime}$/Rh(phi)$_2$(bpy$'$)$^{3+}$ System with Intercalating D/A Groups

Recently Kelley, Holmlin, Stemp, and Barton [67] attempted to meet these conditions by synthesizing a series of duplexes containing a tethered ethidium (EB$^{+\prime}$) intercalating donor with the familiar Rh(phi)$_2$(bpy$'$)$^{3+}$ intercalating acceptor. In this system the ethidium donor was photoexcited rather than the Rh-acceptor yielding a free-energy for ET of -0.87 eV to form the EB$^{2+\prime}$ and Rh(II) photoproducts. Indeed in an associated bimolecular experiment involving non-tethered EB$^+$ and Rh(phi)$_2$(bpy)$^{3+}$ complexes intercalated into CT–DNA, a long-lived absorbance transient with a $\lambda_{max} = 425$ nm was seen. This absorption feature was consistent with the expectation that EB^{2+} should be formed, but corresponded to less than 10% of the quenched donors. Disappointingly, no transient absorbance kinetics measurements have yet been performed on this D/A system with tethered D/A groups.

To date both time-resolved emission and emission quantum yield measurements have been made on four duplexes with tethered EB$^{+\prime}$ and Rh-complexes. The complexity of this seemingly straightforward experiment became immediately apparent. While EB$^+$ exhibited a monoexponential emission decay lifetime of $\tau = 23$ ns when intercalated into CT–DNA, the excited state emission of tethered EB$^{+\prime}$ showed triexponential decay kinetics, $\tau = 1.2$ ns (13%), $\tau = 2.4$ ns (59%), and $\tau = 7.6$ ns (28%). Worse yet, whether this triple-lifetime emission decay corresponded to three discrete EB$^{+\prime}$-binding configurations or a distribution of a large number of binding configurations was not known. When a 10 base-pair duplex, which was covalently labeled with EB$^{+\prime}$ on one 5$'$-end and with Λ-Rh(phi)$_2$(bpy$'$)$^{3+}$ on the other 5$'$-end, was photoexcited, four emission decay processes were observed. $\leq 28\%$ of the emission was quenched in < 150 ps, and the remaining ca. 70% of the emission decayed again with three lifetimes, $\tau = 0.9$ ns (17%), $\tau = 2.3$ ns (65%), and $\tau = 7.8$ ns (18%). Unfortunately, what physical or chemical excited-state relaxations these four emission decay processes represented was not known.

Interestingly, as the length of these end-tethered duplexes was increased from 10 to 14 base pairs, the fraction of $EB^{+\prime}$ emission that was quenched by the Λ-Rh(phi)$_2$(bpy$^\prime$)$^{3+}$ acceptor decreased correspondingly: 29%, 21%, 16%, 11%, and 8%. A similar decrease in emission quenching with increasing duplex length was also found for the Δ-Rh(phi)$_2$(bpy$^\prime$)$^{3+}$ acceptor [67]. These are intriguing results, but this D/A–DNA system is so complex and so incompletely characterized at this time that it is pointless to speculate on their meaning. An obvious and very interesting question is what is the arrangement of D/A configurations that gives rise to the prompt ($\leqslant$ 150 ps) emission quenching in these duplexes? This work by Kelley, Holmlin, Stemp, and Barton [67] represents a major investment of research effort and serves as vivid lesson that constructing sufficient quantities of a series of covalently labeled D/A-DNA duplexes that meet stringent characterization criteria is major chemical and intellectual challenge.

Barton et. al [67] view the above experimental results quite differently. In particular they ignore the complicated emission decay kinetics data and focus instead on the static emission quenching component ($\tau <$ 150 ps). The static quenching decreases from ca. 30% of the ethidium fluorescence for duplexes with 10 base pairs to ca. 10% for duplexes with 14 base pairs. If this $<$ 150-ps quenching were due to ET from $EB^{+\prime}$ to Rh(III) and each of the these groups were rigidly intercalated, it would correspond to an ET rate in excess of 6×10^9 s^{-1} over distances as large as 20–35 Å. This decrease of static quenching with increasing D/A separation is ascribed to an increase in the probability of finding imperfectly stacked base pairs as the number of bridging base pairs is increased. In this model, long-range DNA-mediated ET is highly sensitive to the stacking geometries of the intervening bases and for well-stacked bases can occur at rates in excess of 6×10^9 s^{-1} for D/A separations greater than 30 Å. For comparison it is interesting to note that if the rate of ET were 10^{13} s^{-1} for a D/A pair in contact and β were 1.0 Å^{-1}, at 30 Å separation the ET rate would be 0.9 s^{-1}, and if β were 0.64 Å^{-1} (see below), the ET rate at 30 Å would be 5×10^4 s^{-1}. If the Barton model [67] that ET in DNA has an extremely shallow electronic coupling decay with D/A separation ($\beta < 0.1$ Å^{-1}) and a very high sensitivity to base stacking geometry proves correct, DNA will indeed be a qualitatively different medium for long range ET than are proteins ($\beta =$ 0.9–1.6 Å^{-1}). At present there is insufficient data to know whether or not this is the case. The set of experiments discussed next provides a glimpse of what new results in this area may offer.

B. G/Stilbene ET in Hairpin Duplexes

The second new long-range ET experiment was recently reported by Lewis, Letsinger, and Wasielewski et al. [68]. In this work stilbene was reduced by guanosine nucleotides. The energy of the first excited singlet state of stilbene was 3.4 eV, and its reduction potential was -1.9 V vs. the standard calomel electrode (SCE). Previous investigators had observed that stilbene was quenched strongly by dG and only weakly by dT, whereas the other nucleosides had no quenching effect [118]. Interestingly in fluid solution the first excited state of stilbene lived only 40 ps due to torsion about the central C=C bond [68]. To study long-range DNA-mediated ET, a 4,4'-stilbenedicarboxamide was used as a bridge to connect complementary DNA strands of six or seven bases each. At temperatures below 75°C, hairpin duplexes formed for the complementary strands used in this work. [See Fig. 17 for molecular structures of the stilbene bridge (St) and a seven base-pair hairpin (4GC).] Very interestingly for a T_6-St-A_6 hairpin, restricted motion about the central C=C bond of the St-bridge lengthened the lifetime of the first excited state of stilbene 50-fold to 2.0 ± 0.4 ns.

In this work five hairpins were studied each having zero to four (nGC) A/T base pairs separating the St-bridge and a single G/C base pair: 0GC (G/C base pair adjacent to the bridge) to 4GC. Consistent with ET quenching of the first excited state of stilbene by G and with Eq. (7) (which incorporates electron tunneling through a one-dimensional square barrier as a measure of electronic coupling decay with D/A separation distance), the lifetime of stilbene's first excited state increased as the G/C base pair was moved farther away from the bridge: 1 ps (0GC), 5 ps (1GC), 290 ps (2GC), 1.0 ns (3GC), and 1.4 ns (4GC). Meaningful excited state quenching measurements could not be made beyond four A/T base pairs separating G/C and the bridge. Importantly in this work, both emission kinetics and transient absorbance measurements were made. The above excited state lifetimes are from the latter measurements (300-fs time resolution), and the emission lifetimes for hairpins 2GC to 4GC agreed reasonably well with those measured by absorbance kinetics also lengthening as the distance betwen G/C and the bridge increased. Emission lifetimes for hairpins 0GC and 1GC were too short for emission kinetics measurements (ca. 0.3-ns time resolution).

The transient absorbance spectra of the first excited singlet state of stilbene and of the stilbene radical anion had similar absorbance maxima at 575 nm, but were identified by their different shapes as determined from measurements of the spectra of model compounds. Biexponential decays were found for 0GC and 1GC with the shorter component assigned to the decay of the stilbene excited state and the longer component to the decay of the radical anion product. However, the absorbance decays of 2GC-4GC were more complicated than simply biexponential. In these cases the emission decays were used to assign which absorbance lifetime component corresponded to ET quenching of the stilbene excited state. The ET quenching rates in the nGC hairpins were determined by subtracting the excited state decay rate in the T_6-St-A_6 hairpin (1/2.0 ns) from the observed excited state decay rate for each nGC hairpin (1/observed lifetime). A plot of the logarithm of these ET quenching rates vs. D/A separation distance as determined from molecular models (ca. 3.4 Å per interventing A/T base pair) was reasonably well fit by straight line with a slope of β = 0.64 ± 0.1 Å^{-1}.

This value of the distance decay parameter β for D/A electronic coupling in a DNA duplex is both significantly larger than the ≤ 0.31 Å^{-1} value discussed above for the tethered Ru–Rh duplex and significantly smaller than that found in the EB^{+}–DAP^{2+} and AOH^{+}–DAP^{2+} intercalated duplexes (0.9 Å^{-1}) as well as those generally found for long-range ET in proteins (0.9–1.4 Å^{-1}) [5,53,82–84,105]. Thus this work on ET in DNA hairpins suggests that a DNA duplex can mediate long-range ET somewhat better than proteins but not nearly as well as one would expect for a molecular wire. Interestingly semiempirical calculations of electronic couplings in (dA)$_6$(dT)$_6$ duplex bridges separating D/A groups found that the couplings along either strand decayed monotonically and approximately exponentially as D/A distance increased [β = +2.1 and +1.5 Å^{-1}, respectively, along the (dA)$_6$ and (dT)$_6$ strands]. However, these calculations also suggested that the couplings were very sensitive to the geometry of the DNA duplex. For example, at 20-Å D/A separation same-strand coupling was 150-fold stronger than across-strand coupling [48]. In the hairpin duplexes the stilbene electron acceptor was bound to both strands while the donor was a DNA base. Neither of these conditions was incorporated into previously published calculations. It will be informative to see the results of new calculations on these hairpin duplexes.

A final comment on oxidation potentials of dG and dA nucleosides is in order in light of the numerous studies discussed above that used G or GG nucleotides as electron donors. Two recent papers attempted to measure these values. In the first paper Seidel, Schultz, and Sauer [118] used electrochemical and excited state quenching techniques to correlate aqueous and nonaqueous results. They determined that the deprotonated neutral radicals of the nucleosides dG-(−H)$^{\cdot}$ and dA(−H)$^{\cdot}$ were reduced in water at pH 7 with the addition of a proton at 0.92 and 1.26 V, respectively (vs. SCE). Thus 340 mV separated the oxidation potentials of dG and dA. In the second paper Steenken and Javonovic [116] used pulse radiolysis and dynamic equilibrium measurements between nucleosides and reference couples to determine the nucleoside oxidation potentials. They determined that dG(−H)$^{\cdot}$ and dA(−H)$^{\cdot}$ were reduced in water at pH7 with the addition of a proton at 1.29 and 1.42 V, respectively (vs. SCE). Thus only 130 mV separated the oxidation potentials of dG and dA. These results bear on the hairpin duplex experiments in that if the oxidation potential of dA is close to that of dG, it may not be possible to measure the reduction of stilbene by G in the absence of competitive reduction by A nucleotides. This could become an increasingly significant problem as D/A separations increase because the number of intervening As increases while the ET quenching rate by G drops. Additionally, other researchers have reported that dT nucleosides also quench the stilbene excited state, although only weakly [118]. The above values for the reduction potential and excited state energy of stilbene can be combined according to Eq. (3) with the oxidation potentials of dG and dA to yield estimates of the free energy of ET quenching of stilbene by these two nucleosides. The nucleoside oxidation values from the first paper yield $\Delta G°(ET) = -0.58$ and -0.24 eV, respectively, for dG and dA; while the corresponding values from the second paper yield -0.21 and -0.08 eV, respectively. Thus it seems that weak quenching by intervening A and T nucleotides may have to be considered in the hairpin experiments in conjunction with strong G quenching.

If weak quenching by A and T nucleotides is occuring in the hairpins, the measured 2.0-ns lifetime for the excited state of stilbene in the T_6-St-A_6 hairpin already reflects these processes but not exactly in the same way as in the G/C containing hairpins. The fact that the absorbance spectra of the stilbene excited state and the stilbene radical anion are very similar and that complex kinetics are seen in the 2GC–4GC hairpins might mean that it will be difficult to unravel all of these effects. However, at present is it reasonable to assume that G quenching is the dominant ET quenching path in these stilbene hairpins and that $\beta = 0.64 \pm 0.1$ Å^{-1} is a good description of the distance decay of electronic coupling in these hairpin experiments. While this value of β is the best measurement todate of the decay of electronic coupling in DNA duplexes with increasing D/A separation, its interpretation and use require care. Importantly, this value of β is for ET transfer from a guanosine nucleotide to the excited state of a doubly tethered stilbene acceptor. Whether β would be this low for DNA-mediated ET if DNA bases themselves were not directly involved as electron donors or acceptors is not known.

All of the other kinetics studies of ET in DNA duplexes, in common with all ET studies in proteins, have a large energy difference between the D/A electronic tunneling energy (E_{tun}) and the oxidized and reduced states of the bridge. These virtual states of the DNA bridge (with energy E_N) assist long-range ET [49]. Because H_{AB} is proportional to $1/(E_N - E_{tun})$, DNA-mediated ET experiments in which E_{tun} is very close in energy to the oxidized and reduced bridge states may show enhanced electron transfer rates. These hairpin experiments directly oxidize a guanine base whose oxidation potential is not very different from that of the adenine bridging bases. Thus long-range ET in these experiments is likely to be in the "small-tunneling-gap" regime. If so, the measured value of β may be lower for DNA-mediated ET in these hairpins than those values determined in proteins and in other ET experiments in DNA duplexes not because DNA is ipso facto a better ET mediator than proteins or because the other duplex measurements are wrong, but rather because in these experiments there is a significant decrease in the tunneling gap energy of the ET reaction being studied compared to other studies of long-range ET. Undoubtedly future experiments will refine our understanding of this very interesting topic.

VIII. ACKNOWLEDGMENTS

I thank Drs. David Beratan, Satyam Priyadarshy, and Carolyn Kanagy for providing the molecular modeling figures, donor/accepter distances, and for helpful discussions; Drs. Eimer Tuite, Bengt Norden, Douglas Turner, David Beratan, Bruce Armitage, Gary Schuster, Nicholas Geacintov, Vladimir Shafirovich, Jacqueline Barton, and Fredrick Lewis for sharing their unpublished results; Kirk Schanze for suggestions and corrections to this manuscript; and Mrs. Marla Netzel for invaluable library research assistance. This review was produced at Georgia State University with financial support from the U.S. National Science Foundation (CHE-970918).

REFERENCES

1. Buhks, E.; Jortner, J. *J. Phys. Chem.* 1980, *84*, 3370.
2. DeVault, D. *Q. Rev. Biophys.* 1980, *13*, 387.
3. Kutal, C. *J. Chem. Educ.* 1983, *60*, 882.
4. *Electron Transfer in Biology and the Solid State: Inorganic Compounds with Unusual Properties*; Johnson, M. K.; King, B. R.; Kurtz, D. M., Jr.; Kutal, C.; Norton, M. L.; Scott, R. A., eds.; Advances in Chemistry Series 226; American Chemical Society: Washington, DC, 1990.
5. Moser, C. C.; Keske, J. M.; Warncke, K.; Farid, R. S.; Dutton, P. L. *Nature* 1992, *355*, 796.
6. Fox, M. A. In *Photoinduced Electron Transfer III*; Mattay, J., ed.; Topics in Current Chemistry 159; Springer-Verlag: Berlin, 1991: p. 67.
7. Bowler, B. E.; Meade, T. J.; Mayo, S. L.; Richards, J. H.; Gray, H. B. *J. Am. Chem. Soc.* 1989, *111*, 8757.
8. *Photoinduced Electron Transfer III*; Mattay, J., ed.; Topics in Current Chemistry 159; Springer-Verlag: Berlin, 1991.
9. Bixon, M.; Fajer, J.; Feher, G.; Freed, J. H.; Gamliel, D.; Hoff, A. J.; Levanon, H.; Mobius, K.; Nechushtai, R.; Norris, J. R.; Scherz, A.; Sessler, J. L.; Stechlik, D. *Isr. J. Chem.* 1992, *32*, 369.
10. Gunner, M. R.; Dutton, P. L. *J. Am. Chem. Soc.* 1989, *111*, 3400.
11. Gunner, M. R.; Dutton, P. L. *NATO ASI Ser., Ser. A* 1988, *149*, 259.
12. Giangiacomo, K. M.; Dutton, P. L. *Proc. Natl. Acad. Sci. U.S.A.* 1989, *86*, 2658.
13. Dutton, P. L. *Electron Transfer Mechanisms in Reaction Centers: Final Progress Report*, DOE/ER/13476-1; U.S. Department of Energy: Washington, DC, 1989; DE90001204.
14. Dutton, P. L.; Alegria, G.; Gunner, M. R. *NATO ASI Ser., Ser. A* 1988, *149*, 185.
15. Whitten, D. G. *J. Chem. Educ.* 1983, *60*, 867.
16. Kirschdemesmaeker, A.; Lecomte, J. P.; Kelly, J. M. *Top. Curr. Chem.* 1996, *177*, 25.
17. Fox, M. A.; Jones, W. E. J.; Watkins, D. *Chem. Eng. News* 1993, *March 15*, 38.
18. Gust, D.; Moore, T. A.; Moore, A. L.; Leggett, L.; Lin, S.; DeGraziano, J. M.; Hermant, R. M.; Nicodem, D.; Craig, P.; Seely, G. R.; Nieman, R. A. *J. Phys. Chem.* 1993, *97*, 7926.

19. Moore, T. A.; Gust, D.; Moore, A. L. *NATO ASI Ser., Ser. C* 1992, *371*, 295.
20. Wasielewski, M. R.; O'Neil, M. P.; Gosztola, D.; Niemczyk, M. P.; Svec, W.A. *Pure Appl. Chem.* 1992, *64*, 1319.
21. Wasielewski, M. R. *Chem. Rev.* 1992, *92*, 435.
22. Paddon-Row, M. N.; Oliver, A. M.; Warman, J. M.; Smit, K. J.; De Haas, M. P.; Oevering, H.; Verhoeven, J. W. *J. Phys. Chem.* 1988, *92*, 6958.
23. Gust, D.; Moore, T. A.; Bensasson, R. V.; Mathis, P.; Land, E. J.; Chachaty, C.; Moore, A. L.; Liddle, P. A.; Nemeth, G. A. *J. Am. Chem. Soc.* 1985, *107*, 3631.
24. Gust, D.; Moore, T. A. In *Photoinduced Electron Transfer III*; Mattay, J., ed.; Topics in Current Chemistry 159; Springer-Verlag: Berlin, 1991; p. 257.
25. Candeias, L. P. *Radiation Chemistry of Purines in Aqueous Solution*, 1992; Inst. Super. Tec., Tech. Univ. Lisbon: Lisbon, Portugal, Order No. PB92-229418, Avail. NTIS From Gov. Rep. Announce. Index (U.S.) 1992, *92*(23), Abstr. No. 265,515.
26. Close, D.M. *Radiat. Res.* 1993, *135*, 1.
27. Sevilla, M. D. *Mechanisms for Radiation Damage in DNA*, Comprehensive Report, June 1, 1986-May 31, 1992; U.S. Department of Energy, Division of Energy Research: Washington, DC, 1991; DOE/ER/60455-6.
28. Michalik, V. *Int. J. Radiat. Biol.* 1992, *62*, 9.
29. Schulte-Frohlinde, D.; Von, S. C. *UCLA Symp. Mol. Cell. Biol., New Ser.* 1990, *136*, 31.
30. Schulte-Frohlinde, D. *Chem. Unserer Zeit* 1990, *24*, 37.
31. Bien, M.; Steffen, H.; Schulte-Frohlinde, D. *Mutat. Res.* 1988, *194*, 193.
32. Schulte-Frohlinde, D.; Bothe, E. *NATO ASI Ser., Ser. H* 1991, *54*, 317.
33. Miller, J. H. *NATO ASI Ser., Ser. H* 1992, *54*, 157.
34. Symons, M. C. R. *NATO ASI Ser., Ser. H* 1992, *54*, 111.
35. Sutin, N.; Creutz, C. *J. Chem. Educ.* 1983, *60*, 809.
36. Marcus, R. A.; Sutin, N. *Biochim. Biophys. Acta* 1985, *811*, 265.
37. Sutin, N. *Acc. Chem. Res.* 1982, *15*, 275.
38. Sutin, N. *Prog. Inorg. Chem.* 1983, *30*, 441.
39. Brunschwig, B.; Sutin, N. *Comm. Inorg. Chem.* 1987, *6*, 209.
40. Jortner, J. J. *J. Chem. Phys.* 1976, *64*, 4860.
41. Liang, N.; Miller, J. R.; Closs, G. L. *J. Am. Chem. Soc.* 1989, *111*, 8740.
42. Closs, G. L.; Calcaterra, N. J.; Green, N. J.; Penfield, K. W.; Miller, J. R. *J. Phys. Chem.* 1986, *90*, 3673.
43. Rehm, D.; Weller, A. *Isr. J. Chem.* 1970, *8*, 259.
44. Brunschwig, B. S.; Ehrenson, S.; Sutin, N. *J. Phys. Chem.* 1986, *90*, 3657.
45. Sutin, N.; Brunschwig, B. S.; Creutz, C.; Winkler, J. R. *Pure Appl. Chem.* 1988, *60*, 1817.
46. Geacintov, N. E.; Zhao, R.; Kuzmin, V. A.; Seog, K. K.; Pecora, L. J. *Photochem. Photobiol.* 1993, *58*, 185.
47. Sutin, N. In *Electron Transfer in Inorganic, Organic, and Biological Systems;* Bolton, J. R., Mataga, N. and McLendon, G., eds.; Advances in Chemistry Series 228; American Chemical Society: Washington, DC, 1991; p. 25.
48. Priyadarshy, S.; Beratan, D. N.; Risser, S. M. *Int. J. Quantum Chem.* 1996, *60*, 1789.
49. Priyadarshy, S.; Risser, S. M.; Beratan, D. N. *J. Phys. Chem.* 1996, *100*, 17678.

50. Wofram, S. *The Mathematica Book*; 3rd ed.; Wolfram Media/Cambridge University Press: Cambridge, 1996.
51. Closs, G. L.; Miller, J. R. *Science* 1988, *240*, 440.
52. Liang, N.; Miller, J. R.; Closs, G. L. *J. Am. Chem. Soc.* 1990, *112*, 5353.
53. Gray, H. B.; Winkler, J. R. *Pure Appl. Chem.* 1992, *64*, 1257.
54. Therien, M. J.; Chang, J.; Raphael, A. L.; Bowler, B. E.; Gray, H. B. *Struct. Bonding (Berlin)* 1991, *75*, 109.
55. Gray, H. B. *Aldrichim. Acta* 1990, *23*, 87.
56. Chang, I. J.; Gray, H. B.; Winkler, J. R. *J. Am. Chem. Soc.* 1991, *113*, 7056.
57. Therien, M. J.; Bowler, B. E.; Selman, M. A.; Gray, H.B.; Chang, I-J.; Winkler, J. R. In *Electron Transfer in Inorganic, Organic, and Biological Systems;* Bolton, J. R., Mataga, N. and McLendon, G., eds.; Advances in Chemistry Series 228; American Chemical Society: Washington, DC, 1991; p. 191.
58. Winkler, J. R.; Gray, H. B. *Chem. Rev.* 1992, *92*, 369.
59. Meade, T. J.; Gray, H. B.; Winkler, J. R. *J. Am. Chem. Soc.* 1989, *111*, 4353.
60. Therien, M. J.; Selman, M.; Gray, H. B.; Chang, I. J.; Winkler, J. R. *J. Am. Chem. Soc.* 1990, *112*, 2420.
61. Marcus, R. A. *J. Chem. Phys.* 1956, *24*, 966.
62. Brunschwig, B. S.; Ehernson, S.; Sutin, N. *J. Phys. Chem.* 1987, *91*, 4714.
63. Brun, A. M.; Harriman, A. *J. Am. Chem. Soc.* 1992, *114*, 3656.
64. Murphy, C. J.; Arkin, M.R.; Jenkins, Y.; Ghatlia, N. D.; Bossmann, S. H.; Turro, N. J.; Barton, J. K. *Science* 1993, *262*, 1025.
65. Meade, T. J.; Kayyem, J. F. *Angew. Chem., Int. Ed. Engl.* 1995, *34*, 352.
66. Jenkins, Y.; Barton, J. K. *J. Am. Chem. Soc.* 1992, *114*, 8736.
67. Kelley, S. O.; Holmlin, R. E.; Stemp, E. D. A.; Barton, J. K. *J. Am Chem. Soc.* 1997, *119*, 9861.
68. Lewis, F. D.; Wu, T.; Zhang, Y.; Letsinger, R. L.; Greenfield, S. R.; Wasielewski, M. R. *Science* 1997, *277*, 673.
69. Shafirovich, V. Y.; Levin, P. P.; Kuzmin, V. A.; Thorgeirsson, T. E.; Kliger, D. S.; Geacintov, N. E. *J. Am. Chem. Soc.* 1994, *116*, 63.
70. O'Connor, D.; Shafirovich, V. Y.; Geacintov, N. E. *J. Phys. Chem.* 1994, *98*, 9831.
71. Shafirovich, V. Y.; Courtney, S. H.; Ya, N. Q.; Geacintov, N. E. *J. Am. Chem. Soc.* 1995, *117*, 4920.
72. Netzel, T. L.; Nafisi, K.; Headrick, J.; Eaton, B. E. *J. Phys. Chem.* 1995, *99*, 17948.
73. Netzel, T. L; Zhao, M.; Nafisi, K.; Headrick, J.; Sigman, M. S.; Eaton, B. E. *J. Am. Chem. Soc.* 1995, *117*, 9119.
74. Armitage, B.; Koch, T.; Frydenlund, H.; Orum, H.; Batz, H. G.; Schuster, G. B. *Proc. Natl. Acad. Sci. U.S.A.* 1997, *94*, 12320.
75. Shafirovich, V. Y.; Dourandin, A.; Luneva, N. P.; Geacintov, N. E. *J. Phys. Chem.* 1997, *101 B*, 5863.
76. Hall, D. B.; Holmlin, R. E.; Barton, J. K. *Nature* 1996, *382*, 731.
77. Dandliker, P. J.; Holmlin, R. E.; Barton, J. K. *Science* 1997, *275*, 1465.
78. Stemp, E. D. A.; Arkin, M. R.; Barton, J. K. *J. Amer. Chem. Soc.* 1997, *119*, 2921.
79. Arkin, M. R.; Stemp. E. D. A.; Pulver, S. C.; Barton, J. K. *Chem. Bio.* 1997, *4*, 389.
80. Hall, D. B.; Barton, J. K. *J. Amer. Chem. Soc.* 1997, *119*, 5045.
81. Risser, S. M.; Beratan, D. N.; Meade, T. J. *J. Am. Chem. Soc.* 1993, *115*, 2508.

82. Casimiro, D. R.; Beratan, D. N.; Onuchic, J. N.; Winkler, J. R.; Gray, H. B. In *Mechanistic Bioinorganic Chemistry;* Thorp, H. H., and Pedoraro, V. L., ed.; Advances in Chemistry Series 246; American Chemical Society: Washington, DC, 1995; p. 471.
83. Casimiro, D. R.; Richards, J. H.; Winkler, J. R.; Gray, H. B. *J. Phys. Chem.* 1993, *97*, 13073.
84. Langen, R.; Chang, I.-J.; Germanas, J. P.; Richards, J. H.; Winkler, J. R.; Gray, H. B. *Science* 1995, *268*, 1733.
85. Norden, B.; Lincoln, P.; Akerman, B.; Tuite, E. *Met. Ions Biol. Syst.* 1996, *33*, 177.
86. Tuite, E.; Lincoln, P.; Norden, B. *Photochem. Photobiol.* 1996, *63*, 9S.
87. Tuite, E.; Lincoln, P.; Becker, H.-C.; Norden, B. *J. Am. Chem. Soc.* 1998, *120* (submitted).
88. Tuite, E. in *Organic and Inorganic Photochemistry*; Ramamurthy, V. and Schanze, K. S., eds.; Marcel Dekker, Inc.: New York, 1988; Molecular and Supramolecular Photochemistry Series 2, 55–74.
89. Hartshorn, R. M.; Barton, J. K. *J. Am. Chem. Soc.* 1992, *114*, 5919.
90. Chambron, J.-C.; Sauvage, J.-P.; Amouyal, E.; Koffi, P. *New. J. Chem.* 1985, *9*, 527.
91. Amouyal, E.; Homsl, A.; Chambron, J.-C.; Sauvage, J.-P. *J. Chem. Dalton Trans.* 1990, 1841.
92. Arkin, M. R.; Stemp, E. D. A.; Turro, C.; Turro, N. J.; Barton, J. K. *J. Am. Chem. Soc.* 1996, *118*, 2267.
93. Hiort, C.; Lincoln, P.; Norden, B. *J. Am. Chem. Soc.* 1993, *115*, 3448.
94. Ferraudi, G. *Elements of Inorganic Photochemistry*; Wiley: New York, 1988.
95. Indelli, M. T.; Carioli, A.; Scandola, F. *J. Phys. Chem.* 1984, *27*, 2685.
96. Turro, C.; Evenzahav, A.; Bossmann, S. H.; Barton, J. K.; Turro, N. J. *Inorg. Chim. Acta* 1996, *243*, 101.
97. Stemp, E. D. A.; Arkin, M. R.; Barton, J. K. *J. Am. Chem. Soc.* 1995, *117*, 2375.
98. Murphy, C. J.; Arkin, M. R.; Ghatlia, N. D.; Bossmann, S.; Turro, N. J.; Barton, J. K. *Proc. Natl. Acad. Sci. U.S.A.* 1994, *91*, 5315.
99. Lincoln, P.; Tuite, E.; Norden, B. *J. Am. Chem. Soc.* 1997, *119*, 1454.
100. McGhee, J. D.; von Hippel, P. H. *J. Mol. Biol.* 1976, *103*, 679.
101. McGhee, J. D.; von Hippel, P. H. *J. Mol. Biol.* 1974, *86*, 469.
102. Arkin, M. R.; Stemp, E. D. A.; Holmlin, R. E.; Barton, J. K.; Hormann, A.; Olson, E. J. C.; Barbara, P. F. *Science* 1996, *273*, 475.
103. Olson, E. J. C.; Hu, D. H.; Hormann, A.; Barbara, P. F. *J. Phys. Chem., B* 1997, *101*, 299.
104. Stemp, E. D.; Barton, J. K. *Met. Ions Biol. Syst.* 1996, *33*, 325.
105. Casimiro, D. R.; Wong, L.-L.; Colon, J. L.; Zewert, T. E.; Richards, J. H.; Chang, I.-J.; Winkler, J. R.; Gray, H. B. *J. Am. Chem. Soc.* 1993, *115*, 1485.
106. Brun, A. M.; Harriman, A. *J. Am. Chem. Soc.* 1994, *116*, 10383.
107. Turner, D. H.; Li, Y.; Fountain, M.; Profenno, L.; Bevilacqua, P. C. *Nuc. Acids Molec. Biol.* 1996, *10*, 19.
108. Bevilacqua, P. C.; Li, Y.; Turner, D. H. *Biochemistry* 1994, *33*, 11340.
109. Li, Y.; Bevilacqua, P. C.; Mathews, D.; Turner, D. H. *Biochemistry* 1995, *34*, 14394.
110. Bevilacqua, P. C.; Johnson, K. A.; Turner, D. H. *Proc. Natl. Acad. Sci. U.S.A.* 1993, *90*, 8357.

111. Bevilacqua, P. C.; Kierzek, R.; Johnson, K. A.; Turner, D. H. *Science* 1992, *258*, 1355.

112. Geacintov, N. E.; Mao, B.; France, L. L.; Zhao, R.; Chen, J.; Liu, T. M.; Ya, N. Q.; Margulis, L. A.; Sutherland, J. C. *Proc. SPIE-Int. Soc. Opt. Eng.* 1992, *1640*, 774.

113. Telser, J.; Cruickshank, K. A.; Morrison, L. E.; Netzel, T. L. *J. Am. Chem. Soc.* 1989, *111*, 6966.

114. Manoharan, M.; Tivel, K. L.; Zhao, M.; Nafisi, K.; Netzel, T. L. *J. Phys. Chem.* 1995, *99*, 17461.

115. Taubes, G. *Science* 1997, *275*, 1420.

116. Steenken, S.; Jovanovic, S. V. *J. Am. Chem. Soc.* 1997, *119*, 617.

117. Kittler, L.; Lober, G.; Gollmick, F.; Berg, H. *J. Electroanal. Chem.* 1980, *116*, 503.

118. Seidel, C. A. M.; Schulz, A.; Sauer, M. H. M. *J. Phys. Chem.* 1996, *100*, 5541.

119. Steenken, S.; Telo, J. P.; Novais, H. M.; Candeis, L. P. *J. Am. Chem. Soc.* 1992, *114*, 4701.

120. Faraggi, M.; Klapper, M. H. *J. Chim. Phys.* 1994, *91*, 1054.

121. Saito, I.; Takayama, M.; Sugiyama, H.; Nakatani, K. *J. Am. Chem. Soc.* 1995, *117*, 6406.

122. Ly, D.; Kan, Y. Z.; Armitage, B.; Schuster, G. B. *J. Am. Chem. Soc.* 1996, *118*, 8747.

123. Sugiyama, H.; Saito, I. *J. Am. Chem. Soc.* 1996, *118*, 7063.

124. Hall, D. B.; Holmlin, R. E.; Barton, J. K. *Nature* 1996, *382*, 731.

125. Rosen, M. A.; Shapiro, L.; Patel, D. J. *Biochemistry* 1992, *31*, 4015.

126. Breslin, D. T.; Schuster, G. B. *J. Am. Chem. Soc.* 1996, *118*, 2311.

127. Egholm, M.; Buchardt, O.; Nielsen, P. E.; Berg, R. H. *J. Am. Chem. Soc.* 1992, *114*, 1895.

128. Egholm, M.; Buchardt, O.; Christensen, L.; Behrens, C.; Freier, S. M.; Driver, D. A.; Berg, R. H.; Kim, S. K.; Norden, B.; Nielsen, P. E. *Nature* 1993, *365*, 566.

129. Hyrup, B.; Egholm, M.; Nielsen, P. E.; Wittung, P.; Norden, B.; Buchardt, O. *J. Am. Chem. Soc.* 1994, *116*, 7964.

130. Leijon, M.; Graslund, A.; Nielsen, P. E.; Buchardt, O.; Norden, Bl; Kristensen, S. M.; Eriksson, M. *Biochemistry* 1994, *33*, 9820.

131. Wittung, P.; Kim, S. K.; Buchardt, O.; Nielsen, P.; Norden, B. *Nuc. Acids Res.* 1994, *22*, 5371.

132. Wittung, P.; Nielsen, P. E.; Buchardt, O.; Egholm, M.; Norden, B. *Nature* 1994, *368*, 561.

133. The free energies of ET quenching in aqueous solution at pH 7 were estimated using Eq. (3) and the following parameters: $E_{0,0}$ (^{3}AQ) = 2.76 eV, $E°(AQ/AQ^{·-})$ = -0.80 V (vs. SCE), and the oxidation potentials for dG and dA given in ref. [116]. If the oxidation potentials for dG and dA given in Ref 118 are used, $\Delta G°(ET)$ = -1.04 eV and -0.70 eV, respectively.

134. Sheu, C.; Foote, C. S. *J. Am. Chem. Soc.* 1995, *117*, 6439.

135. Eriksson, M; Neilsen, P. E. *Nature Struct. Biol.* 1996, *3*, 410.

136. Sancar, A. *Annu. Rev. Biochem,* 1996, *65*, 43.

137. Yeh, S.-R.; Flavey, D. E. *J. Am. Chem. Soc.* 1991, *113*, 8557.

138. Stramel, R. D.; Nguyen, C.; Webber, S. E.; Rodgers, M. A. J. *J. Phys. Chem.* 1988, *92*, 2934.

139. Shand, M. A.; Rodgers, M. A. J.; Webber, S. E. *Chem. Phys. Lett.* 1991, *177*, 11.

2

Coordination Complexes and Nucleic Acids. Perspectives on Electron Transfer, Binding Mode, and Cooperativity

Eimer Tuite
Chalmers University of Technology, Göteborg, Sweden

I. INTRODUCTION

There has been long-term interest in how far and how fast, and by what mechanism DNA can disperse radiative damage by mediating electron and hole migration along the helix. These questions have come to the fore again recently, spurred by reports that DNA can mediate rapid long-range electron transfer between bound intercalators or between an intercalator and a nucleobase. The relevance lies not only in the biologically important areas of DNA damage and repair but also in the development of new types of sensors for DNA sequencing and mutation analysis.

There are several distinct but related systems that have been studied in this context which are classified, for the sake of this review, into three groups. The first is electron transfer between two DNA-bound ligands, and the second is electron transfer between a DNA-bound ligand and DNA bases. In these two groups, where the mechanism may be quite different, the electron transfer reaction is mainly photoinduced in experimental studies. In the third group no DNA-bound ligands are involved and what is studied is transfer (migration) within a DNA molecule of radicals (electrons and/or holes) introduced by ionizing radiation (e.g., UV, γ-radiation, pulse radiolysis). Within each group there are different types of experimental conditions such as whether ligands are physically bound to the DNA or covalently tethered, and whether they are intercalated or externally bound, whether the DNA is prepared as fibres or in solution, whether the solvent is liquid or frozen, and so forth. The interpretations of these diverse experimental systems lead to conclusions that range from DNA being highly insulating to highly conducting. Rather than attempting to correlate all the information available on this topic, this review focuses on the area of electron transfer reactions between donor and acceptor molecules that are bound, but not covalently attached, to DNA since special attention must then be paid to the binding properties of the molecules which can be of equal or greater importance than their redox properties. Readers are referred to Chapter 1 of this volume by Netzel for a broader review and a discussion of electron transfer involving covalently bound photosensitizers.

II. PHOTOINDUCED ELECTRON TRANSFER BETWEEN DNA-BOUND COMPOUNDS

A. Quenching of Intercalated Ethidium Bromide

The first reports in the literature about electron transfer between two compounds bound to DNA were made by Baguley and coworkers [1–3]. Ethidium bromide (**I**, EB$^+$) is a well-characterized DNA binder that intercalates with little sequence selectivity. Its lifetimes in solution and when bound to DNA are well separated (ca. 1.8 ns and 23 ns, respectively), so free and bound dye can be distinguished. It was first found that amsacine (**II**, Am$^+$) but not 9-aminoacridine (**IIIa**, 9AAH$^+$) quenched the fluorescence of ethidium bromide when both were bound to DNA [1] and further studies indicated that the degree of quenching varied widely for different 9-anilinoacridine derivatives [4,5]. An electron transfer mechanism (to *EB$^+$ from acridine) was suggested through elimination of all other mechanisms and later by correlation of the acridine redox properties with quenching efficiency [3]. In a detailed study of the quenching, Am$^+$ was found to quench EB$^+$ steady-state fluorescence in a variety of nucleic acids with efficiencies in the order [poly(dA-dT)]$_2$>DNA>[poly(dG-dC)]$_2$ and the quenching was virtually identical using different types and lengths of DNA such as high molecular weight and sheared calf-thymus DNA (CT–DNA) and T7–DNA (39 kbp) [2]. Moreover, the changes in fluorescence were largely independent of the ethidium binding ratio [4] which indicates an absence of cooperativity in the binding of the two reagents (vide infra). Since EB$^+$ was reported to have a preference for binding to certain basepairs and to have different lifetimes depending on the DNA composition,

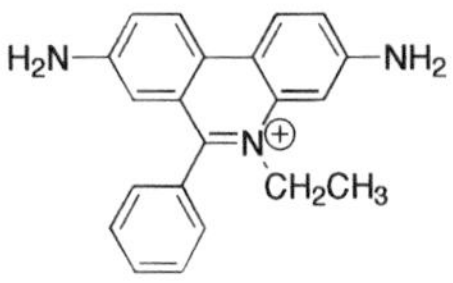

I: ethidium bromide (EB$^+$)

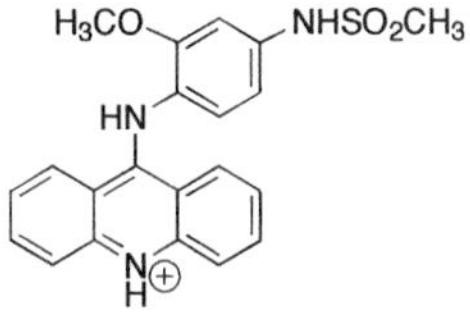

II: amsacrine (Am$^+$)

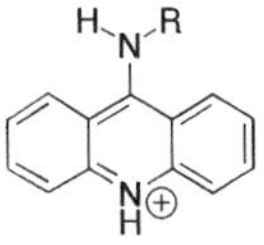

IIIa: R=H; 9-aminoacridine (9AAH$^+$)
IIIb: R=CH$_3$; 9-methylaminoacridine (9MAAH$^+$)

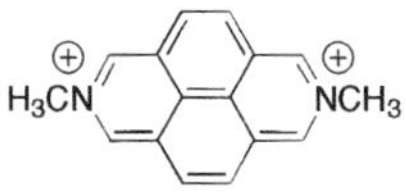

IV: *N,N*-dimethyl-2,7-diazapyrene (Me$_2$DAP^{2+})

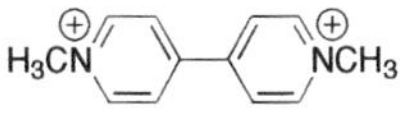

V: methyl viologen (MV^{2+})

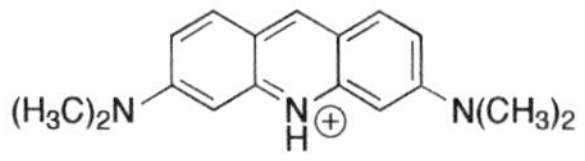

VI: acridine orange (AOH$^+$)

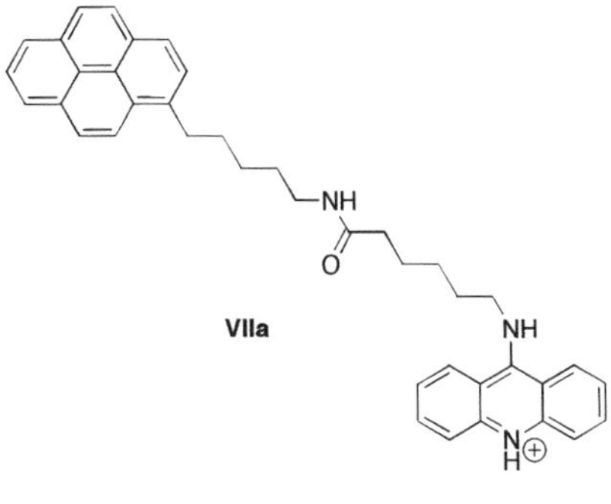

VIIa

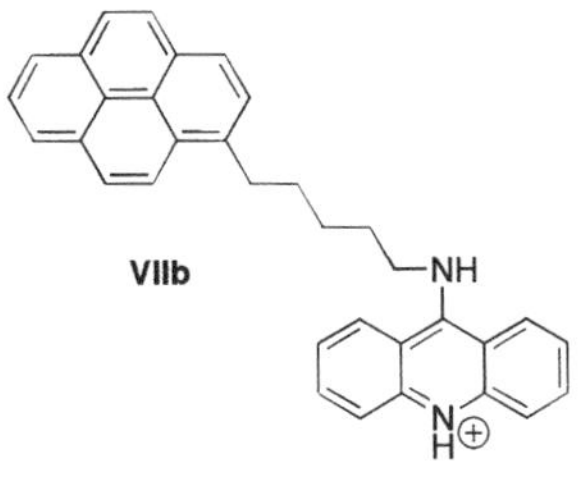

VIIb

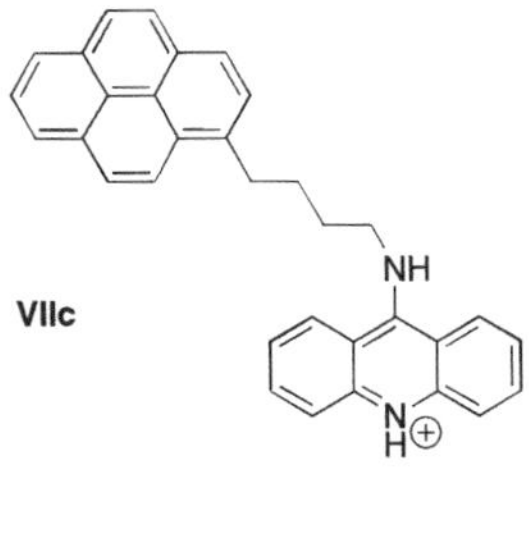

VIIc

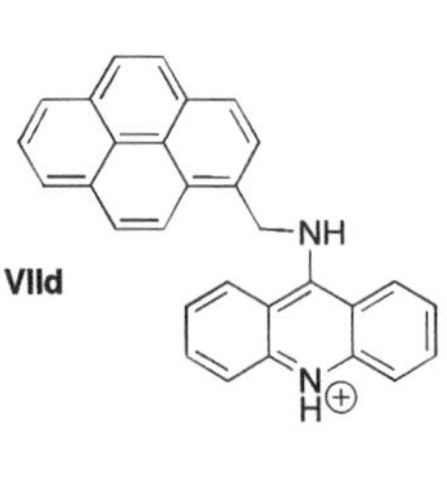

VIId

time-resolved fluorescence studies were carried out using [poly(dA-dT)]$_2$ [3]. In the absence of Am$^+$, the fluorescence decay of EB$^+$ had two components; a small amount (ca. 4%) of fast decay could not be distinguished from that of free dye (1.6 ns) while the slow decay was due to intercalated dye (25 ns). Addition of Am$^+$ led to a shortening of the lifetime of the slower component and also to its proportion being reduced relative to that of the fast component. The lifetime of the fast component became ca. 3 ns in the presence of Am$^+$ and the 1.63 ns component could not be resolved from this. One possible explanation was that the 3-ns component was due to quenching of EB$^+$ by a nearest neighbor Am$^+$ and the reduction in lifetime of the long-lived component resulted from electron transfer at longer distances, or less favorable orientations. However, the proportion of the short lifetime was much less than the proportion of nearest neighbors calculated (Monte Carlo method) for randomly distributed molecules—70% nearest neighbors calculated for 0.3 amsacrine molecules per base pair where the short lifetime component was only 41%. Ruling out anticooperativity from dialysis experiments, the interpretation was that nearest-neighbor binding was necessary but not sufficient for the production of the 3-ns component. Another explanation not considered in the original treatise is that there is an even faster decay component that is unresolved, and support for this comes from a comparison of steady-state quenching data and the integrated emission decay which we have extracted from the original papers [2,3] and plotted in Fig. 1. From this plot it seems clear that there is more fluorescence quenching that can be accounted for by the two time-resolved components. The steady-state fluorescence data matches very well to the quenching profile that we have calculated (solid line, Fig. 1) using conditional probabilities for noncooperative interactions between EB$^+$ and Am$^+$ molecules (using binding constants and site sizes from the original paper) assuming a quenching sphere of action of 3.4 Å, i.e., total quenching of only nearest neighbors.

Brun and Harriman later studied the quenching of intercalated EB$^+$ by an intercalating viologen analog N,N-dimethyl-2,7-diazapyrene (**IV**, Me$_2$DAP^{2+}) in CT–DNA using time-resolved emission [6]. Strong steady-state fluorescence quenching was observed when Me$_2$DAP^{2+} was added to intercalated EB$^+$ (Fig. 2), due to electron transfer from *EB$^+$ to Me$_2$DAP^{2+}, and in this case the integrated time-resolved decays match the steady-state quenching indicating the absence of any unresolved short-lived component in the decay profiles. The fluorescence decays in this case were best analyzed with a three-exponential fit, the three different lifetimes—ca. 20 ns (cf. unquenched dye), 8.3 ns, 0.72 ns—remaining approximately constant with increasing concentrations of Me$_2$DAP^{2+} but their amplitudes changing significantly. The two shorter lifetimes increased in amplitude with increasing quencher concentration, while the third lifetime decreased in both amplitude and magnitude (from 21 ns to 19.4 ns). The different components were explained by Me$_2$DAP^{2+} molecules bound at different distances from EB$^+$ molecules having different rate constants for electron transfer, so the 0.72-ns

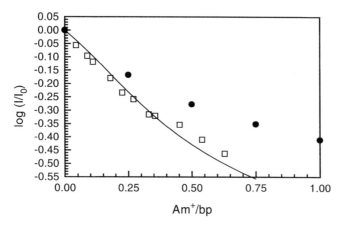

FIGURE 1 Quenching of ethidium bromide fluorescence by amsacrine in the presence of poly(dA-dT)·poly(dA-dT). (□) Steady-state intensity and (●) integrated fluorescence decay ($\Sigma\alpha\tau$). (Data compiled from Refs. 2 and 3, respectively.) $[EB^+] = 2\ \mu M$; $[bp] = 20\ \mu M$; buffer is 0.01 M SHE (9.4 mM NaCl, 2 mM N-(2-hydroxylethyl)piperazine-N-2-ethan-sulfonate (pH 7), 20 μM EDTA). The solid line is quenching calculated using statistical probabilities to model the binding distribution and allowing complete quenching of only nearest neighbors (sphere of action of radius 3.4 Å) [28].

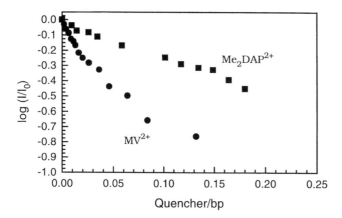

FIGURE 2 Steady-state quenching of ethidium bromide fluorescence by viologens in the presence of CT–DNA. (■) Me_2DAP^{2+} and (●) MV^{2+}. (Data compiled from Refs. 6 and 10, respectively). For Me_2DAP^{2+}, $[EB^+] = 70\ \mu M$; $[bp] = 700\ \mu M$; buffer is 5 mM phosphate. For MV^{2+}, $[EB^+] = 1\ \mu M$; $[bp] = 50\ \mu M$; buffer is 1.5 mM cacodylate (pH 6).

component was assigned to three-base-pair (10.2 Å) separation, the 8.3-ns component to four-base-pair (13.6 Å) separation and, at high Me_2DAP^{2+} loading, the 19.4-ns component to five-base-pair (17.0 Å) separation. Fitting the forward electron transfer data (R = separation distance; $k_{et} = 1/\tau - 1/\tau_0$) to Eq. (1) gave a β value of 0.91 ± 0.04 Å^{-1}, which is comparable to values obtained for modified proteins and, assuming that this attenuation factor referred to electron tunnelling through the intervening basepairs, it was suggested that electron transfer over considerable distances might be possible for a sensitizer possessing a long-lived excited state.

$$k_{et} = A\exp[-\beta R] \tag{1}$$

The electron transfer products were monitored by transient absorption spectroscopy and rates of charge recombination were also measured, giving a short-lived component of 20 ns dominating the decay at high loadings of Me_2DAP^{2+}, and longer-lived components with lifetimes of 2 μs and 0.62 ms. From these data an attenuation factor for charge recombination was calculated to be β = 1.49 ± 0.07 Å^{-1}, although because of the long lifetimes (possible diffusion) and the stated complexity of the decay profiles this value should not be overinterpreted.

There have also been several reports of electron transfer quenching of *EB$^+$ using externally bound quenchers. Atherton and Beaumont studied oxidation of photoexcited intercalated ethidium bromide by externally bound transition metal ions [7,8]. In the presence of the metal ions Cu^{2+}, Ni^{2+}, or Co^{2+} (which bind to the DNA phosphate groups under the conditions employed and do not alter the DNA conformation) the fluorescence decay of *EB$^+$ became biexponential, indicating two noninterchangeable populations of EB$^+$ excited states on the timescale of fluorescence. As the metal ion concentration was increased the amplitude of the faster component increased at the expense of the slower component and the rates of both decays increased with increasing metal ion concentration: nonlinearly with ion concentration for the fast decay (ca. 8–4 ns), and linearly for the slow decay (ca. 23–14 ns) in a manner that depended on the metal ion. The interpretation offered was that strong quenching (larger rate) occurred when there was a metal ion bound close to an intercalated EB$^+$ and weak quenching (small rate) was due to diffusion of metal ions into the quenching sphere—as the quencher concentration is increased, its concentration within the quenching sphere should also increase giving a higher proportion of fast quenching, as observed. Analysis of the quenching data indicated that there were ca. 4–6 phosphate groups within the quenching radius of intercalated ethidium bromide. The mobilities of the metal ions were judged to be about 20-fold lower than in solution and the rate of phosphate-to-phosphate motion was estimated as $4–20 \times 10^8$ s^{-1} given a residence time per phosphate of 5–20 ns. It was concluded that electron transfer could occur to metal ions bound to phosphates up to three bases away from the intercalated ethidium over distances of ca. 1 nm with a rate of ca. 2×10^8 s^{-1}. A more recent study from the same authors on reverse electron transfer following quenching of $*EB^+$ by Cu^{2+} bound to polynucleotides showed evidence from transient absorption spectroscopy for the electron transfer products and led to the calculation of an attenuation factor β of 0.73 ± 0.05 Å^{-1} for this system [9].

In short communication, very efficient electron transfer from photoexcited intercalated ethidium bromide to externally bound methyl viologen (**V**, MV^{2+}) studied with steady-state emission (Fig. 2) was reported by Fromherz and Rieger [10]. Electron transfer was assigned as the quenching mechanism considering the redox potentials (Table 1) and by transient absorption spectroscopy. The apparent enhancement of the rate of $*EB^+$ quenching in the presence of CT–DNA (Fig. 2) was attributed mainly to the polyelectrolyte effect of the nucleic acid which caused an increase in the local concentrations of the reactants. In fact, when this was taken into account, there were indications that the electron transfer reaction itself might be slowed down compared to aqueous solution and assuming that the reaction is diffusion-controlled in the limit of low occupancy, the mobility of MV^{2+} along the DNA helix appeared to be restricted. Preliminary time-resolved studies indicated nonexponential decay of $*ET^+$ within 1 ns, faster than observed for quenching by transition metal ions (vide supra), and recombination on the ms timescale. Unfortunately, a complete time-resolved analysis of the forward electron transfer in this system was never reported although transient absorption

TABLE 1 Oxidation and Reduction Potentials for the Donors and Acceptors Reviewed

Redox couple	$E°$ vs. NHE	Reference
EB^{2+}/EB^+	1.68 V	48
$EB^{2+}/*EB^+$	−0.52 V	48
AOH^{2+}/AOH^+	1.40 V	48
$AOH^{2+}/*AOH^+$	−1.00 V	48
$MV^{2+}/MV^{•+}$	−0.44 V	10
$Me_2DAP^{2+}/Me_2DAP^{•+}$	−0.26 V	49
$Py^{•+}/Py$	1.40 V	12
$9AAH^+/9AAH^•$	−0.97 V	12
$*9AAH^+/9AAH^•$	1.95 V	12
$EB^+/EB^•$	−1.12 V	48
$*EB^+/EB^•)$	1.08 V	48
$[Ru(phen)_2dppz]^{3+}/[Ru(phen)_2dppz]^{2+}$	1.52 to 1.63 V	22, 23, 25, 26
$[Ru(phen)_2dppz]^{3+}/*[Ru(phen)_2dppz]^{2+}$	−0.61 to −0.77 V	22, 23, 25, 26
$[Rh(phi)_2phen]^{3+}/[Rh(phi)_2phen]^{2+}$	−0.05 to 0.02 V	22, 25
$[Rh(phi)_2bpy]^{3+}/[Rh(phi)_2bpy]^{2+}$	−0.05 to −0.1 V	23, 25
$[Rh(NH_3)_6]^{3+}/[Rh(NH_3)_6]^{2+}$	0.04 V	22

studies on the back electron transfer by Atherton and Beaumont [11] agreed that in the presence of MV^{2+} the singlet state absorption and emission of $*EB^+$ both decay within ca. 1 ns. Recombination on the ms timescale was attributed to back reaction with $MV^{\cdot +}$ that had escaped the helix (ca. 2% yield) whereas recombination of the redox products on the helix could not be separated from the decay of the EB^+ singlet state.

B. Quenching of Intercalated Acridines

Brun and Harriman also investigated the quenching of the intercalator acridine orange (**VI**, AOH^+) by Me_2DAP^{2+} [6] using time-resolved methods. As for ethidium bromide (vide supra) the fluorescence decay could best be fit by a three-exponential function with three almost invariant lifetimes—ca. 4.6 ns (cf. 5 ns for unquenched AOH^+), 2.1 ns, 0.21 ns—where the amplitudes of the short-lived components increase with increasing quencher concentration. Compared to the experiments with EB^+, lower concentrations of Me_2DAP^{2+} were required to give the same amount of quenching because of the greater driving force for electron transfer with AOH^+. Nonetheless, fitting the data to Eq. (1) gave a rather similar value for the attenuation factor of $\beta = 0.86 \pm 0.04$ Å^{-1}, again in line with values obtained from studies of electron transfer in proteins.

Recently, Bassani et al. [12] described the synthesis of a series of pyrene-acridine bis-intercalators (**VIIa–d**), in which the pyrene can quench the acridine moiety both in acetonitrile solution and when bound to DNA. This construction allows a more defined separation of the reactants compared to the situation where they are bound statistically, as long as the linker length is optimized to allow bis-intercalation only at designated, e.g., nearest-neighbor binding sites. The four compounds described have different linker lengths and characterization of the binding modes implied that **VIIa** bis-intercalates, **VIIc** and **VIId** intercalate only the acridine chromophore, and **VIIb** binds by partial bis-intercalation without nearest-neighbor exclusion (i.e., at contiguous sites). In acetonitrile solution, the linked pyrene quenches the fluorescence of the acridine moiety compared with 9-methylaminoacridine (**IIIb**, $9MAAH^+$) by electron transfer and the degree of quenching increases with decreasing linker length. Compared with $9MAAH^+$ bound to sonicated CT–DNA (300 bp), **VIIb** and **IIIc** had enhanced fluorescence while **VIIa** and **VIId** had reduced fluorescence. However, the authors were unable to separate sequence-dependent binding effects or base quenching effects from intramolecular quenching by pyrene, and further studies with defined sequences are required to assign rates to electron transfer between the intercalated moieties.

C. Quenching of Groove-Bound Metal Complexes

The first report regarding quenching of DNA-bound metal complexes was of enantiomeric ruthenium(II)tris-diammine complexes by racemic cobalt(III)tris-diammine complexes [13]. As with organic donors and acceptors, increased electron transfer rates were observed in the presence of DNA, in a manner that depended on the binding constants of the complexes, consistent with DNA playing a concentrating polyelectrolyte role. However, the efficiency of electron transfer was found to depend on the length of the DNA strands and was significantly more efficient with native (ca. 11 kbp) than with sonicated (ca. 290 bp) CT–DNA. This could be rationalized in terms of diffusion of the quencher, if diffusion along a strand is much faster than jumping from strand to strand. An alternative explanation, mooted already at this point, was that the π-framework of DNA might be involved in mediating long-range electron transfer through extended donor-acceptor coupling.

This idea was expounded upon in a later paper [14] considering the same systems. The quenching of $[Ru(phen)_3]^{2+}$ by $[Co(phen)_3]^{3+}$ and $[Co(bpy)_3]^{3+}$ (**VIII**, phen = 1,10-phenanthroline; **IX**, bpy = 2,2'-bipyridyl) was studied as a

VIII: 1,10-phenanthroline (phen)   IX: 2,2'-bipyridine (bpy)

X: dipyrido[3,2-*a*:2',3'-*c*]phenazine (dppz)   XI: 9,10-phenanthrenequinone diimine (phi)

function of temperature and viscosity. Much of the interpretation was based on tris-phenanthroline complexes having two binding modes—intercalation and groove binding—although it was subsequently shown that these compounds do not intercalate [15–18]. This binding model arose from the observation of two lifetimes for $[Ru(phen)_3]^{2+}$ with DNA, one about 4-fold greater than that of free dye in solution and the other very similar to free dye [19]. However, the similarity of the second lifetime to that of free dye prompted others to ask whether it did not simply arise from unbound dye and subsequent studies indicated the existence of only one bound form in addition to free dye [16,20]. However, it is clear from this study [14] and a subsequent one [21] that the short as well as the long lifetime is really reduced by added quenchers, and the quenching is much more efficient than the quenching of the metal complex in the absence of DNA. Hence, the shorter lifetime probably represents a mixture of unbound metal complex and complexes that are loosely bound to DNA, while the long lifetime represent complexes that are intimately bound to DNA in the minor groove, and the finding that the short lifetime is quenched more efficiently than the long lifetime could be attributed to the increased mobility of the loosely bound complexes. However, experiments in glycerol at low temperature to eliminate diffusion indicated that the rate constant for quenching of the short lifetime actually increased with increasing viscosity, while that for the long lifetime was decreased. Since cooperative clustering of donors and acceptors was ruled out, this result suggests that at low temperature the solvent around DNA may act as a better medium for electron transfer than DNA itself. Although the original report [14] suggested the possibility that DNA acts as a special mediator of electron transfer between metal complexes bound to DNA, later work on similar systems [21] comparing the effects of DNA and the hydrophobic polyanion polystyrenesulphonate indicated that the primary effect of the nucleic acid was increase the local donor and acceptor concentrations and to reduce reactant mobility.

D. Quenching of Intercalated Metal Complexes

The recent interest in DNA-mediated electron transfer has been largely stimulated by the report of Barton and coworkers of extraordinary quenching of intercalated Δ-*[Ru(phen)$_2$dppz]$^{2+}$ by intercalated Δ-[Rh(phi)$_2$bpy]$^{3+}$ [22] (X, dppz = dipyrido[3,2-a:2′,3′-c]phenazine; XI, phi = 9,10-phenanthrenequinone diimine) with CT–DNA. Compared to quenching by externally bound [Ru(NH$_3$)$_6$]$^{3+}$ (with almost the same driving force), quenching by intercalated [Rh(phi)$_2$bpy]$^{3+}$ was more efficient (Fig. 3) and more notably was extremely fast. Differences were observed in the luminescence decays (on the nanosecond timescale)—while both quenchers decreased the lifetimes, only quenching by [Rh(phi)$_2$bpy]$^{3+}$ decreased the initial intensity indicating a faster unresolved quenching component, estimated at 120 ps. A series of publications followed [23–26], using different metal complexes containing dppz and phi ligands, different enantiomeric donor/acceptor pairs, and different polynucleotides and heterogeneous media to reach the conclusion that in order for DNA to efficiently mediate long-range electron transfer, both reactants must be not only intercalators but also be well stacked between the base pairs to ensure good overlap with the stacked π-system of the bases—the so-called "π-way." In a fast time-resolved study, a single electron transfer rate was found regardless of loading which suggested either a single distance or a shallow distance-dependence of the rate [26]. The possibility of cooperative binding of the donor and acceptor was more or less discounted from photocleavage studies and by comparison with the system in which the reactants are attached to oligonucleotides where fast electron transfer over 40 Å is reported [27]. However, by comparison with other literature reports of electron transfer between DNA-bound molecules, the indications of ultrafast long-range electron transfer in the *[Ru(phen)$_2$dppz]$^{2+}$/[Rh(phi)$_2$bpy]$^{3+}$ and related systems present

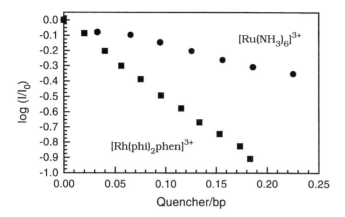

FIGURE 3 Steady-state quenching of Δ-[Ru(phen)$_2$dppz]$^{2+}$ luminescence by metal complexes in the presence of CT–DNA. (Data compiled from Ref. 27.) (■) Δ-[Rh(phi)$_2$phen]$^{3+}$ and (●) [Ru(NH$_3$)$_6$]$^{3+}$·[Ru(phen)$_2$dppz]$^{2+}$ = 10 μM; [bp] = 500 μM; buffer is 5 mM Tris-HCl/50 mM NaCl (pH 7.2).

an anomaly rather than a generality. An examination of the data prompts considerations of cooperativity, i.e., the possibility that for some reason the donor and acceptor have a preference to bind very close to each other on the DNA strand.

In order to investigate the possibility of cooperativity in this system, a simple model to analyze the emission quenching which focused on binding rather than electron transfer rates was constructed [28]. Rather than modeling in terms of a fall-off of electron transfer rate with distance, the model assumed all-or-none quenching within a given quenching sphere of action, which is justified if β is in the normal range of 1–1.5 Å^{-1}. The binding was modeled by developing an extension to the McGhee–von Hippel binding analysis [29] to accommodate heterocooperativity; this models binding to a one-dimensional lattice using conditional probabilities. Steady-state quenching experiments were also carried out to provide raw data for the modeling. Because of the complex photophysics of *[Ru(phen)$_2$dppz]$^{2+}$ which depend on enantiomer, binding ratio, and DNA sequence [30–34], the experiments were carried out using the polynucleotide poly(dA-dT)·poly(dA-dT) and Δ-[Ru(phen)$_2$dppz]$^{2+}$, the combination with the greatest fluorescence enhancement and the longest lifetimes, and Δ-[Rh(phi)$_2$bpy]$^{3+}$ was used as the quencher. Any heterocooperativity was expected to be manifest in the quenching curves (as a function of quencher loading) not being independent of the initial loading of the sensitizer, so quenching was measured at several bp/[Ru(phen)$_2$dppz]$^{2+}$ ratios as shown in Fig. 4. By comparison with Fig. 3, this polynucleotide allows more efficient quenching than CT–DNA, a feature also reported by Barton and coworkers [26]. It was found that the differences between

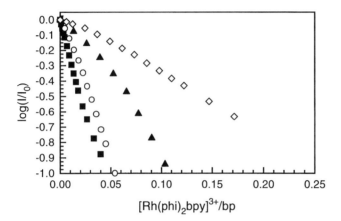

FIGURE 4 Steady-state quenching of Δ-[Ru(phen)$_2$dppz]$^{2+}$ luminescence by Δ-[Rh(phi)$_2$-bpy]$^{3+}$ in the presence of poly(dA-dT)·poly(dA-dT). (Data from Ref. 28.) (■) 75 bp/Ru (8.5 μM Ru + 635 μM bp), (○) 25 bp/Ru (2.5 μM Ru + 61 μM bp), (▲) 8.5 bp/Ru (2.5 μM Ru + 21.25 μM bp), and (◇) 2.5 bp/Ru (27.5 μM Ru + 68.5 μM AT). Buffer is 5 mM sodium phosphate/45 mM NaCl (pH 6.9).

the quenching curves could best be modeled in terms of strong cooperativity of binding between the [Ru(phen)$_2$dppz]$^{2+}$ and [Rh(phi)$_2$bpy]$^{3+}$ complexes with complete quenching of only nearest neighbors [28]. No model that used only random binding provided satisfactory global fits, even including possible weak [Ru(phen)$_2$dppz]$^{2+}$ homocooperativity. Assuming a random distribution of complexes, each curve gave a different "sphere of action" for the quencher with the implication of a remarkable radius of > 100 Å at bp/[Ru(phen)$_2$dppz]$^{2+}$ = 75! The conclusion of strong heterocooperativity was supported by circular dichroism (CD) measurements which even at low loadings of [Ru(phen)$_2$dppz]$^{2+}$ and [Rh(phi)$_2$bpy]$^{3+}$ showed a perturbation that was characteristic of the CD observed when the complexes are forced to bind adjacently at high binding ratios [28].

A similar conclusion of [Ru(phen)$_2$dppz]$^{2+}$/[Rh(phi)$_2$bpy]$^{3+}$ clustering was arrived at by Olson, Barbara et al. using Monte Carlo simulations to model the equilibrium distribution of donor and acceptor molecules and subsequently modeling quenching assuming an exponential distance dependence of the excited-state electron transfer and recombination rates [35]. Comparison of the simulations with experimental data indicated a typical distance dependence of the electron transfer rate ($\beta \approx 1$ Å^{-1}), and furthermore the simulation including clustering was able to account for the random photocleavage of DNA by *[Rh(phi)$_2$bpy]$^{3+}$ in the presence of [Ru(phen)$_2$dppz]$^{2+}$. Turro and co-workers also developed a phenomenological model for describing DNA-mediated electron transfer in a nonequilibrium system [36]. Their model was used to calculate for the

*[Ru(phen)$_2$dppz]$^{2+}$/[Rh(phi)$_2$bpy]$^{3+}$ system at a single initial bp/[Ru(phen)$_2$dppz]$^{2+}$ ratio (using the data from Ref. 22) and assuming random binding suggested that electron transfer could occur over a distance of 24 Å. However, as Fig. 4 shows, the quenching efficiency is very dependent on the bp/[Ru(phen)$_2$dppz]$^{2+}$ ratio and fitting each curve individually will give different distances [28]. In order to globally model all data for this system, heterocooperativity must be invoked.

We suggest therefore that the different efficiencies observed for quenching of dppz–metal complexes by phi–rhodium complexes as a function of enantiomer, complex, and base sequence may simply be due to different degrees of binding heterocooperativity depending on the precise system studied. However, since for the *EB$^+$/AM$^+$ system which seems to have completely random binding [2], the same dependence of quenching efficiency on base sequence was observed as for the Ru/Rh system [26]—poly(dA-dT)·poly(dA-dT) > CT–DNA > poly(dG-dC)·poly(dG-dC)—the DNA sequence probably does have some role to play in the mediation of electron transfer. Whether this is due to different sequence dependencies of tunneling, or to base-dependent structural and dynamic fluctuations is another matter which theoretical calculations may be able to sort out.

E. Comparison of Intercalated and Externally Bound Viologens as Quenchers of an Intercalated Metal Complex

In order to examine how the binding mode of a quenching molecule affects its ability to quench the excited state of an intercalated molecule, we have recently embarked on a systematic study of the quenching of the intercalated [Ru(phen)$_2$dppz]$^{2+}$ by various intercalating and nonintercalating quenchers [37]. Initially the comparison has been of oxidative quenching of the excited state ruthenium complex by methyl viologen (MV^{2+}) which binds externally to DNA and dimethyldiazapyrene (Me$_2$DAP^{2+}) which we have recently confirmed by linear dichroism as a DNA intercalator [38]. Both these viologens have been used previously in quenching studies with DNA and have similar reduction potentials, but a comparison of their abilities to quench the same sensitizer under the same conditions has never been made before. The luminescence decay of each [Ru(phen)$_2$dppz]$^{2+}$ enantiomer is biexponential with all types of nucleic acid: With an alternating polynucleotide the lifetimes remain constant while their proportions vary depending on the binding ratio [33]. In this study the homopolymeric alternating polynucleotide poly(dA-dT)·poly(dA-dT) was used to simplify interpretation of the photophysics and to maximize the unquenched emission [30,33], and the quenching was studied at several different bp/[Ru(phen)$_2$dppz]$^{2+}$ ratios to test for cooperativity.

Figure 5 shows a comparison of the quenching of [Ru(phen)$_2$dppz]$^{2+}$ by MV^{2+} (panel a) and Me$_2$DAP^{2+} (panel b) at a bp/[Ru(phen)$_2$dppz]$^{2+}$ ratio of

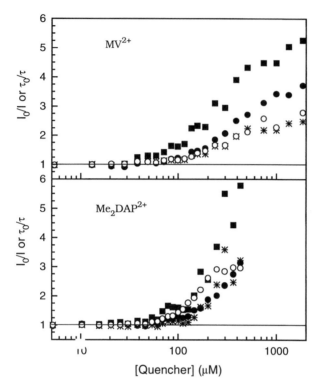

FIGURE 5 Quenching of Δ-[Ru(phen)$_2$dppz]$^{2+}$ luminescence by viologens in the presence of poly(dA-dT)·poly(dA-dT)—comparison of steady-state and lifetime quenching. (●) Short lifetime, (■) long lifetime, (*) steady-state intensity, and (○) intensity calculated from lifetimes ($\Sigma\alpha\tau$). Upper panel MV^{2+} and lower panel Me$_2$DAP^{2+}. [Ru(phen)$_2$dppz]$^{2+}$ = 20 μM; bp = 500 μM; buffer is 5 mM sodium phosphate (pH 7).

25 (20 μM Ru). The quenching profiles are quite similar and the most notable feature is that there is very little quenching of either lifetime or of the emission at low concentrations of added Me$_2$DAP^{2+}, where most donor and acceptor molecules would be separated by >5–6 bp (17–20 Å) on average if binding is random. In fact, quenching by Me$_2$DAP^{2+} only becomes very efficient at higher added quencher concentrations when the average separation should be <5 bp (17 Å). However, under these conditions, there are spectroscopic indications that Me$_2$DAP^{2+} also starts to bind in its second binding mode which is non-intercalative and may resemble the electrostatic binding of MV^{2+}.

Figure 6 shows quenching of the emission intensity by MV^{2+} (panel a) and Me$_2$DAP^{2+} (panel b) at different bp/[Ru(phen)$_2$dppz]$^{2+}$ ratios. For each quencher, the curves are similar when log (I/I_0) is plotted against the ratio of quencher to

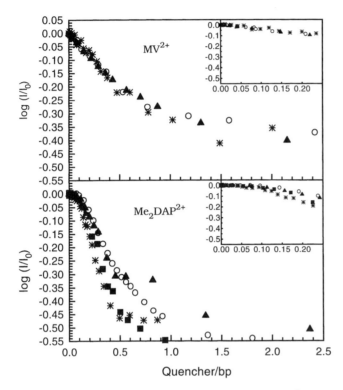

FIGURE 6   Steady-state quenching of Δ-[Ru(phen)$_2$dppz]$^{2+}$ luminescence by viologens in the presence of poly(dA-dT)·poly(dA-dT). (■) 75 bp/Ru (5 μM Ru + 375 μM bp), (○) 25 bp/Ru (5 μM Ru + 125 μM bp), (▲) 10 bp/Ru (5 μM Ru + 50 μM bp), and (*) 25 bp/Ru (20 μM Ru + 500 μM bp). Upper panel MV^{2+} and lower panel Me$_2$DAP^{2+}. [Ru(phen)$_2$dppz]$^{2+}$ = 5 μM except for * [Ru(phen)$_2$dppz]$^{2+}$ = 20 μM; buffer is 5 mM sodium phosphate (pH 7). The insets are re-scaled to the same X-scale as Fig. 2 to show how much less efficiently these viologens quench *[Ru(phen)$_2$dppz]$^{2+}$ compared to *EB$^+$ emission.

polynucleotide concentrations. This is an indication that there is no significant cooperative donor-acceptor binding in either of these systems. With Me$_2$DAP^{2+}, the curves deviate downward with little quenching at low quencher concentrations. This behavior could indicate anticooperative binding of the reactants which would give curves of this shape but, in that case the curves should differ and that for the highest binding density should lie below the others (the curves should have the opposite trend to that for cooperative binding, vide supra). A more likely explanation is that there are two parts to the quenching curve; at low concentrations, quenching is only by intercalated Me$_2$DAP^{2+} molecules and at higher concentrations externally bound Me_2DAP^{2+} molecules also causes quenching. The second phase of Me_2DAP^{2+} quenching is much more efficient than the first and the curves resemble those observed with MV^{2+}, the stronger quenching with the former perhaps partially being due to the larger driving force for electron transfer but also there should be a contribution from nearest-neighbor quenchers. At higher Me_2DAP^{2+} concentrations the quenching of the steady-state emission is stronger than that calculated from quenching of the lifetimes and this indicates that there is a fast unresolved quenching component which may be attributed to quenching of $[Ru(phen)_2dppz]^{2+}$ by adjacently intercalated Me_2DAP^{2+} molecules (ca. 10 Å separation). For both quenchers quenching saturates at about 1 quencher molecule per base pair which is the point at which electrostatic binding of a di-cation should become saturated. This is further indication that it is external, electrostatically bound Me_2DAP^{2+} that causes the strong quenching. Intercalated Me_2DAP^{2+}, on the other hand, seems to be a rather poor quencher of the intercalated metal complex, suggesting that in this case the π-stacked DNA basepairs do not mediate electron transfer very efficiently. One notable point is that the quenching of $[Ru(phen)_2dppz]^{2+}$ by these two viologens is less efficient than the quenching of $*EB^+$ by the same compounds (compare Fig. 6 with Fig. 2) when conditions and driving forces (Table 1) are reasonably comparable. The reason for this is not clear, but it is probably related to different types of binding of the two intercalated photosensitizers. Notably, in both cases there is a trend for the external binding MV^{2+} to quench more efficiently than the intercalating Me_2DAP^{2+}. There is a possibility that $[Ru(phen)_2dppz]^{2+}$ may favor self-clustering with a weak cooperativity and, even with no positive or negative heterocooperativity, if only molecules bound at the ends of a contiguous sequence are available for quenching then the quenching efficiency will be lower than expected for a randomly binding donor such as ethidium bromide. Inclusion of such weak homocooperativity in our model for quenching of $*[Ru(phen)_2dppz]^{2+}$ by $[Rh(phi)_2bpy]^{3+}$ did not significantly affect the calculated quenching because of the strong heterocooperativity in that case [28].

II. THEORETICAL CONSIDERATIONS

The recent startling results suggesting long-range electron transfer mediated by the "π-way" of the stacked DNA bases have prompted renewed theoretical consideration [39,40] of whether the extended π-system could efficiently mediate electron transfer. In a series of papers, Beratan and coworkers have performed various calculations to investigate the ability of DNA to mediate electron transfer [41–44]. From the calculations, the π-electrons were found to dominate the long-range coupling, which is the basis for the π-way hypothesis. However, despite this, the prediction is that the distance decay constant of the electron transfer rate is protein-like ($\beta \sim 1.0$–1.5 Å^{-1}) for transfer between two DNA-bound compounds (intercalated or not). The source of this large β is the through-space gap of 3.4 Å between the stacked basepairs. Experimental data for the quenching of intercalated *EB$^+$ by Me$_2$DAP^{2+} [6] ($\beta \approx 0.9$ Å^{-1}), for electron transfer between two nonintercalating ruthenium complexes covalently attached to DNA oligomers [45] ($\beta \approx 1.2$ Å^{-1}), and reassessed data considering cooperativity for the quenching of *[Ru(phen)$_2$dppz]$^{2+}$ by [Rh(phi)$_2$bpy]$^{3+}$ [35] ($\beta \approx 1.0$ Å^{-1}) were found to give distance dependencies falling in the expected range of values. The only anomolous result is the value estimated for quenching of *[Ru(phen)$_2$dppz]$^{2+}$ by [Rh(phi)$_2$bpy]$^{3+}$ when they are tethered to oligonucleotides [27] ($\beta \approx 0.2$ Å^{-1}); hence, further studies on these kind of tethered systems are motivated in order to resolve the issue of whether DNA can mediate electron transfer with little distance dependence, even if only in selected systems.

In a theoretical study proposing a model that could explain rapid long-range DNA-mediated electron transfer, it was concluded that for transfer through a delocalized bridge system, a distance-independent mechanism could occur when the bridge is weakly populated [46], via electron injection into a DNA conduction band. However, Friesner and co-workers clearly point out that if this is the mechanism, then fluorescence quenching should depend exponentially on temperature— such an experiment remains to be carried out with any of the systems for which long-range electron transfer is claimed.

III. CONCLUSIONS REGARDING DNA-ASSISTED ELECTRON TRANSFER

The data in the figures throughout this review have been collected from different studies of electron transfer between an intercalated photosensitizer and either an intercalated or externally bound quencher, and are all plotted in the same way to allow direct comparison. There is clearly quite a lot of variation in the efficiencies but in almost all cases electron transfer between intercalators appears to be possible only over a small and reasonable range of separations (up to maximally 5 bp, or 17 Å). There is no evidence that intercalators are much more efficient than external binders, and indeed extremely efficient quenching of *EB$^+$ is observed with MV^{2+} although this is attributed mainly to the concentrating polyelectrolyte effects of the DNA [10]. Only in the *[Ru(phen)$_2$dppz]$^{2+}$/[Rh(phi)$_2$bpy]$^{3+}$ case was there evidence for extremely strong and fast quenching that might be attributed to efficient long-range electron transfer [26,27], but by varying the initial binding ratio of photosensitizer it was possible to show that this was not a system where the quencher had an extremely large sphere-of-action (> 100 Å could be estimated from the quenching at the lowest binding ratio used [28]), but rather that there was cooperative binding of the donor and acceptor molecules on DNA [28,35]. In the other cases discussed, the sphere of action of the quencher is not unreasonably large, 3–5 basepairs for either intercalators or externally bound quenchers, which seems to agree with the theoretical predictions about the fall-off of rate with distance [43]. In the most detailed analysis of distance dependence of quenching [6], the rate is found to fall off with an attenuation factor ($\beta \approx 0.86 \text{ Å}^{-1}$) that is similar to that found for proteins, so that quenching over a separation of 5 bp (17 Å) is extremely weak. Considering these data, the hypothesis that intercalators quench other intercalators most efficiently because of their insertion into the π-way does not hold up. In fact the most inefficient quenching is that of intercalated *[Ru(phen)$_2$dppz]$^{2+}$ by intercalated Me$_2$DAP^{2+}. By contrast, when the Me$_2$DAP^{2+} loading is increased so that it also binds externally, the quenching increases enormously and resembles that of the related MV^{2+} which is exclusively an external binder. Moreover, much of the quenching enhancement observed in the presence of DNA can be attributed to the increased local concentrations of cationic reactants in the polyelectrolyte field of the anionic DNA. Indeed two studies have suggested that when this factor and also the reduced mobility of reactants is accounted for, the electron transfer reaction itself might be slowed down in the DNA matrix compared to solution [10,21]. The proponents of efficient, long-range DNA-mediated electron transfer have cited such conclusions, which are based on experiments where one or both reactants are not intercalated, as further evidence that intercalation of both reactants is required to access the π-way [47]. However, a comparison of all pertinent data in this review shows that intercalators are not in any way exceptional in this sense.

ACKNOWLEDGMENTS

Financial support of the EU TMR program, the Wenner-Gren and Carl Trygger Foundations is gratefully acknowledged. Per Lincoln is thanked for many insightful discussions and for carrying out modeling of quenching using statistical probabilities.

REFERENCES

1. Baguley, B. C.; Falkenhaug, E. M. *Nucleic Acids Research* 1978, *5*, 161.
2. Baguley, B. C.; Le Bret, M. *Biochemistry* 1984, *23*, 937.
3. Davis, L. M.; Harvey, J. D.; Baguley, B., C. *Chem.-Biol. Interactions* 1987, *62*, 45.
4. Baguley, B. C.; Denny, W. A.; Atwell, G. J.; Cain, B. F. *J. Med. Chem.* 1981, *24*, 170.
5. Baguley, B. C.; Cain, B. F. *Mol. Pharmacol.* 1982, *22*, 486.
6. Brun, A. M.; Harriman, A. *J. Am. Chem. Soc.* 1992, *114*, 3656.
7. Atherton, S. J.; Beaumont, P. C. *Photobiochem. Photobiophys.* 1984, *8*, 103.
8. Atherton, S. J.; Beaumont, P. C. *J. Phys. Chem.* 1986, *90*, 2252.
9. Atherton, S. J.; Beaumont, P. C. *J. Phys. Chem.* 1995, *99*, 12025.
10. Fromherz, P.; Rieger, B. *J. Am. Chem. Soc.* 1986, *108*, 5361.
11. Atherton, S. J.; Beaumont, P. C. *J. Phys. Chem.* 1987, *91*, 3993.

12. Bassani, D.; Wirz, J.; Hochstrasser, R.; Leupin, W. *J. Photochem. Photobiol. A: Chem.* 1996, *100*, 65.
13. Barton, J. K.; Kumar, C. V.; Turro, N. J. *J. Am. Chem. Soc.* 1986, *108*, 6391.
14. Purugganan, M. D.; Kumar, C. V.; Turro, N. J.; Barton, J. K. *Science* 1988, *241*, 1645.
15. Hiort, C.; Nordén, B.; Rodger, A. *J. Am. Chem. Soc.* 1990, *112*, 1971.
16. Satyanarayana, S.; Dabrowiak, J. C.; Chaires, J. B. *Biochemistry* 1993, *32*, 2573.
17. Eriksson, M.; Leijon, M.; Hiort, C.; Nordén, B.; Gräslund, A. *Biochemistry* 1994, *33*, 5031.
18. Coury, J. E.; Anderson, J. R.; McFail-Isom, L.; Williams, L. D.; Bottomley, L. *J. Am. Chem. Soc.* 1997, *119*, 3792.
19. Barton, J. K.; Goldberg, J. M.; Kumar, C. V.; Turro, N. J. *J. Am. Chem. Soc.* 1986, *108*, 2081.
20. Satyanarayana, S.; Dabrowiak, J. C.; Chaires, J. B. *Biochemistry* 1992, *31*, 9319.
21. Orellana, G.; Kirsch-De Mesmaeker, A.; Barton, J. K.; Turro, N. J. *Photochem. Photobiol.* 1991, *54*, 499.
22. Murphy, C. J.; Arkin, M. R.; Ghatlia, N. D.; Bossmann, S.; Turro, N. J.; Barton, J. K. *Proc. Natl. Acad. Sci. USA* 1994, *91*, 5315.
23. Stemp, E. D. A.; Arkin, M. R.; Barton, J. K. *J. Am. Chem. Soc.* 1995, *117*, 2375.
24. Holmlin, R. E.; Stemp, E. D. A.; Barton, J. K. *J. Am. Chem. Soc.* 1996, *118*, 5236.
25. Arkin, M. R.; Stemp, E. D. A.; Turro, C.; Turro, N. J.; Barton, J. K. *J. Am. Chem. Soc.* 1996, *118*, 2267.
26. Arkin, M. R.; Stemp, E. D. A.; Holmlin, R. E.; Barton, J. K.; Hörmann, A.; Olson, E. J. C.; Barbara, P. F. *Science* 1996, *273*, 475.
27. Murphy, C. J.; Arkin, M. R.; Jenkins, Y.; Ghatlia, N. D.; Bossmann, S. H.; Turro, N. J.; Barton, J. K. *Science* 1993, *262*, 1025.
28. Lincoln, P.; Tuite, E.; Nordén, B. *J. Am. Chem. Soc.* 1997, *119*, 1454.
29. McGhee, J. D.; von Hippel, P. H. *J. Mol. Biol.* 1974, *86*, 469.
30. Hiort, C.; Lincoln, P.; Nordén, B. *J. Am. Chem. Soc.* 1993, *115*, 3448.
31. Lincoln, P.; Broo, A.; Nordén, B. *J. Am. Chem. Soc.* 1996, *118*, 2644.
32. Nordén, B.; Lincoln, P.; Åkerman, B.; Tuite, E. *DNA Interactions with Substitution-Inert Transition Metal Ion Complexes*; Nordén, B.; Lincoln, P.; Åkerman, B.; Tuite, E.; eds.; Marcel Dekker, Inc.: New York, 1996; vol. 33, p. 177.
33. Tuite, E.; Lincoln, P.; Nordén, B. *J. Am. Chem. Soc.* 1997, *119*, 239.
34. Choi, S.-D.; Kim, M.-S.; Kim, S. K.; Lincoln, P.; Tuite, E.; Nordén, B. *Biochemistry* 1997, *36*, 214.
35. Olson, E. J. C.; Hu, D.; Hörmann, A.; Barbara, P. F. *J. Phys. Chem. B* 1997, *101*, 299.
36. Schulman, L. S.; Bossmann, S. H.; Turro, N. J. *J. Phys. Chem.* 1995, *99*, 9283.
37. Tuite, E. *Photochem. Photobiol.* 1996, *63S*, 9.
38. Becker, H.-C.; Norden, B. *J. Am. Chem. Soc.* 1997, in press.
39. Hoffmann, T. A.; Ladik, J. *Adv. Chem. Phys.* 1964, *7*, 84.
40. Dee, D.; Baur, M. E. *J. Chem. Phys.* 1974, *60*, 541.
41. Risser, S. M.; Beratan, D. N.; Meade, T. J. *J. Am. Chem. Soc.* 1993, *115*, 2508.
42. Priyadarshy, S.; Beratan, D. N.; Risser, S. M. *Int. J. Quant. Chem: Quant. Biol. Symp.* 1996, *23*, 65.
43. Priyadarshy, S.; Risser, S. M.; Beratan, D. N. *J. Phys. Chem.* 1996, *100*, 17678.
44. Beratan, D. N.; Priyadarshy, S.; Risser, S. M. *Chemistry and Biology* 1997, *4*, 3.

45. Meade, T. J.; Kayyem, J. F. *Angew. Chem., Int. Ed. Engl.* 1995, *34*, 352.
46. Felts, A. K.; Pollard, W. T.; Freisner, R. A. *J. Phys. Chem.* 1995, *99*, 2929.
47. Stemp, E. D. A.; Barton, J. K. *Electron Transfer Between Metal Complexes Bound to DNA: Is DNA a Wire?*; Stemp, E. D. A.; Barton, J. K., eds.; Marcel Dekker, Inc.; New York, 1996; vol. 33, p. 325.
48. Kittler, L.; Löber, G.; Gollmich, F. A.; Berg, H. *J. Electroanal. Chem.* 1980, *116*, 503.
49. Brun, A. M.; Harriman, A. *J. Am. Chem. Soc.* 1991, *113*, 8153.

3

Photoinduced Electron Transfer in Metal-Organic Dyads

Kirk S. Schanze and Keith A. Walters
University of Florida, Gainesville, Florida

I. INTRODUCTION

Photoinduced electron transfer has played a pivotal role in the development of an understanding of factors that control electron transfer (ET) reactions [1–15]. Early studies of photoinduced ET focused on organic chromophore/organic electron acceptor (or donor) systems [16,17]. It was not long, however, before well-defined examples of photoinduced ET in transition metal-based systems were discovered [18–23]. Following these initial studies, a substantial number of investigations were carried out using transition metal chromophores that provided a wealth of data concerning the effect of thermodynamics, molecular structure, and the medium upon the electron transfer rate [4].

Early work on the kinetics of photoinduced ET in transition metal complex systems focused exclusively on bimolecular reactions between transition metal chromophores and electron donors or acceptors. However, concomitant with the advances in rapid photochemical kinetic methods and chemical synthetic methodology, emphasis shifted to photoinduced ET in "chromophore–quencher" assemblies that comprise a metal complex chromophore covalently linked to an organic electron donor or acceptor [24]. These supramolecular compounds afford several advantages for kinetic studies, including: (1) the ability to directly determine the rates of forward and back ET (k_{FET} and k_{BET}, respectively) without concern for the dynamics of diffusion; (2) control of the relative orientation of and separation distance between the donor–acceptor pair; and (3) control over the structure of the molecular spacer between the donor–acceptor pair. In addition to the obvious advantages associated with ET rate studies on covalently linked systems, the transition metal-based chromophore–quencher systems possess features that are particularly amenable for the study of long-range electron transfer including long-lived and luminescent metal-to-ligand charge transfer (MLCT) excited states, tunable excited state energies and redox potentials, and a versatile synthetic chemistry [25].

The study of photoinduced ET in covalently linked donor-acceptor assemblies began with comparatively simple "dyad" systems which contain a transition metal center covalently linked to a single electron donor or acceptor unit [26]. However, work in this area has naturally progressed and in recent years complex supramolecular assemblies comprised of one or more metal complexes that are covalently linked to one or more organic electron donors or acceptors have been synthesized and studied [27–36]. Furthermore, several groups have utilized the useful photoredox properties of transition metal complexes to probe electron and energy transfer across spacers comprised of biological macromolecules such as peptides [37,38], proteins [39,40], and polynucleic acids [41].

Although studies of the comparatively more complex transition metal based supramolecular assemblies have led to many useful findings, this review is limited to studies of photoinduced intramolecular ET in *metal-organic dyads*. A metal-organic dyad is defined as a supramolecular assembly wherein a transition metal complex serves as the light absorbing chromophore (and as the excited state electron donor or acceptor) and a single organic electron donor or acceptor is covalently linked to the metal complex chromophore. Furthermore, this chapter will consider in detail only work done on metal-organic dyads wherein the driving force for photoinduced forward and back electron transfer (ΔG_{FET} and ΔG_{BET}, respectively) are known and in which the kinetics of forward and/or back electron transfer in the assemblies have been determined by excited state kinetic methods. The rationale for this specificity is motivated by the objective of the review, which is to survey relationships that exist between the dynamics of ET and ΔG_{ET}, donor-acceptor distance, spacer structure and spin-state factors in structurally well-defined transition metal dyad systems. Work in this area that was published through early 1997 is covered, although the chapter is by no means a comprehensive review. (We apologize in advance for any oversights that we make in reviewing the literature in this area.) Several other recent papers are available that review work on photoinduced intramolecular ET in organic-based dyads and multichromophoric systems [42–43].

II. PHOTOINDUCED ELECTRON TRANSFER AND ELECTRON TRANSFER THEORY

A. Electron Transfer in Metal-Organic Dyads

In most of the metal-organic dyads described in this review the metal center has a
d^6 electronic configuration. Further, the lowest excited state typically has a metal-
to-ligand charge transfer (MLCT) configuration arising from promotion of a metal
centered d-electron into a ligand based π^* level, e.g.,

$$L - M(d^6) \xrightarrow{h\nu} {}^*[\bar{L} - M(d^5)] \tag{1}$$

The MLCT basis for the reactive excited state leads to interesting consequences with respect to the orbitals involved in photoinduced forward and back
ET in metal complex dyads. In order to categorize this difference we define two
categories of dyad systems: Type 1 dyads contain an electron acceptor covalently
attached to the d^6 metal chromophore and type 2 dyads contain an electron donor
covalently attached to the d^6 metal chromophore (see Fig. 1). In the type 1 dyads,
photoinduced forward ET involves transfer of an electron from a π^* orbital
localized on the acceptor ligand, L, to a π^* orbital on the organic electron
acceptor, A. Back ET involves transfer of an electron from a π^* orbital of the
organic electron acceptor, A, to the d-shell of the transition metal center. By
contrast, in the type 2 dyads photoinduced forward ET involves transfer of an
electron from a π orbital on the organic donor, D, into the hole in the d-shell of the

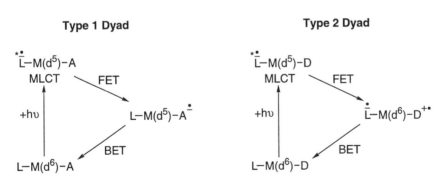

FIGURE 1 Photoinduced electron transfer schemes for type 1 and type 2 metal-organic
dyads. Key: L is a diimine ligand such as 2,2'-bipyridine; M is a transition metal; d^n
indicates the electron count in the valence shell d-orbitals of M; A is an organic electron
acceptor; D is an organic electron donor; FET is forward ET; BET is back ET.

excited metal. Back ET involves transfer of an electron from a π^* orbital localized on the acceptor ligand, L, to a π orbital localized on the organic donor, D. Thus, in the type 2 dyads forward ET involves donor to metal ET, while back ET is essentially a charge recombination reaction between an organic radical anion (L$\bar{}$) and an organic radical cation (D$\overset{+}{}$).

B. Thermodynamics of Photoinduced Electron Transfer

A detailed discussion of the energetics of photoinduced intramolecular ET in organic dyads is available in the literature [44]. Therefore, only a brief discussion of the energetic features which are unique to transition metal dyads are presented herein. The free energy for photoinduced forward and back ET (ΔG_{FET} and ΔG_{BET}, respectively) between an electron donor and an electron acceptor (D and A, respectively) is given by

$$\Delta G_{FET} = E_{1/2}(D/\overset{+\cdot}{D}) - E_{1/2}(A/\overset{-\cdot}{A}) - E_{00} - CSE \tag{2a}$$
$$\Delta G_{BET} = E_{1/2}(A/\overset{-\cdot}{A}) - E_{1/2}(D/\overset{+\cdot}{D}) + CSE \tag{2b}$$

where $E_{1/2}(A/\overset{-\cdot}{A})$ and $E_{1/2}(D/\overset{+\cdot}{D})$ are, respectively, the half-wave electrochemical potentials for reduction of the acceptor and oxidation of the donor (usually obtained by cyclic voltammetry), E_{00} is the 0–0 energy of the excited state reactant (either *A or *D) and CSE is the coulombic stabilization energy of the charge separated state (CSE = $e^2/\varepsilon r_{DA}$, where e = electron charge, ε = solvent dielectric constant, and r = center-to-center donor-acceptor separation distance) [44].

Equations (2a) and (2b) have been applied widely to organic-based dyads where forward ET involves charge separation (e.g., *D-A $\rightarrow$ D$^+$-A$^-$) and back ET charge annihilation (e.g., D$^+\cdot$-A$^-\cdot$ $\rightarrow$ D-A). In this situation the CSE term provides an estimate of the stabilization energy of the charge-separated state which arises due to coulombic attraction of the (+) and (−) charges on the acceptor and donor. However, the situation is often more complicated (or simpler!) in metal-organic dyads. In the simple cases, forward and back ET are charge-shift reactions and the coulombic stabilization energy is exactly zero. Alternatively, in a host of metal-organic dyads the metal complex and/or the organic electron acceptor (or donor) are charged in the ground state. In these cases, accurate calculation of the coulombic stabilization energy is difficult and most often this term is neglected.

C. Electron Transfer Theory

Two different theoretical approaches have been used to relate the ET rate constant (k_{ET}) to the thermodynamic driving force (ΔG_{ET}) and parameters related to molecular structure. The first approach is the semiclassical treatment derived from the early work by Marcus on ET theory [2–4,45].

$$k_{ET} = \nu_n \kappa_{el} \kappa_n \tag{3a}$$

where

$$\nu_n \kappa_{el} = \left[\frac{2H_{DA}^2}{h}\right]\sqrt{\frac{\pi^3}{\lambda k_B T}} \quad \text{and} \quad \kappa_n = \exp\left[-\frac{(\Delta G_{ET} + \lambda)^2}{4\lambda k_B T}\right] \tag{3b}$$

In Eq. (3a), ν_n is an average vibrational frequency ($\nu_n \approx 6 \times 10^{12}$ s^{-1}) and κ_n and κ_{el} are the nuclear and electronic transmission coefficients (the latter parameters vary from 0 to 1). In Eq. (3b), H_{DA} is the donor-acceptor electronic coupling matrix element (proportional to the donor-acceptor orbital overlap), λ is the total reorganization energy ($\lambda = \lambda_i + \lambda_s$, where λ_i and λ_s are the inner- and outer-sphere reorganization energies), ΔG_{ET} is the driving force for electron transfer and h, k_B, and T have their usual meanings. Note that the electronic transmission coefficient (κ_{el}) is proportional to the square of the donor-acceptor electronic coupling matrix element (H_{DA}^2). When the donor–acceptor electronic coupling is large ($H_{DA} \geq 50$ cm^{-1}), $\kappa_{el} = 1.0$ and the ET reaction is adiabatic; by contrast, when electronic coupling is small ($H_{DA} \ll 50$ cm^{-1}), $\kappa_{el} \ll 1.0$ and the ET reaction is nonadiabatic [5,6].

Many experimentalists use a simplified version of Eq. (3) to analyze experimental rate data [45]

$$k_{ET} = k_{ET}^{\circ} \exp\left\{-\frac{(\Delta G + \lambda)^2}{4\lambda k_B T}\right\} \tag{4}$$

where k_{ET}° is the "maximum" rate constant that is observed when $\Delta G = \lambda$. Comparison of Eqs. (3) and (4) reveals that $k_{ET}^{\circ} = \nu_n \kappa_{el}$; note that when $\kappa_{el} = 1$, k_{ET}° attains its maximum value which is equal to the vibrational frequency. In other words, the maximum ET rate corresponds to the frequency of the vibrational mode(s) that is (are) coupled to the ET reaction.

The second theoretical approach is quantum mechanical in nature and is based on the Fermi Golden Rule expression for nonradiative decay processes [45,46]

$$k_{ET} = \frac{4\pi^2}{h} H_{DA}^2 \langle FC \rangle \tag{5a}$$

In this approach k_{ET} is proportional to the square of the donor–acceptor electronic coupling matrix element (H_{DA}) and a Franck–Condon term that contains the dependence of the ET rate on ΔG_{ET}, λ and factors related to the molecular structure,

$$\langle FC \rangle = \left(\frac{1}{4\pi\lambda_s k_B T}\right)\sum_{w=0}^{\infty}\left(\frac{e^{-S}S^w}{w!}\right)\exp\left\{-\left[\frac{(\lambda_s + \Delta G + wh\omega)^2}{4\lambda_s k_B T}\right]\right\} \tag{5b}$$

In Eq. (5b), $h\omega$ is the frequency and S is the unitless displacement of the high-frequency intramolecular vibrational mode which is coupled to the ET process,

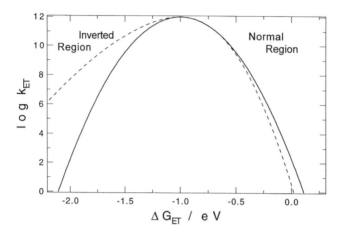

FIGURE 2   Plots of $\log k_{ET}$ vs. ΔG_{ET} calculated by using theory. Solid line calculated by using semiclassical theory [Eq. (4), see text] and dashed line calculated by using quantum theory [Eq. (5)]. Rate maximum occurs when the total reorganization energy (λ) is equal to the driving force (e.g., when $\lambda = |\Delta G_{ET}|$).

and the summation is taken over the quantum levels (ω) of the high-frequency mode.

Figure 2 illustrates plots of $\log (k_{ET})$ vs. ΔG_{ET} calculated by using the semiclassical and quantum mechanical expressions with parameters that are typical for ET between small organic or inorganic molecules in a polar solvent in the adiabatic limit (e.g., H_{DA} is sufficiently large so that at the optimal driving force k_{ET} is limited by the vibrational frequency). Several points are clear from this comparison of the two ET rate theories. First, both theories predict that for endothermic or weakly exothermic ET reactions (e.g., where $|\Delta G| < \lambda$) k_{ET} increases as the driving force increases. This is the Marcus "normal region." Second, both theories predict that for highly exothermic reactions (e.g., where $|\Delta G| > \lambda$) k_{ET} *decreases* as the driving force increases. This is the Marcus "inverted region." Third, the two theories give qualitatively similar rate/free-energy correlations in the normal region. As a result, the relatively simpler semiclassical ET theory is adequate and used most often to model ET processes in the normal region. Finally, for ET in the inverted region the semiclassical theory predicts a quadratic decrease in $\log (k_{ET})$ with increasing driving force, while the quantum theory predicts a nearly linear (and much weaker) dependence of $\log (k_{ET})$ on ΔG_{ET}.

A significant body of experimental and theoretical work has explored the distance dependence of ET [37–39,44,47–59]. On an empirical level it has generally been found that k_{ET} decreases exponentially with increasing distance,

$$k_{ET} = k_{ET}^{\circ} \exp\{-\beta(r - r_0)\} \tag{6}$$

where k_{ET}° is the rate of electron transfer when the donor and acceptor are separated by a distance equal to the sum of their van der Waals radii (r_0), r is the center-to-center donor–acceptor separation distance, and β is an empirical parameter. Electron transfer rate studies on a wide range of model donor–acceptor compounds, modified proteins and nucleic acids generally show that k_{ET} decreases in accord with Eq. (6) with β ranging from 0.8 to 1.5 Å^{-1} [38–44,47–49]. To a first-order approximation the electronic coupling matrix element [H_{DA} in Eqs. (3) and (5)] is the term that dominates the distance dependence and following Eq. (6) we write,

$$H_{DA} = H_{DA}^{\circ} \exp\left\{-\frac{\beta}{2}(r - r_0)\right\} \tag{7}$$

where H_{DA}° is the electronic coupling matrix element when the donor and acceptor are separated by a distance equal to r_0. A thorough examination of the theory shows that the overall distance dependence of k_{ET} is determined not only by changes in H_{DA}, but also by the distance dependence of the outer-sphere reorganization energy (λ_s), which increases with distance according to Eq. (8) [60]:

$$\lambda_s = e^2 \left[\frac{1}{2r_D} + \frac{1}{2r_A} - \frac{1}{r_{DA}}\right]\left[\frac{1}{\varepsilon_{op}} - \frac{1}{\varepsilon_s}\right] \tag{8}$$

In Eq. (8), e is the electron charge, r_D and r_A are the van der Waals' radii of the donor and acceptor, r_{DA} is the center-to-center separation distance, and ε_{op} and ε_s are, respectively, the optical and static dielectric constants of the medium (e.g., the solvent or protein matrix).

III. EXPERIMENTAL DETERMINATION OF RATES OF PHOTOINDUCED INTRAMOLECULAR ELECTRON TRANSFER

Scheme 1 illustrates a kinetic diagram for photoinduced intramolecular ET in a type 2 metal-organic dyad. In this example the ground state metal complex bears a positive charge and is covalently linked to a neutral donor. Photoexcitation of the metal chromophore produces the relaxed excited state, $*M^+$-D. In the dyad, the excited state metal complex can decay by two competing channels: (1) "normal" radiative and nonradiative decay to the ground state with rate constant $k_d^{\circ} = k_r + k_{nr}$, where k_r and k_{nr} are, respectively, the rates for radiative and nonradiative decay; (2) photoinduced forward ET with rate constant k_{FET}. The total decay rate of $*M^+$-D is given by the sum of the rates of the competing decay channels,

$$k_d = k_d^{\circ} + k_{FET} \tag{9}$$

Solving Eq. (9) for the forward ET rate constant leads to Eq. (10), where τ and

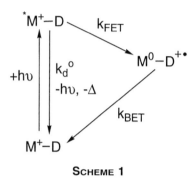

SCHEME 1

τ_{model} are, respectively, excited state lifetimes of the dyad and a "model" complex which contains the metal complex chromophore but not the electron donor.

$$k_{FET} = (k_d - k_d^\circ) = \left(\frac{1}{\tau} - \frac{1}{\tau_{model}}\right) \tag{10}$$

The product of forward ET, M-D$^{+\cdot}$, decays by back ET with rate constant k_{BET}.

Experimentally, forward ET rates are usually measured by using time-resolved emission spectroscopy to determine the lifetime of *M in the dyad, M$^+$-D (τ), and in a suitable model compound, M$^+$ (τ_{model}). Although this is a useful method for determining forward ET rates, there are some important limitations and assumptions that are implicit when this method is applied. First, in order for Eqs. (9) and (10) to hold, the rates of radiative and "normal" nonradiative decay in *M must be the same in the dyad and the model complex. This assumption is likely to be valid when the molecular structures of the model and dyad are very similar. However, the authors know of at least one system where this assumption breaks down, so caution is always advised in selecting a suitable model compound [61]. A second limitation inherent in deriving forward ET rates from time-resolved emission studies of dyads and model complexes is that the uncertainty in k_{FET} increases substantially as k_{FET} becomes small compared to k_d°. Given the errors intrinsic in determining excited state lifetimes, a practical limit is that $k_{FET} \geq 0.05 \times k_d^\circ$ to enable the ET rate to be determined with reasonable accuracy.

Experimental determination of k_{BET} is typically accomplished by using transient absorption spectroscopy to monitor the decay of the charge separated state, M-D$^{+\cdot}$. Given sufficient time resolution, it is often possible to detect the charge separated state since the radical ions that comprise this species usually have relatively strong absorption bands in the near-UV and visible spectral regions. One situation which makes direct determination of k_{BET} very difficult is when $k_{BET} > k_{FET}$. In this case the concentration of the charge separated state does not achieve a level sufficient for its detection. This situation has been encountered frequently in metal-organic dyads which consist of a d^6 metal covalently linked to organic acceptors.

IV. HISTORICAL PERSPECTIVE OF ELECTRON TRANSFER IN METAL-ORGANIC SYSTEMS

The transition metal complex $Ru(bpy)_3^{2+}$ (**1**, Scheme 2) has played a key role in the development of inorganic photochemistry and photophysics and, in particular, in

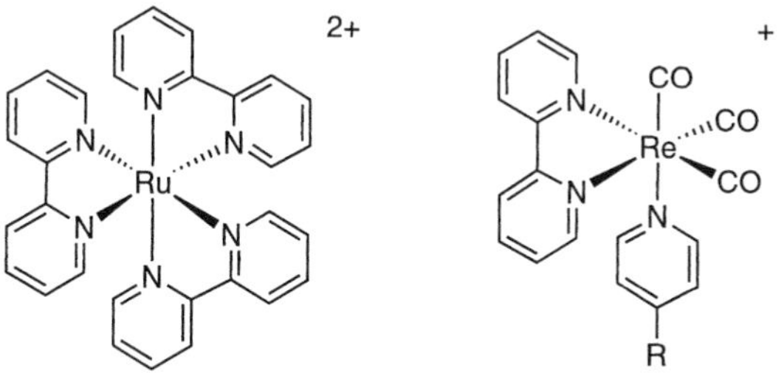

SCHEME 2

the study of photoinduced ET in inorganic systems. The significance of $Ru(bpy)_3^{2+}$ to the development of photoinduced ET in metal-organic systems can be traced to its interesting photophysical and photochemical properties [25,62–71]. (1) $Ru(bpy)_3^{2+}$ absorbs strongly in the visible region ($\lambda_{max} \approx 450$ nm) and therefore can be selectively excited in the presence of UV-absorbing quenchers. The strong visible absorption also makes $Ru(bpy)_3^{2+}$ attractive for applications in solar energy conversion schemes [72,73]. (2) The lowest excited state of $Ru(bpy)_3^{2+}$ is based on a $d\pi$ (Ru) $\rightarrow \pi^*$ (bpy) MLCT transition. The MLCT state has a relatively long lifetime (ca. 1 μs) and is luminescent. (3) The energy and lifetime of the MLCT state can be "tuned" by making simple substitutions on the bipyridyl ligands [25]. (4) $Ru(bpy)_3^{2+}$ and related complexes are relatively inert to photosubstitution and therefore can withstand long-term photolysis without undergoing significant photodecomposition.

By the mid-1960s it was clear that photoinduced ET is an important mechanism for fluorescence quenching in organic donor-acceptor systems [16,17]. The empirical Rehm and Weller equation provided a quantitative basis for predicting the relationship between redox potentials, excited state energy, and the rate of photoinduced ET in organic donor–acceptor pairs [17]. The first suggestion that photoinduced ET was the mechanism for emission quenching in metal complex systems was made in 1972 by Gafney and Adamson [18]. These authors observed that $Co(NH_3)_5Br^{2+}$ quenches the MLCT emission of $Ru(bpy)_3^{2+}$ and suggested that photoinduced ET was the likely mechanism for the quenching. The ET quenching theory was supported by the observation that $Ru(bpy)_3^{2+}$ sensitizes photoreduction of $Co(NH_3)_5Br^{2+}$ to $Co(II)$. Gafney and Adamson proposed that photoinduced ET was the primary step of the photoreduction reaction, e.g.,

$$Ru^{II}(bpy)_3^{2+*} + Co^{III}(NH_3)_5Br^{2+} \rightarrow Ru^{III}(bpy)_3^{3+} + Co^{II}(NH_3)_5Br^{1+} \rightarrow \rightarrow$$
$$Co(II)^{2+} + 5NH_3 + Br^- \quad (11)$$

In a series of papers that appeared shortly after Gafney and Adamson's report, Meyer, Whitten, and co-workers explored bimolecular photoinduced ET reactions between $Ru(bpy)_3^{2+*}$ and a series of mild organic oxidants and reductants [19–23]. Their first communication reported flash photolysis studies which demonstrated that $Ru(bpy)_3^{2+*}$ could efficiently reduce pyridinium ions such as **3** and **4** (Scheme 3), e.g. [19],

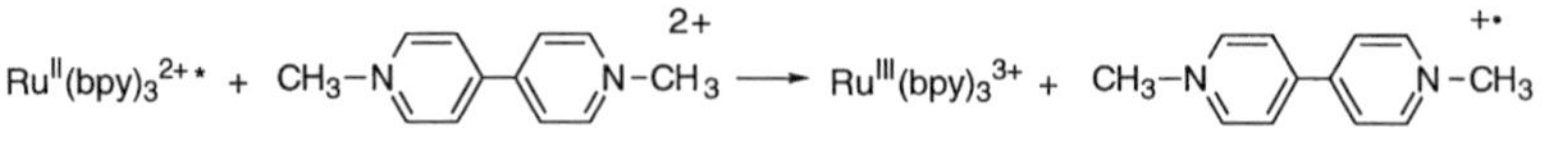

(12)

Shortly thereafter these authors demonstrated (again by using flash photolysis) that $Ru(bpy)_3^{2+*}$ could efficiently oxidize mild reducing agents such as N,N-

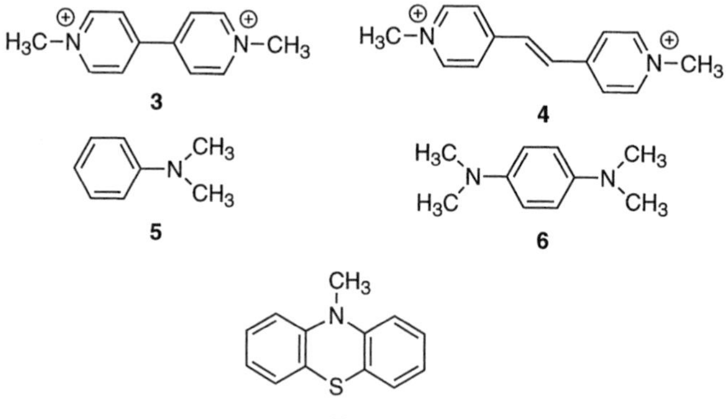

SCHEME 3

dimethylaniline (**5**), N,N,N',N'-tetramethylphenylene diamine (**6**) and 10-methyl-phenothiazine (**7**) [22]. Finally, the series of publications from the Meyer and Whitten collaboration culminated with a full paper in 1979 showing that the free-energy relationship for ET quenching of $Ru(bpy)_3^{2+}$ by organic electron donors and acceptors such as **3–7** is modeled well by Marcus's semiclassical ET theory [23]. By this point in time the field of photoinduced ET in inorganic chemistry was well developed. Significantly, during the 1970s inorganic photochemists clearly demonstrated that the Marcus free-energy relationship was valid for photoinduced ET processes in metal-organic systems.

Another development that was key in determining the course of investigations concerning intramolecular ET in metal-organic dyads was the evolution of the fac-(bpy)Re^I(CO)$_3$(L) chromophore (**2**, Scheme 2). The key synthetic and photophysical work on this family of complexes was accomplished by Wrighton and co-workers during the mid-1970s [74–77]. The Wrighton group's early papers established that the lowest excited state of these complexes is based on a $d\pi$ (Re) → π^* (bpy) MLCT transition, in direct analogy to the MLCT state of Ru(bpy)$_3^{2+}$. The MLCT state of the (bpy)Re^I(CO)$_3$(L) chromophore was demonstrated to be long-lived (ca. 1 μs), luminescent, and quenched by electron donors and acceptors, presumably by an ET mechanism [77].

A significant feature of the (bpy)Re^I(CO)$_3$(L) unit which has facilitated its application in studies of intramolecular ET is the ability to covalently link electron donor and acceptor moieties to the complex via the monodentate "L" ligand [78]. For example, complexes of the type (bpy)Re^I(CO)$_3$(py-X)$^+$, where py-X is a substituted pyridine, are coordinatively stable. Thus, it is possible to use the pyridyl ligand as a "lead-in" to a variety of organic spacers and/or electron donor and acceptor groups. For synthetic reasons this approach to covalent attachment of functional groups to a photoactive metal center is preferable compared to functionalization of bidentate ligands such as 2,2'-bipyridine [79]. Unfortunately, the latter approach is the strategy that must be taken when covalently attaching functional groups to a Ru(II) center, owing to the fact that complexes of the type (bpy)$_2$Ru(L)$_2^{2+}$ typically undergo efficient photosubstitution of the "L" ligands.

V. EARLY EXAMPLES OF METAL-ORGANIC DYADS

The first investigations of intramolecular ET in metal-organic dyads involve systems which use synthetic inorganic or organic chemistry to covalently link together a redox-active metal complex chromophore and an organic quencher. It is logical that these first covalently linked assemblies consist of the "building blocks" that had been used widely in the extensive studies of bimolecular photoinduced ET reactions, i.e., Ru(bpy)$_3^{2+}$ and electron donors and acceptors such as paraquat (**3**) and phenothiazine (**7**).

The first example of a covalently linked metal-organic dyad is reported by Meyer and co-workers in a communication that appeared in 1978 [80]. This manuscript describe the synthesis, electrochemistry, and photophysics of **8** (Scheme 4), which consists of a pair of N-methyl-4,4'-bipyridinium (monoquat) electron acceptors coordinated directly to the (bpy)$_2$Ru^{II} chromophore. This complex is an "inner-sphere" analog of the Ru(bpy)$_3^{2+}$/paraquat (**3**) system which has received significant attention in studies of bimolecular photoinduced ET. Complex **8** displays electrochemical features typical of both the Ru-bpy unit and the monoquat electron acceptor. The photophysical results indicate that in fluid solution the lowest excited state is based on a Ru → monoquat MLCT transition.

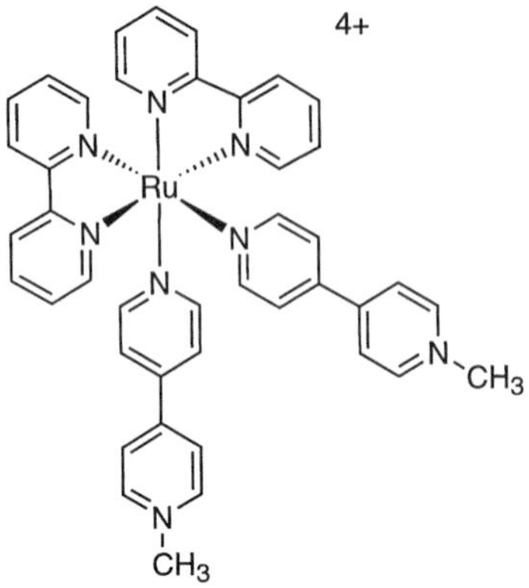

8

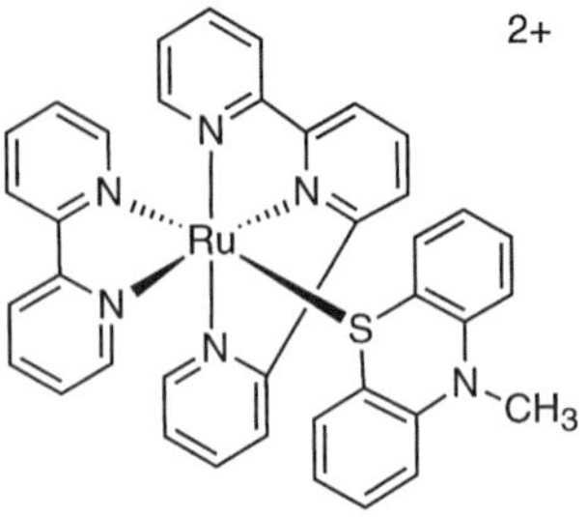

9

SCHEME 4

Moreover, this excited state is produced *directly* by photoexcitation, i.e.,

$$(bpy)_2 Ru^{II}(MQ^+)_2 \xrightarrow{h\nu} {}^*(bpy)_2 Ru^{III}(MQ^{\cdot})(MQ^+) \tag{13}$$

(bpy = 2,2'-bipyridine and MQ^+ = N-methyl-4,4'-bipyridinium)

and decays with a lifetime of approximately 100 ns. Thus, complex **8** is the first example of a metal-organic dyad in which photoexcitation produces a charge separated state having an electronic structure analogous to that achieved by bimolecular photoinduced ET between the separated molecular redox units. Unfortunately, analysis of the dynamics of forward and back ET in **8** is complicated by the fact that electronic coupling between the Ru and monoquat sites is comparatively large. As a result, the charge-separated state is best considered as a charge transfer excited state (with partial rather than complete ET between the metal and the acceptor ligand).

The second report of a covalently linked metal-organic dyad appeared in 1983 [81]. This study reports the synthesis, electrochemistry, and photophysics of complex **9** which consists of a N-methylphenothiazine electron donor coordinated directly to the Ru^{II}(bpy)(trpy) (trpy = 2,2',2''-terpyridine) chromophore. The structure of **9** is not established unequivocally; however, there is a strong indication that the phenothiazine unit is coordinated to Ru through the heterocyclic sulfur as shown in Scheme 4. Complex **9** is an "inner sphere" analog of the bimolecular photoinduced ET system consisting of $Ru(bpy)_3^{2+}/N$-methylphenothiazine (**7**). Although the electrochemistry of **9** features waves characteristic of the phenothiazine and Ru(II) components and indicates that photoinduced ET from phenothiazine to the excited Ru(bpy)(trpy) chromophore should be exothermic, the luminescence data suggests that photoinduced ET does not occur. Despite the ambiguity of the photophysical results obtained with **9**, this complex is the forerunner of a substantial number of metal-organic dyad systems that appeared shortly after this initial report. Many of these more refined studies are described in the ensuing sections of the chapter.

VI. PHOTOINDUCED INTRAMOLECULAR ELECTRON TRANSFER IN METAL-ORGANIC DYADS: "INNER-SPHERE" ACCEPTORS AND DONORS

A. Dyads with Pyridinium Acceptors

A significant amount of work has been carried out on complexes that are analogous to **8** in that they contain pyridinium acceptors directly coordinated to a photoactive metal center [82–85]. As noted above, in these complexes electronic coupling between the metal center and the pyridinium acceptor is comparatively large, and as a result the dynamics of photoinduced forward and back ET are best considered by using excited state decay theory [86]. In any event, these complexes have figured prominently in the study of ET in metal-organic dyads and some of the important discoveries made with them are briefly reviewed in this section.

The Re(I) complex **10a** (Scheme 5) has been featured in much of this work

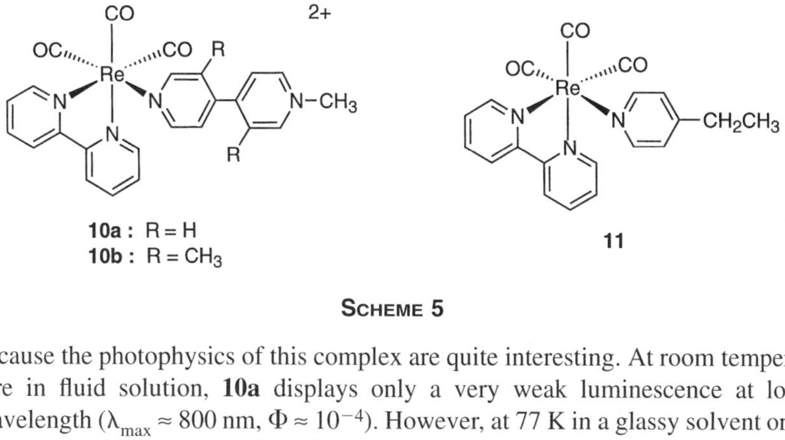

10a : R = H
10b : R = CH$_3$

11

SCHEME 5

because the photophysics of this complex are quite interesting. At room temperature in fluid solution, **10a** displays only a very weak luminescence at long wavelength ($\lambda_{max} \approx 800$ nm, $\Phi \approx 10^{-4}$). However, at 77 K in a glassy solvent or at room temperature in a rigid polycarbonate glass the complex displays strong luminescence ($\lambda_{max} \approx 560$ nm) that is superimposable with the emission of **11** (Scheme 5) which does not contain the N-methyl-4,4'-bipyridinium (MQ$^+$) electron acceptor ligand. The luminescence observed when **10a** is dissolved in rigid environments is characteristic of the $d\pi$ (Re) $\rightarrow \pi^*$ (bpy) MLCT excited state; thus the photophysical data implies that the MQ$^+$ ligand is only able to "quench" this excited state in fluid media [82,83].

Other studies shed light on the mechanism of the unusual luminescence properties of **10a** [84]. A cyclic voltammogram of **10a** features reversible waves at

−0.68 and −1.17 V that are due, respectively, to reduction of the N-methyl-4,4-bipyridinium (MQ^+) and 2,2′-bipyridine ligands. Given this electrochemical result, two sets of MLCT excitations are anticipated (in order of increasing energy): the first based on a $d\pi$ (Re) → π^* (MQ^+) transition and the second on a $d\pi$ (Re) → π^* (bpy) transition. However, the energy of the $d\pi$ (Re) → π^* (MQ^+) MLCT state is a strong function of the dihedral (twist) angle (Θ) between the planes defined by the two pyridyl rings of the MQ^+ ligand. Interestingly, the energy of the $d\pi$ (Re) → π^* (MQ^+) MLCT state is at a minimum when $\Theta = 0°$ (e.g., when MQ^+ is planar); this effect is due to the increased delocalization of the odd electron on MQ˙ which is imparted by Re → MQ^+ MLCT excitation. Another important piece of information comes from the X-ray crystal structure of **10a** which indicates that in the ground-state complex the inter-ring dihedral angle is approximately 45° [84].

   With this information in hand the photophysical properties of the complex are explained. In the relaxed ground state, the MQ^+ ligand is twisted and the $d\pi$ (Re) → π^* (bpy) MLCT state is lowest in energy. In fluid solution, near UV photoexcitation produces the Re → bpy MLCT state and rapidly thereafter intramolecular bpy → MQ^+ ET occurs to produce the $d\pi$ (Re) → π^* (MQ^+) MLCT state (Scheme 6). By contrast, in rigid environments (77 K solvent glass or 298 K

polycarbonate matrix) rotation around the inter-ring bond in MQ^+ is slow or completely inhibited and intramolecular bpy → MQ^+ ET (the second step in Scheme 6) does not occur. The very different electronic structure of the two MLCT configurations is underscored by comparison of the transient absorption spectra of **10a** and model complex **11** taken 40 ns after excitation in fluid solution (Fig. 3) [83]. The model complex (Fig. 3a) features a strong absorption band with $\lambda_{max} \approx 370$ nm that is very characteristic of the 2,2′-bipyridine radical anion (bpy⁻˙) that is produced by Re → bpy MLCT excitation. By contrast, the transient absorption of **10a** (Fig. 3b) exhibits a sharp, narrow band at $\lambda_{max} \approx 365$ nm and a very strong, broad band with $\lambda_{max} \approx 610$ nm; these features are associated with the reduced MQ ligand. Final convincing evidence for the importance of the inter-ring torsional mode in controlling the intramolecular bpy → MQ^+ ET process is provided by the observation that in complex **10b** strong luminescence is observed from the Re → bpy MLCT state at ambient temperature in fluid solution [84]. In this complex the energy of the Re → MQ^+ MLCT state is "pinned" above that

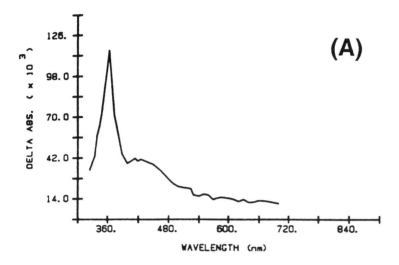

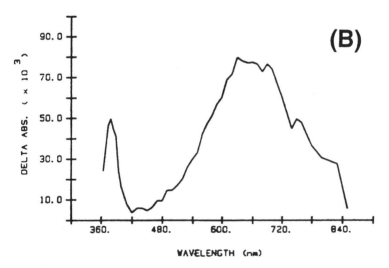

FIGURE 3   Transient absorption difference spectra obtained 40 ns following 355-nm pulsed excitation (10-ns pulse width). (A) Model complex **11**; (B) metal-organic dyad **10a**. Reprinted with permission from Ref. 83.

of the Re → bpy MLCT state because the MQ^+ acceptor ligand cannot attain a planar conformation.

B. "Inner-Sphere" Complexes with Aromatic Amine Electron Donors

Reports have appeared concerning the photophysics of several families of complexes that contain either Re(I) or Os(II) metal centers directly coordinated to aromatic amine electron donors [87,88]. These complexes display unusual, albeit complex, photophysics associated with photoinduced ET from the amine donor to the metal. By analogy with the systems discussed in the preceding section, in these complexes electronic coupling between the metal and amine donor is strong and consequently the dynamics of ET are not easily analyzed in the context of conventional nonadiabatic ET theories. However, the complexes display interesting and rather unusual properties and therefore are briefly discussed herein.

Perkins et al. reported a time-resolved emission and absorption study of the series of Os(II) complexes (**12a–d**, Scheme 7) that contain several different

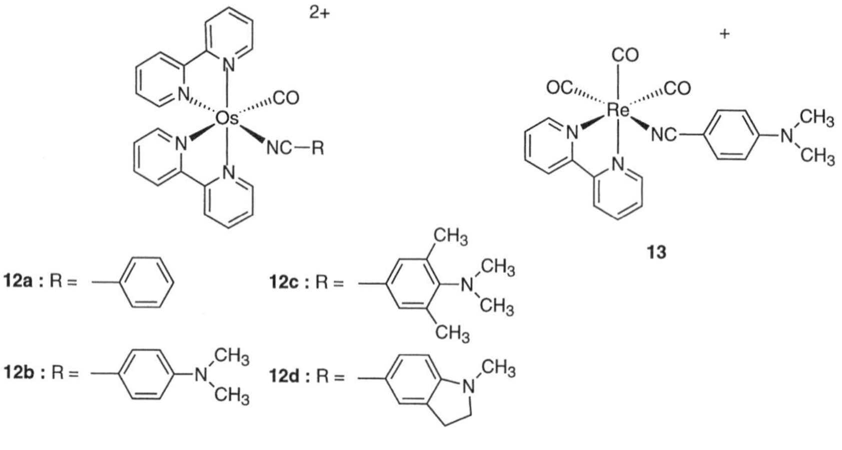

SCHEME 7

aromatic amine donors [87]. The low-energy region in the absorption spectra of **12a–d** is dominated by transitions based on $d\pi$ (Os) → π^* (bpy) MLCT. Model complex **12a**, which does not feature an electron donor, exhibits strong and long-lived luminescence from the MLCT state. By contrast, MLCT emission from each of the complexes that contain amine donors (**12b–d**) is strongly suppressed. Time-resolved emission studies of **12b–d** in acetonitrile show a short-lived ($\tau < 5$ ns) luminescence assigned to the MLCT state. The lifetime of the short-lived emission component increases with the oxidation potential of the amine donor, suggesting that emission quenching is due to ET from the amine to the metal center (e.g., **14** → **15**, Scheme 8). Picosecond transient absorption spectra of the amine

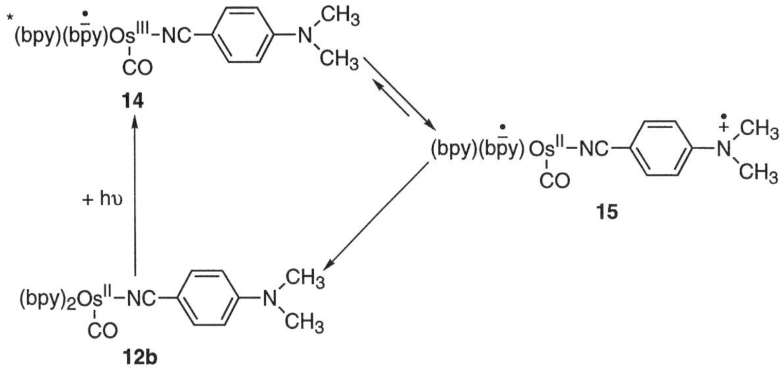

SCHEME 8

substituted complexes feature visible absorption bands characteristic of the aromatic amine radical cations, supporting the photoinduced ET mechanism (Fig. 4).

An unusual feature is that the luminescence intensity and lifetime of **12b** increases substantially in nonpolar solvents relative to the lifetime in polar solvents. Moreover, as the emission lifetime increases, the decay kinetics become distinctly nonexponential. The explanation for the unusual solvent-dependent luminescence properties of **12b** is that in a low dielectric solvent the MLCT and charge separated states (**14** and **15**, respectively, in Scheme 8) are similar in energy and equilibrium is established between the two states. In polar solvents **15** is stabilized with respect to **14**, and photoinduced ET is irreversible. Similar, but attenuated, solvent dependent luminescence is observed for **12c** and **12d**. The effect is attenuated in these complexes because the amine donors are easier to oxidize and thus the charge separated state is stabilized with respect to the MLCT state.

The study of the Os(II) complexes **12a–d** was followed shortly by a report concerning the photophysics of the Re(I) analog **13** (Scheme 7) [88]. This complex exhibits luminescence properties similar to Os(II) complex **12b**, but the solvent dependence is more exaggerated for **13**. Thus, in a low dielectric solvent **13** is strongly luminescent ($\Phi_{em} \approx 0.1$) but in a high dielectric solvent the emission is completely quenched ($\Phi_{em} < 0.001$). The electrochemistry of **13** in solvent mixtures of varying dielectric constant shows that the oxidation potential of the amine electron donor decreases as solvent polarity increases. The authors suggest that the solvent-induced quenching of the luminescence results because the charge separated state (analogous to **15** in Scheme 8) is stabilized in high dielectric solvents.

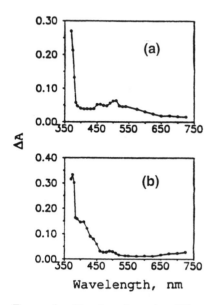

FIGURE 4; Transient absorption difference spectra obtained 500 ps after 355-nm excitation pulse (30-ps pulse width). (a) Model complex **12a**; (b) metal-organic dyad **12b**. (Reprinted with permission from Ref. 87.)

VII. PHOTOINDUCED INTRAMOLECULAR ELECTRON TRANSFER IN METAL-ORGANIC DYADS: DRIVING FORCE DEPENDENCE

A. Type 1 Dyads

Several systematic studies of the driving force dependence of the rate of forward and back ET in type 1 dyads (see Fig. 1) were carried out during the past decade. As might be expected, the type 1 dyads used in these investigations consist of covalently linked assemblies of metal complexes and organic quenchers used in early studies of bimolecular photoinduced ET reactions. Thus, the type 1 dyads consist of polypyridine Ru(II) complexes linked to pyridinium acceptors such as paraquat and diquat (quaternized 2,2′-bipyridine).

Scheme 9 illustrates the sequence of events that occur when these Ru(II)-pyridinium type 1 dyads (**16**) are photoexcited. Visible light excitation produces the MLCT excited state, **17**. Forward ET occurs via transfer of an electron from the bipyridine acceptor ligand to the covalently linked pyridinium acceptor to produce charge separated state **18**, which features a d^5 Ru(III) ion linked to the reduced pyridinium acceptor. Finally, back ET occurs by transfer of the odd electron from the pyridinium radical to the Ru(III) center.

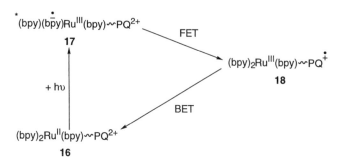

SCHEME 9

The first detailed study of the driving force dependence for photoinduced ET was reported by Elliott and co-workers [89]. The dyads used in Elliott's work consist of a tris-diimine Ru(II) chromophore covalently linked to a series of quaternized 2,2'-bipyridine (diquat) electron acceptors, **19a–c** (Scheme 10, Table

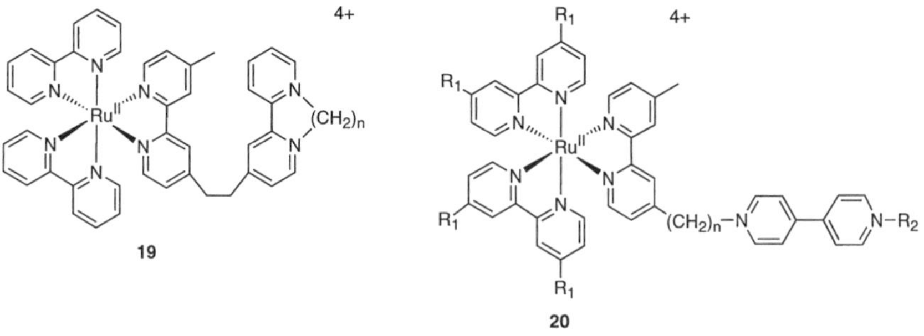

SCHEME 10

1). The driving force dependence for ET in this series of complexes arises from the effect of the length of the $-(CH_2)_n-$ bridge that connects the pyridine nitrogens on the reduction potential of the diquat. As the length of the $-(CH_2)_n-$ bridge increases, the diquat becomes harder to reduce [90]. This change in reduction potential is due to the effect of the bridge length on the ability of the diquat to become planar [90]. The driving force for photoinduced forward ET (ΔG_{FET}) in **19a–c** varies from -0.24 to -0.57 eV (Marcus normal region), and rates for forward ET were determined by monitoring the decay of the emission from the Ru $\rightarrow$ bpy MLCT excited state by using a streak camera. Forward ET rates in **19a–c** determined from the MLCT emission decays are listed in Table 1. In general, intramolecular ET in **19a–c** is very rapid, and the rates increase as ΔG_{FET} becomes more negative. Analysis of the rate-free energy correlation for **19a–c** with Marcus semiclassical theory [Eqs. (3) and (4)] suggests that $k° \approx 1.4 \times 10^{11}$ s^{-1} and $\lambda \approx 0.85$ eV [89]. Picosecond transient absorption spectroscopy is used to examine

TABLE 1 Photoinduced Electron Transfer in Type 1 Metal-Organic Dyads[a]

Compound	n	Substituents		ΔG_{FET}/eV	r_{DA}/Å	$k_{FET}/10^9$ s^{-1}	ΔG_{BET}/eV	$k_{BET}/10^9$ s^{-1}	Ref.
		R_1	R_2						
19a	4			−0.24	8.5	0.16		c	89
19b	3			−0.36	8.5	0.59		c	89
19c	2			−0.57	8.5	4.0		c	89
20a	2	−H	−CH$_3$	−0.42	10.2	7.9	−1.65	16	91
20b	2	−CH$_3$	−CH$_3$	−0.52	10.2	39	−1.54	29	91
20c	2	−H	−CH$_2$CN	−0.55	10.2	11	−1.51	34	91
20d	2	−CH$_3$	−CH$_2$CN	−0.63	10.2	29	−1.42	51	91
20e	1	−H	−CH$_3$	−0.42	9.4	53	−1.65	40	91
20f	1	−CH$_3$	−CH$_3$	−0.52	9.4	140	−1.54	84	91
20g	1	−H	−CH$_2$CN	−0.55	9.4	120	−1.51	100	91
20h	1	−CH$_3$	−CH$_2$CN	−0.63	9.4	160	−1.42	120	91
21[b]				−0.35	10	1.0		c	94
22				−0.44	12–15	<1.0		c	31
23				−0.07	12–18	0.0055		c	32

[a]Ambient temperature in CH$_3$CN solvent.
[b]CH$_2$Cl$_2$ solvent.
[c]Charge-separated state not detectable by flash photolysis. Conclude that $k_{BET} \gg k_{FET}$

19a–c in an effort to detect the charge separated state that is produced by forward ET. However, the only transient detected is the MLCT excited state, and the authors conclude that the rate of back ET is equal to or greater than the rate of forward ET (e.g., $k_{BET} \geqslant k_{FET}$).

More recently, Mallouk and co-workers examined the driving force dependence of photoinduced ET in another series of type 1 dyads which feature a tris-diimine Ru(II) chromophore covalently linked to a series of paraquat derivatives (structures **20**, Scheme 10 and Table 1) [91]. In this series of complexes, ΔG for forward and back ET is varied by (1) changing the substituents on the diimine ligands, and (2) changing the N-alkyl substituent on the paraquat acceptor. The (average) separation distance between the Ru(II) ion and the paraquat acceptor is also varied slightly by using one or two methylene groups in the linker chain. The dynamics of photoinduced ET in complexes **20** was studied by using picosecond transient absorption spectroscopy. Figure 5 illustrates transient absorption spectra of complex **20b** taken at delay times ranging from 0 to 30 ps following excitation. The difference spectra at the earliest delay times are characteristic of the Ru $\rightarrow$ bpy MLCT excited state (moderate absorption at $\lambda = 370$ nm and bleaching at $\lambda = 450$ nm) [92], and those at later times exhibit absorption bands characteristic of the reduced paraquat acceptor (sharp absorption at $\lambda = 390$ nm and broad absorption in the red) [93]. Kinetic plots of transient absorption data reveal that both the MLCT and charge separated states decay at nearly the same rate, indicating that the rates of photoinduced forward and back ET are comparable ($k_{FET} \approx k_{BET}$). Analysis of the time-resolved data leads to estimates for k_{FET} and k_{BET} for the entire series of complexes **20** (Table 1). The kinetic data for **20** indicates that k_{FET} generally increases as ΔG_{FET} becomes more negative, consistent with Marcus normal region behavior. Analysis of the forward ET rate data using the semiclassical theory [Eqs. (3) and (4)] implies that for the series with a single methylene spacer, $k° \approx 8 \times 10^{10}$ s^{-1} and $\lambda \approx 0.7$ eV, and for the series with two methylene spacers, $k° \approx 7.4 \times 10^{10}$ s^{-1} and $\lambda \approx 1.0$ eV. This analysis, albeit qualitative, suggests that the generally faster rates observed for the complexes with the shorter methylene spacer can be attributed to a smaller reorganization energy.

The Mallouk study stands apart from all other studies of type 1 dyads in that the charge separated state (e.g., **18** in Scheme 9) is detected by picosecond transient absorption spectroscopy. Analysis of the back ET rate data for complexes **20** indicate that k_{BET} increases as ΔG_{BET} becomes less negative, consistent with Marcus inverted region behavior (see Fig 2). A more detailed discussion of the back ET rate data is provided below.

Inspection of the literature uncovers ET rate data for three other type 1 metal-organic dyad systems (Scheme 11). The first system comprises complex **21**, which features a p-benzoquinone acceptor linked to a polypyridine-Ru(II) chromophore via a 3-atom amide-based spacer [94]. Photoinduced forward ET occurs

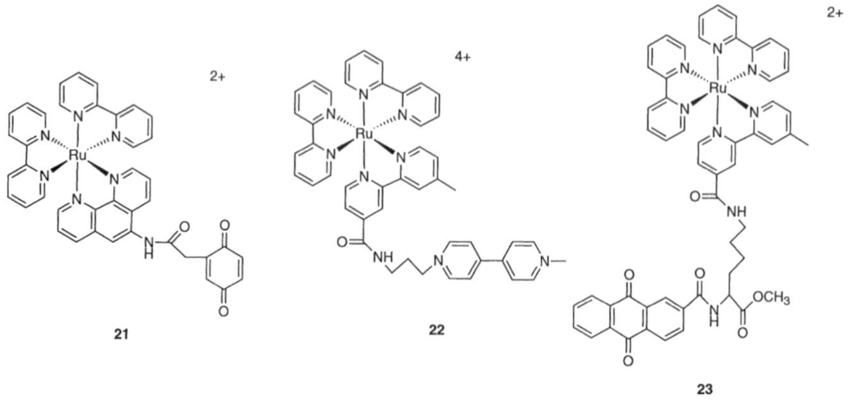

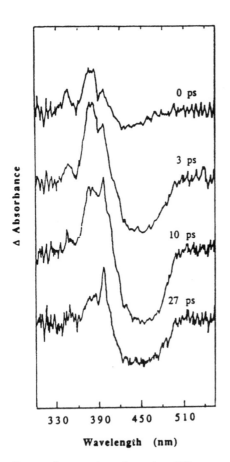

SCHEME 11

FIGURE 5 Transient absorption difference spectra of **20b** obtained following 295-nm excitation (800-fs pulse width). Delay times are listed above each spectrum. (Reprinted with permission from Ref. 91.)

in **21** with a driving force of ≈ -0.35 eV and with a rate of $k_{FET} = 1.0 \times 10^9$ s^{-1}. In two separate reports Meyer and co-workers examined photoinduced ET in type 1 dyads **22** and **23** [31,32]. Dyad **22** is very similar in structure to the complexes reported by Mallouk and co-workers, and not too surprisingly it was discovered that k_{FET} was too fast to be measured by using nanosecond instrumentation ($k_{FET} > 1 \times 10^9$ s^{-1}) [31]. Dyad **23** is unusual in that the driving force for forward ET is small, and this fact coupled with the relatively long spacer yields a relatively slow forward ET rate (Table 1) [32]. The significant feature with respect to dyads **21–23** is that they all have the common feature that the rate of back ET is so fast that the charge separated states are very difficult to detect by transient absorption spectroscopy (e.g., k_{BET} is greater than or equal to k_{FET}). Importantly, this general property is shared by all Type 1 metal-organic dyads that have been examined to date.

B. Type 2 Dyads

Most type 2 metal-organic dyads consist of the (diimine)Re^I(CO)$_3$(py)$^+$ chromophore covalently attached to an organic electron donor (**24**, Scheme 12). The Re(I)

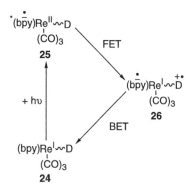

SCHEME 12

chromophore has been used in these studies for several reasons. (1) The $d\pi$ (Re) $\rightarrow \pi^*$ (diimine) MLCT state is a strong and tunable oxidizing agent; indeed, the excited state reduction potential can be varied from +1.0 to +1.5 V vs. SCE [78,95]. (2) The rhenium chromophore is inert to substitution under both thermal and photochemical conditions. (3) The synthetic chemistry of the system is straightforward. Scheme 12 illustrates the sequence of events that occur when type 2 dyads **24** are photoexcited. Near-UV excitation populates the $d\pi$ (Re) $\rightarrow \pi^*$ (bpy) MLCT excited state (**25**). Then forward ET occurs via transfer of an electron from the covalently attached donor (D in Scheme 12) to the Re center, which is formally in a d^5 configuration in the MLCT state. ET produces charge-separated state **26**, which ultimately decays by back ET from the diimine anion radical to the donor cation radical. As noted above, in a one-electron approximation back ET in the type 2 dyads involves ET from an organic anion radical to a cation radical which are held in proximity by a transition metal ion.

The first successful and detailed study of photoinduced ET in a metal-organic dyad was carried out on type 2 dyad **27c** (Scheme 13, Table 2) which

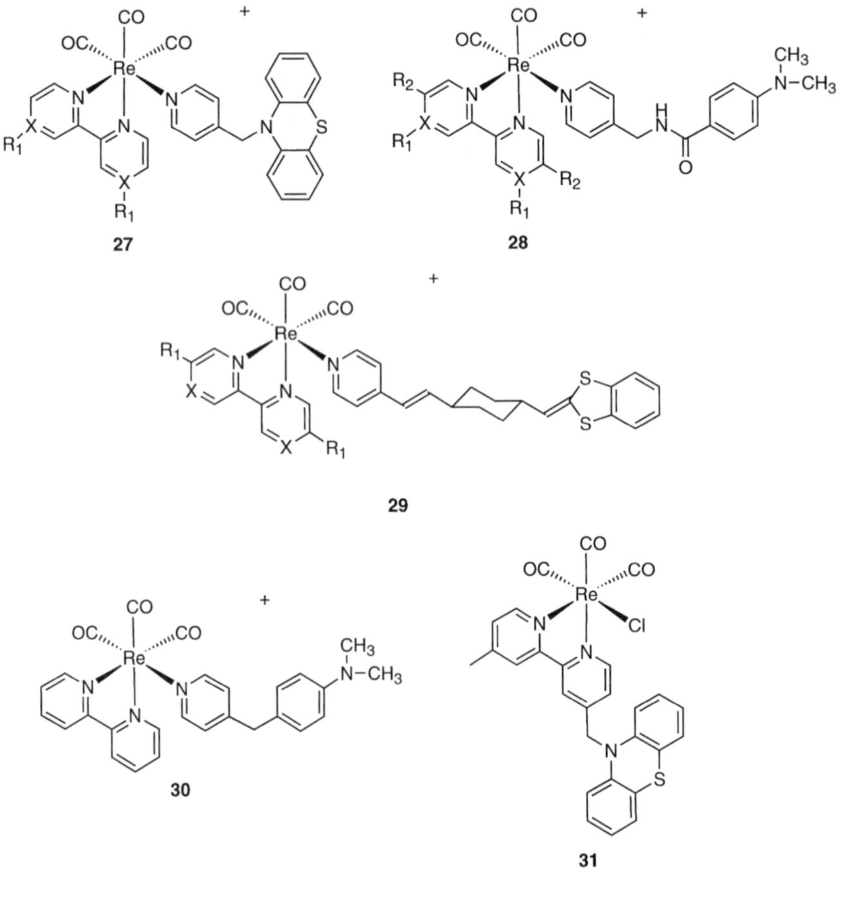

SCHEME 13

consists of the $(bpy)Re^I(CO)_3(pyridine)^+$ chromophore covalently linked to a phenothiazine (PTZ) electron donor. The properties of **27c** were first delineated in a communication [26] which was followed up by a full paper several years later [96]. Electrochemical measurements on **27c** in CH_3CN solution reveal that for-

TABLE 2 Photoinduced Electron Transfer in Type 2 Metal-Organic Dyads[a,b]

Compound	Substituents			r_{DA}/Å	ΔG_{FET}/eV	$k_{FET}/10^8$	ΔG_{BET}/eV	$k_{BET}/10^6$	Ref.
	X	R_1	R_2						
27a	C	OCH₃		6–9	c	≫2[d]	−2.15[d]	9.8[d]	98
27b	C	CH₃		6–9	c	≫2[d]	−2.09[d]	10[d]	98
27c	C	H		6–9	c	>200[d]	−1.99[d]	17[d]	98
27d	C	CONEt₂		6–9	c	≫2[d]	−1.80[d]	30[d]	98
27e	C	CO₂Et		6–9	c	≫2[d]	−1.59[d]	67[d]	98
27f	N	—		6–9	c	≫2[d]	−1.40[d]	>100[d]	98
28a	C	CH₃	CH₃	9–10.5	−0.03 (−0.19)	1.8 (0.21)	−2.57 (−2.38)	2.2 (4.7)	101
28b	C	CH₃	H	9–10.5	−0.07 (−0.20)	5.2 (0.50)	−2.44 (−2.25)	3.6 (8.3)	101
28c	C	H	H	9–10.5	−0.17 (−0.22)	7.1 (0.86)	−2.28 (−2.13)	5.3 (13)	101
28d	C	CONEt₂	H	9–10.5	−0.27 (−0.35)	12 (1.9)	−2.05 (−1.95)	9.1 (19)	101
28e	C	CO₂Et	H	9–10.5	−0.48 (−0.49)	>20 (25)	−1.75 (−1.66)	25 (36)	101
28f	N	—	H	9–10.5	−0.53 (−0.37)	>20 (9.6)	−1.59 (−1.71)	37 (—)	101
29a	C	CONEt₂		16	−0.35 (−0.26)	5 (0.23)	−1.93 (−1.96)	10 (13)	102
29b	N	H		16	−0.54 (−0.43)	8 (0.91)	−1.56 (−1.62)	31 (36)	102
30				7			(−1.99)	(100)	104
31				c	c	c	c (−2.21)	4.7[d] (6.5)	99
32				≈15	−0.12[e] (−0.10)	0.054[e] (0.027)	c	c	105
33				≈17	(−0.18)	(≫2)	(−1.97)	(30)	31

[a]Data for solutions in CH₂Cl₂/0.1 M tetrabutylammonium hexafluorophosphate unless noted otherwise.
[b]Data in parenthesis for CH₃CN solutions.
[c]Not given in reference.
[d]Solutions in CH₂ClCH₂Cl/0.1 M tetrabutylammonium hexafluorophosphate.
[e]Solutions in CH₂ClCH₂Cl.

ward ET is exothermic by ≈ 0.4 eV ($\Delta G_{FET} \approx -0.4$ eV) and back ET is exothermic by ≈ 1.95 eV ($\Delta G_{BET} \approx -1.95$ eV). Steady-state luminescence spectroscopy on **27c** demonstrates that the MLCT state is strongly quenched, presumably via intramolecular ET as shown in Scheme 12. Nanosecond transient absorption spectroscopy confirms that forward ET occurs, as evidenced by the observation of a strong absorption attributed to the cation radical of the donor, PTZ$^{+\cdot}$ (Fig. 6). The decay kinetics in the nanosecond transient absorption experiment indicate that back ET in **27c** occurs with a rate of $k_{BET} \approx 1.7 \times 10^7$ s^{-1}. Picosecond time-resolved experiments on **27c** reveal that in CH$_2$Cl$_2$ solution the absorption attrib-

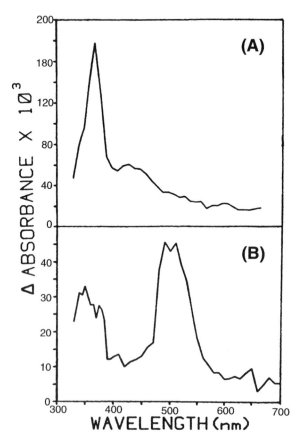

FIGURE 6   Transient absorption difference spectra obtained following 355-nm excitation (10-ns pulse width). (A) Model complex **11**, delay time 50 ns; (B) metal-organic dyad **27c**, delay time 25 ns. (Reprinted with permission from Ref. 26.)

uted to PTZ$^{+ \cdot}$ at 500 nm rises during the 30-ps laser pulse, suggesting that forward ET occurs with $k_{FET} > 2 \times 10^{10}$ s^{-1}. This ultrafast rate of forward ET is consistent with the relatively short donor–acceptor separation and the high driving force for forward ET.

Meyer and co-workers later extended their work on the rhenium–phenothiazine system by measuring the rate of back ET as a function of driving force in the entire series of complexes **27a–f** [97–99]. In this series, the driving force for back ET was varied by 0.8 eV by changing the substituents on the diimine ligand. The rates of back ET in **27a–f** were determined by monitoring the decay of the charge separated state (**26** in Scheme 12) by using nanosecond transient absorption spectroscopy (Table 2 lists the rate data). Interestingly, the rate data clearly reveals Marcus inverted region behavior, in that k_{BET} decreases as $|\Delta G_{BET}|$ increases. This is consistent with expectation, given that in each case the reactions are highly exothermic. However, although a Marcus inverted region dependence is observed, the rates of back ET in **27a–f** only vary from 10^7 to 10^8 s^{-1} over the 0.8 eV range of driving force. This is a rather weak driving force dependence (ca. 1 decade eV^{-1}) and is irregular compared to other systems in which inverted region behavior has been observed [49]. Unfortunately, the rates of forward ET were not determined for the series **27a–f**.

MacQueen and Schanze reported an extensive study of the driving force and solvent dependence of the rates of forward and back ET in the series of rhenium-based type 2 dyads **28a–f** (Scheme 13 and Table 2) [100,101]. In **28a–f** the driving force for both forward and back ET is varied by changing the substituents on the diimine ligands in a manner similar to that used by Meyer for the rhenium–phenothiazine series **27a–f**. The most significant difference between these two rhenium-based type 2 dyad systems lies in the oxidation potentials of the organic donors. Thus, the dimethylaminobenzoate donor, which is in **28a–f**, is more difficult to oxidize by approximately 300 mV compared with the phenothiazine donor in **27a–f**. The larger oxidation potential of the dimethylaminobenzoate donor makes ΔG_{FET} generally less exothermic in **28a–f** compared to **27a–f**. Importantly, the driving force for forward ET was generally low enough in all of **28a–f** such that it is feasible to use time-resolved emission to effect detailed measurements on the solvent and free energy dependence of the rate of forward ET [101]. In these experiments, time-correlated single photon counting is used to monitor the decay kinetics of the luminescence from the MLCT excited state of the (diimine)Re(CO)$_3$(py)$^+$ chromophore. The lifetimes of the MLCT state in **28a–f** are significantly shorter compared to analogous model complexes that do not contain the dimethylaminobenzoate donor; therefore, forward ET rates are given directly by the inverse of the emission decay lifetimes.

Nanosecond transient absorption experiments on **28a–f** provide clear evidence for the charge separated state (**26** in Scheme 12) that is the product of forward ET [101]. For example, Fig. 7 compares the transient absorption spectra

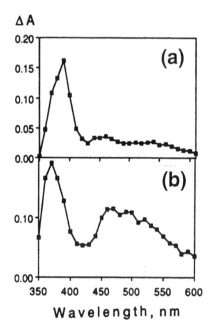

ΔA

Wavelength, nm

FIGURE 7   Transient absorption difference spectra obtained 20 ns after 355-nm excitation pulse (10-ns pulse width). (a) Model complex [(2,2′-bipyrazine)Re^{I}(CO)$_3$(4-N-benzoyl)-aminomethylpyridine]$^+$; (b) metal-organic dyad **28f**. (Reprinted with permission from Ref. 101.)

of donor substituted complex **28f** with that of a "model" complex which does not contain the electron donor. The spectrum of the model complex is typical of the Re → diimine MLCT state; the strong absorption feature at 380–390 nm is due to the diimine radical ion. By contrast, the spectrum of donor substituted complex **28f** features a strong and broad band in the mid-visible region which is attributed to the dimethylaminobenzoate radical cation that is produced by forward ET. MacQueen and Schanze used nanosecond transient absorption spectroscopy to determine the rate of back ET in **28a–f** by monitoring the decay kinetics of the charge separated state in each complex; a listing of the rate data for complexes **28a–f** is provided in Table 2.

    A variety of interesting conclusions are drawn from the study of photo-induced ET in **28a–f**. First, the rate of forward ET in this series varies from 10^8 to 10^{10} s^{-1} and increases with driving force over the range $0 < \Delta G_{FET} < -0.5$ eV. This observation is consistent with Marcus "normal" region behavior. When comparing reactions at a similar driving force, the rates of forward ET in **28a–f** are generally similar to those for the ruthenium–pyridinium type 1 dyads, suggesting that similar electronic (and spin state) factors are operating in the two types of systems. Second, for reactions at a similar driving force, the rate of forward ET in **28a–f** increases according to the trend, $k_{FET}(CH_3CN) < k_{FET}(DMF) < k_{FET}$ (CH_2Cl_2). MacQueen and Schanze explain the solvent dependence in the context of the Marcus semiclassical theory by proposing that the outer sphere reorganization energy decreases along the series $\lambda_s(CH_2Cl_2) = 0.8$ eV $< \lambda_s(DMF) = 0.9$ eV $<$ $\lambda_s(CH_3CN) = 1.0$ eV. Finally, the rates of back ET for the series **28a–f** vary from 10^6 to 5×10^7 s^{-1} over a 0.8 eV range of driving force. Importantly, k_{BET} decreases as the driving force increases, consistent with Marcus inverted region behavior. An interesting point is that a plot of log k_{BET} vs. ΔG_{BET} for the two series **27a–f** and **28a–f** follows the same correlation, which at first glance seems to imply that the electronic coupling matrix element (H_{DA}) is similar for the two systems. However, as detailed below in Section VII.D, it is possible that terms other than the Franck Condon factors and electronic coupling are important in determining the driving force dependence for charge recombination in **27a–f** and **28a–f**.

Another detailed study of a rhenium-based type 2 metal-organic dyad focused on complexes **29a,b** (Scheme 13) [100,102]. In this system, the (diimine)-Re^I(CO)$_3$(py) chromophore is covalently linked to a dithiafulvene electron donor via a rigid *trans*-cyclohexane spacer. The driving force for forward and back ET in **29** is varied by changing the diimine ligand on the rhenium chromophore. The occurrence of photoinduced ET in **29** is confirmed by the fact that the lifetime of the MLCT state is suppressed in the donor-substituted complexes compared to model complexes that contain the cyclohexane spacer but not the dithiafulvene donor. Rates of forward ET determined from the luminescence decay data are listed in Table 2. Forward ET is faster in **29b** than in **29a**, consistent with the fact that the driving force is greater in the former complex. Comparison of the forward ET rate data for complexes **29** and **28** indicates that a similar driving force forward ET is slower by nearly an order of magnitude in **29**. The lower rate of ET is likely due to weaker electronic coupling in the cyclohexane bridged system.

The charge separated state produced by forward ET in **29** is observed by nanosecond transient absorption spectroscopy [102]. The charge separated state features a strong absorption band in the mid-visible region that is due to the dithiafulvene radical cation. The lifetime of the charge separated state is determined from the decay kinetics of the transient absorption (Table 2). Interestingly, the rate of back ET in **29** is not significantly slower than in **27a–f** or **28a–f**. This is surprising, given that the cyclohexane spacer is expected to make electronic coupling weaker in **29**. This fact again points to the possibility that terms other than the Franck Condon factors and electronic coupling are important in determining the driving force dependence for charge recombination in the type 2 metal-organic dyads.

Two other rhenium–donor dyad systems have been examined as components of investigations of a broad scope. First, as part of a study of photochemically reactive metal-organic dyads, Schanze and co-workers determined the rate of back ET in **30**, which features the (bpy)Re(CO)$_3$(py)$^+$ chromophore linked to dimethylaniline via a single methylene spacer [103,104]. Back ET in **30** occurs at a driving force of approximately 2.0 eV and with a rate of $k_{BET} = 1.0 \times 10^8$ s^{-1}. The Meyer group examined dyad **31**, which comprises a (diimine)Re^I(CO)$_3$Cl chromophore linked to phenothiazine via a single methylene linker [99]. Dyad **31** is unique among the large collection of rhenium–donor based dyads (Scheme 13) because the donor is linked to the chromophore via the diimine ligand, as opposed to being attached via the monodentate ligand (pyridine in most cases). The rate of forward ET in **31** was reported to be too fast to resolve using nanosecond laser excitation; however, the charge separated state is readily detected by transient absorption and the rate of back ET was determined to be approximately 5×10^6 s^{-1}. Note that the rate of back ET is very similar in **31** and pyridine linked dyad **27a**; this correspondence is remarkable given the significant difference in connectivity between the metal and donor sites.

Inspection of the literature reveals rate data for two Ru(diimine)$_3^{2+}$ based type 2 dyads. As part of a larger study, Elliott determined the rate of photoinduced forward ET in **32** (Scheme 14) which consists of a *tris*-dimethylbipyridine

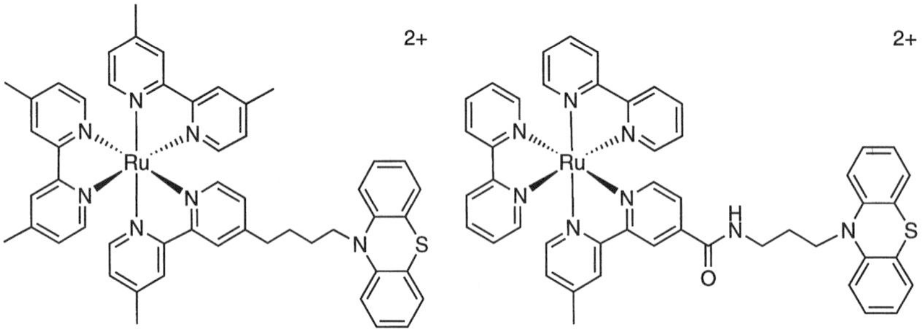

SCHEME 14

ruthenium(II) chromophore covalently linked to a putative phenothiazine donor by a $-(CH_2)_4-$ spacer [105]. Forward ET in this system occurs with a driving force of approximately 100 mV and at a rate of $k_{FET} = 5 \times 10^6$ s^{-1}. Since the Elliott study relied exclusively on luminescence techniques, no information is available concerning k_{BET} in **32**. Note that k_{FET} is considerably smaller in **32** compared to Meyer's rhenium–phenothiazine dyads (**27**), wherein forward ET was fast on the picosecond timescale. The slower rate of forward ET in **32** is likely due to the small driving force and the relatively large donor–acceptor separation (and concomitant small electronic coupling).

Finally, Meyer and co-workers examined the ruthenium-phenothiazine dyad **33** (Scheme 14) [31]. The structure of this dyad is very similar to that of **32**, except that the donor is connected to the metal center via a 5 atom tether that contains an amide link. The rate of forward ET has not been determined in **32**. The charge-separated state was detected by nanosecond transient absorption, however, and by monitoring its decay rate k_{BET} was found to occur relatively slowly with $k_{BET} \approx 3 \times 10^7$ s^{-1}. The finding that back ET is slow in this ruthenium–phenothiazine dyad is significant because it demonstrates that the slow back ET which is observed in type 2 dyads is not limited only to rhenium-based dyads.

C. Other Metal-Organic Dyads

Several recent studies examined photoinduced ET in dyads featuring transition metal chromophores that are dissimilar to the d^6 transition metal polypyridine complexes used in the type 1 and type 2 dyads that have been discussed in the preceding sections. Since the molecular and electronic structure of the excited states involved in these systems is unique from the type 1 and type 2 dyads, results on these systems are discussed separately.

Gray and co-workers investigated the metal-organic dyad system **34** (Scheme 15 and Table 3) which comprises a dimeric iridium(I) complex (Ir$_2$) as the

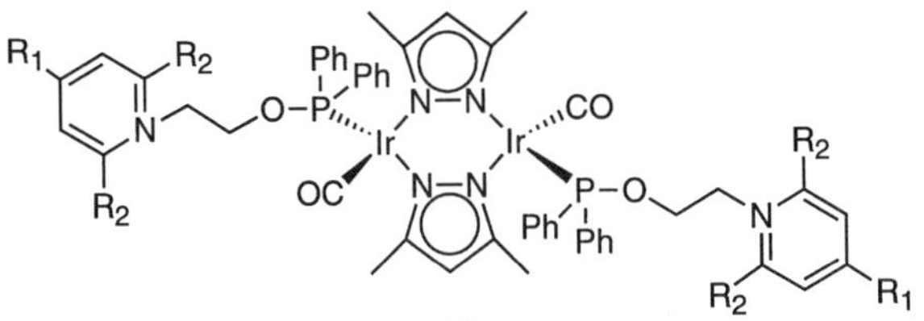

34

SCHEME 15

photoactive metal chromophore and a series of pyridinium ions as putative electron acceptors [106,107]. Before describing the ET results obtained on **34**, however, it is necessary to briefly delineate the photophysics of the Ir$_2$ chromophore [108,109]. The chromophore in **34** consists of two square planar d^8 Ir(I) ions held in close proximity by the pyrazole bridging ligands. Because of the short separation distance between the metals, there is an interaction that affords a manifold of excited states based on the transition from the highest filled d_{z^2} orbital (which is formally an antibonding orbital, $d\sigma^*$) and the lowest empty $6p_z$ orbital (which is formally a bonding orbital, $p\sigma$) [108,109]. A set of low-lying singlet and triplet states arise from the lowest $d\sigma^* \rightarrow p\sigma$ excitation; these two states are labeled ^{3}B and ^{1}B in the C$_2$ symmetry of **34**. Several features are unique with respect to the $d\sigma^* \rightarrow p\sigma$ excited states. First, $d\sigma^* \rightarrow p\sigma$ excitation formally moves an electron from an antibonding into a bonding orbital; thus, there is a bonding interaction between the two iridium centers in the $d\sigma^* \rightarrow p\sigma$ excited state. As a result, low-

TABLE 3 Photoinduced Electron Transfer in Other Metal-Organic Dyads[a]

| Compound | Substituents | | r_{DA}/Å | Reactive excited state | ΔG_{FET}/eV | k_{FET}/10^8 s^{-1} | ΔG_{BET}/eV | k_{BET}/10^8 s^{-1} | Ref. |
	R_1	R_2							
34a	–H	–H	6–8	singlet	–0.89	1100	–1.79	33	106
34b	–CH$_3$	–H	6–8	singlet	–0.71	500	–1.61	67	106
				triplet	–0.21	1.7			106
34c	–Ph	–H	6–8	singlet	–0.97	>1100	–1.53	200	106
34d	–CH$_3$	–CH$_3$	6–8	singlet	–0.58	270	–1.92	0.67	106
				triplet	–0.08	0.035			106
39			≈15				–0.5	0.22 (0.012)[b]	110
40			≈15				–1.1	0.34 (0.0056)[b]	110

[a]Data for acetonitrile solutions.
[b]Data in parentheses for DMSO solution.

frequency metal–metal modes are likely to be strongly coupled to the excitation. A corollary to this point is that one-electron oxidation of the iridium dimer (e.g., $Ir_2 \rightarrow Ir_2^+ + e^-$) removes an electron from the $d\sigma^*$ (antibonding) orbital. This introduces a bonding interaction between the metals and therefore low-frequency Ir–Ir vibrational modes are also expected to be strongly coupled to this oxidation. The next point which is unique to the Ir_2 chromophore is that the low-lying ^3B and ^1B states are energetically far removed from upper charge transfer excited states. As a consequence of this energetic isolation from the CT states and the symmetry of the orbitals involved in the $d\sigma^* \rightarrow p\sigma$ states, spin-orbit coupling is small and therefore the two lowest electronic states are nearly pure singlet and triplet states. Quite interestingly, fluorescence and phosphorescence emission is observed in fluid solution from the ^1B and ^3B states, respectively (see Fig. 8).

Gray and co-workers examined the driving force dependence of ET in the Ir_2 dyads (**34**) by synthesizing a series of complexes in which the reduction potential of the pyridinium acceptor is varied [106,107]. Scheme 16 illustrates the

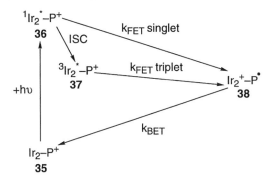

SCHEME 16

events that arise upon photoexcitation of dyads **34**. Excitation produces the singlet excited state **36**, which can either undergo forward ET to the pyridinium acceptor to afford charge separated state **38**, or intersystem crossing to the triplet state **37**. The triplet state can also undergo forward ET to produce charge separated state **38**. The charge-separated state **38** subsequently decays by back ET. By examining the rate of forward ET from both the singlet and triplet excited states and the rate of back ET Gray and co-workers compiled rate data at 10 driving forces using four Ir_2 dyads (**34a–d**, Table 3). The ET reactions in the Ir_2 dyads span a driving force regime from $\Delta G \approx 0$ to $\Delta G \approx -2.0$ eV and the rates vary by over four orders of magnitude. Inspection of the data reveals that for the weakly exergonic reactions (Marcus normal region) the rates increase strongly with driving force, while for the strongly exothermic cases (Marcus inverted region), the rates decrease with increasing driving force.

In an entirely new approach to the design of metal-organic dyads, G. Meyer

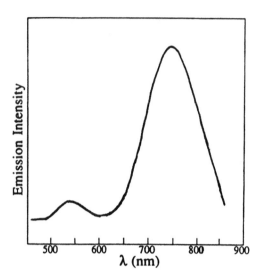

FIGURE 8 Corrected emission spectrum of pyrazolyl-bridged iridium(I) dimer, $Ir_2(\mu\text{-}pz)_2(CO)_4$, where pz = pyrazole. Fluorescence maximum is at 540 nm and phosphorescence maximum is at 750 nm. (Reprinted with permission from Ref. 109.)

developed a system in which a d^{10} polypyridine Cu(I) complex is covalently tethered to one or two paraquat acceptors (**39** and **40**, Scheme 17) [110]. The unique features of the Cu(I) dyads arise largely because of the unusual photophysics of the $d\pi$ (Cu) → π^* (bpy) MLCT excited state, which have been mapped out by McMillin and co-workers [111–115]. In the ground state, the Cu(I) ion prefers a four-coordinate tetrahedral ligand environment. However, consequent to $d\pi$ (Cu) → π^* (bpy) MLCT excitation the metal is formally Cu(II); in this oxidation state the ion prefers to be five coordinate with a square pyramidal ligand environment. As a result, the Cu → bpy MLCT excited state is strongly distorted from the ground state geometry and is a strong Lewis acid. McMillin and co-workers have demonstrated that nucleophiles such as CH_3OH and phosphines quench the weakly luminescent Cu → bpy MLCT state, presumably via nucleophilic attack on the coordinatively unsaturated Cu(II) center [114,115]. By analogy to the MLCT excited state, single-electron oxidation of the tetrahedral Cu(I) complex to afford the Cu(II) oxidation state is accompanied by geometric rearrangement and coordination to fifth (axial) ligand.

With this background in mind it is now possible to consider the results obtained by Meyer and co-workers on the Cu(I) dyads **39** and **40** [110]. In both of these dyads, the Cu → bpy MLCT state is strongly reducing; therefore, forward ET is strongly exoergic ($\Delta G \approx -1.0$ eV in both systems). Consequently, the rate

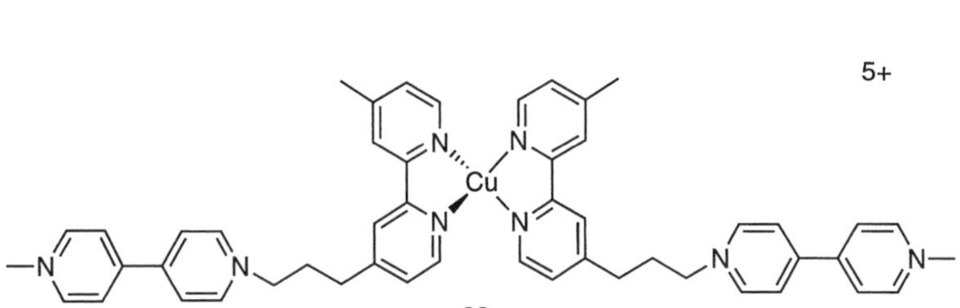

SCHEME 17

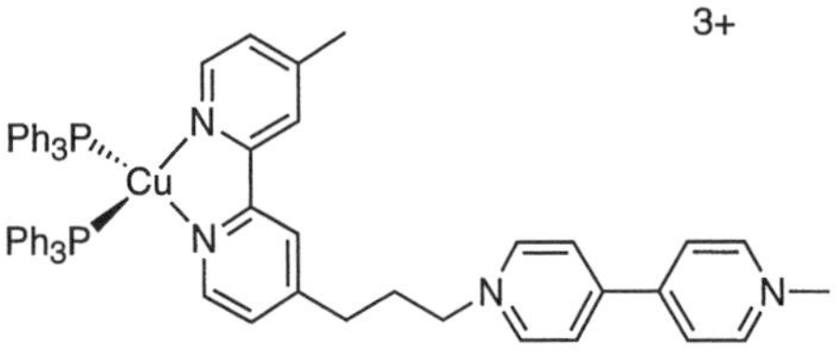

of forward ET is too fast to observe in **39** and **40**. However, nanosecond transient absorption spectroscopy provides clear evidence for the occurrence of intra-molecular ET in the dyads. As illustrated in Fig. 9, the transient absorption of **40** exhibits features characteristic of the paraquat radical cation, which is present in the charge separated state. A remarkable feature is that the charge separated state in both Cu(I) dyads is significantly longer-lived compared to the lifetime of the analogous type 1 dyads studied by Mallouk and Elliott. A second remarkable feature is that the lifetime of dyad **40**, which contains two phosphine ligands in place of one of the bipyridyl ligands present in **39**, is increased by 60-fold in DMSO solvent compared to the lifetime in CH_3CN. Meyer suggests that the increased lifetime of the charge separated state in DMSO may be due to nucleo-philic attack by the solvent on the Cu(II) center, see Scheme 18 [110].

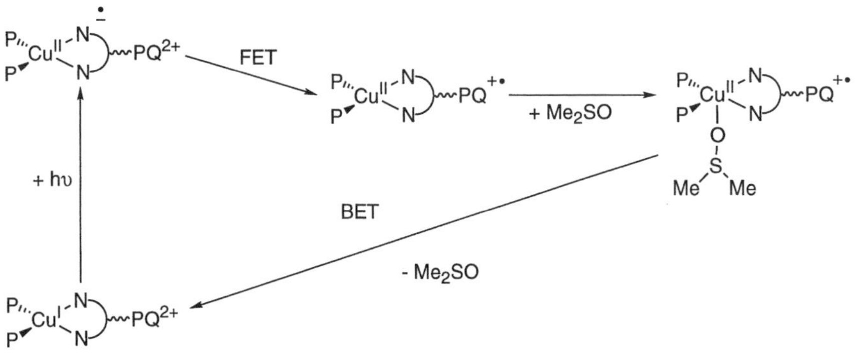

SCHEME 18

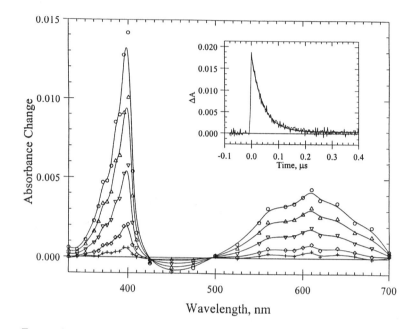

FIGURE 9 Transient absorption difference spectrum of metal-organic dyad **40** obtained 20 ns after 417-nm excitation pulse (10-ns pulse width). (Figure provided by Professor G. J. Meyer, Johns Hopkins University [110].)

D. Analysis of Driving Force Dependence in Metal-Organic Dyads

It is useful and interesting to compare the driving force dependence of photo-induced ET in all of the metal-organic complexes discussed in Sections VIIA to VIIC. First we focus on the rate data for photoinduced forward ET. Figure 10 illustrates a plot of $\log k_{FET}$ vs. ΔG_{FET} for photoinduced forward ET in the type 1 and type 2 dyads (all data for CH_3CN solutions, data extracted from Table 1). The solid line drawn through the data points was calculated by using the Marcus semiclassical model [Eq. (4)] with $k° = 1 \times 10^{11}$ s^{-1} and $\lambda = 1.0$ eV. The correlation shown in this figure clearly illustrates that despite significant structural differences among the metal-organic dyads, the forward ET rate data for all of the complexes is very good agreement with the semiclassical theory of ET [Eq. (4)]. Note that there is some scatter in the data; for example, forward ET in the ruthenium-paraquat dyads **20e–h** is faster than predicted by the generalized correlation. The faster than average rates observed in these dyads is likely due to the relatively short spacer and concomitantly larger than average electronic coupling. It is also evident that k_{FET} for rhenium–dithiafulvene dyads **29** falls below the generalized

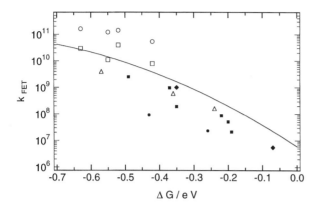

FIGURE 10 Plot of $\log k_{FET}$ vs. ΔG_{FET} for metal-organic dyads. Data from Tables 1 and 2. Figure legend: (○) **20e–h**; (□) **20a–d**; (△) **19a–c**; (■) **28a–f**; (◆) **21, 23**; (●) 29a,b.

plot. The slower rates for these compounds are likely due to the fact that electronic coupling is weaker than for the other compounds owing to the cyclohexyl spacer.

Figure 11 illustrates a plot of $\log k_{BET}$ vs. ΔG_{BET} for back ET in the type 1 and type 2 dyads. For comparison, back ET rate data for a series of porphyrin–quinone dyads (**41a–c**, Scheme 19) studied by Wasielewski is also included on the

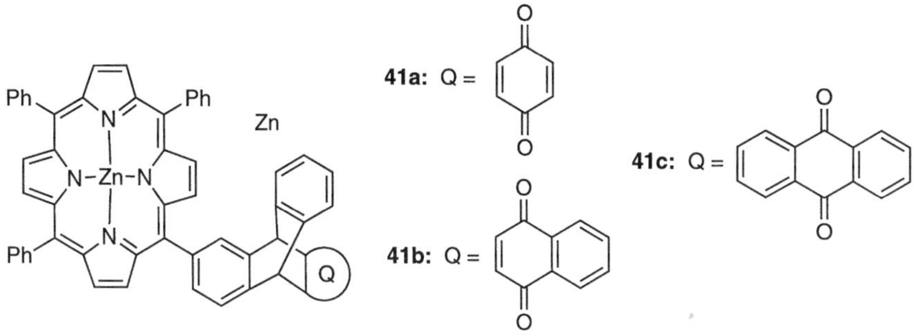

Ph, Ph
Zn
N–Zn–N
Ph, N
Q

41a: Q =

41b: Q =

41c: Q =

SCHEME 19

plot [116]. The correlation in Fig. 11 uncovers several noteworthy features with respect to back ET in the metal-organic dyads. First, for all of the metal-organic systems that have been studied, k_{BET} increases as ΔG_{BET} becomes less exothermic, a pattern which is the hallmark of the Marcus inverted region. The data presented in this figure represent some of the best examples of inverted region behavior that have been uncovered in studies of photoinduced ET.

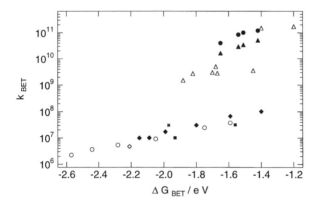

FIGURE 11   Plot of $\log k_{BET}$ vs. $\triangle G_{BET}$ for metal-organic dyads. Data from Tables 1 and 2. Figure legend: (○) **28a–f**; (◆) **27a–f**; (■) **29a,b**; (▲) **20e–h**; (●) **20a–d**; (△) **41a–c**; (◇) **31**; (✳) **33**.

    A second remarkable feature is that the back ET rate data for the type 1 and type 2 dyads follow very different correlations. Thus, rates of back ET in a wide variety of type 2 dyads adhere to a remarkably good (and nearly linear semi-logarithmic) correlation over a 1.2-eV range of driving force. By contrast, for similar driving force, back ET occurs 100–1000 times faster in the type 1 dyads than in the type 2 dyads. Interestingly, the back ET rate data for the type 1 dyads runs closely parallel to the correlation defined by the porphyrin–quinone dyads (**41**) studied by Wasielewski [116]. Another less noticeable difference between the correlations for the type 1 and type 2 dyad systems is that the slopes of the log (rate) vs. driving force correlations are quite different. Thus, for the Type 1 dyads, the rates vary by more than 2 decades per eV, while for the type 2 dyads the rates vary by only 1 decade per eV.
    An important question concerns the origin of the exceptional difference in the rates of back ET for the type 1 and type 2 dyads. One possible explanation for the considerably larger k_{BET}'s in the type 1 dyads is that electronic coupling is much larger in these systems. Indeed, Mallouk and co-workers fit back ET rate data for dyads **20** by using the quantum mechanical expression [Eq. (5)] with $H_{DA} \approx 100 \text{ cm}^{-1}$ [91]. By contrast, MacQueen and Schanze fit back ET rates for dyads **28** by using the same quantum expression with $H_{DA} \approx 0.7 \text{ cm}^{-1}$ [101]. This striking apparent difference in H_{DA} for back ET is surprising in light of several factors. First, the comparison of the forward ET rate data illustrated in Fig. 10 strongly implies that H_{DA} is *not* significantly different in the type 1 and type 2 systems (in the context of forward ET), despite the large difference in the nature of

the donors, acceptors, and spacers. Furthermore, the general fit of the data in Fig. 10 implies that for forward ET in the various type 1 and type 2 systems, H_{DA} values ranging from 10 to 50 cm^{-1} are typical. These electronic coupling values are completely in accord with values observed in a wide range of all-organic donor–acceptor assemblies. For example, Wasielewski fit forward and back ET rates for porphyrin–quinone dyads **41** using $H_{DA} \approx 75$ cm^{-1} [116]. Thus, we conclude that the apparent H_{DA} observed for the type 2 dyads (e.g., 0.7 cm^{-1}) is unreasonably small. Another remarkable feature is that the magnitude of the apparent electronic coupling in the various type 2 dyads is markedly insensitive to the structure of the spacer. For example, k_{BET} is almost the same in dyads **27** and **31**, despite the significant difference in connectivity between the acceptor and donor ligands. Moreover, back ET is not much slower in dyad **29** compared to dyads **28**, despite the presence of the cyclohexane spacer in the former.

Taken together the data strongly imply that back ET occurs at the "expected" rates in the type 1 dyads and that back ET is "retarded" by some additional factor in the type 2 dyads. That back ET occurs at "expected" rates in the type 1 dyads is underscored by the comparison of the type 1 rate data with data from Wasielewski's porphyrin–quinone dyads **41** shown in Fig. 11. Moreover, note that the data on dyads **20** is not unusual compared to the other type 1 dyads that have been studied. Recall that in every type 1 dyad that has been examined, the charge separated state was too short lived to be observed.

We suggest that electronic spin-factors are the origin of the exceptionally low electronic coupling in the type 2 dyads [101,117]. Schemes 20 and 21

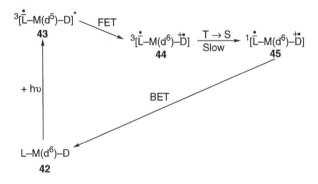

SCHEME 20

illustrate schemes for photoinduced ET in the type 1 and type 2 dyads; these diagrams are more complete than those presented in Fig. 1 by virtue of the fact that they contain the spin states of the reactive excited states and the charge separated states. Studies of ruthenium and rhenium diimine complexes indicate that photo-

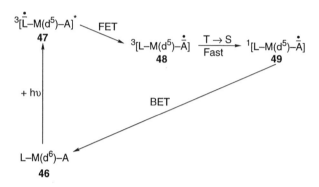

SCHEME 21

excitation affords the lowest $d\pi$ (M) $\rightarrow \pi^*$ (diimine) MLCT state within several picoseconds [118]. Although for the ruthenium and rhenium systems there is strong spin-orbit coupling, the lowest MLCT state has predominantly triplet spin character (**43** and **47**) [69–71,119]. Because of the triplet spin character of the MLCT state, triplet charge separated states are formed by forward ET (**44** and **48**) [92,120–123]. Since the ground state is a singlet, the charge separated state must undergo a triplet to singlet spin transition (T $\rightarrow$ S) prior to back ET [92,120–123]. For reasons described below, we believe that the T $\rightarrow$ S transition is rapid for the type 1 dyads, but is relatively slow in the type 2 dyads. Since the T $\rightarrow$ S transition is rapid in the type 1 dyads it does not significantly retard back ET and these systems undergo back ET at rates that are typical for covalently linked dyads where photoinduced ET involves singlet excited states (e.g., porphyrin–quinones, **41**). However, since the T $\rightarrow$ S spin transition is slower in the type 2 dyads, the rates of back ET are significantly lower; indeed, the observed back ET rates in the type 2 systems may correspond to the rate of the T $\rightarrow$ S conversion.

The dramatically different rate of the T $\rightarrow$ S conversion in the type 1 and type 2 dyads arises because the electronic structure of the charge separated state is significantly different in the two systems. Thus, as described earlier, in the type 1 dyads the charge separated state consists of a d^5 metal ion linked to an organic radical ion (**48**, Scheme 21). In this system the presence of the d^5 open shell metal [e.g., Ru(III)] introduces substantial spin-orbit coupling into the charge separated state [123]. The large degree of spin-orbit coupling renders the T $\rightarrow$ S transition (e.g., **48** $\rightarrow$ **49**) fast in the charge separated state. (Indeed, one study suggests that the T $\rightarrow$ S transition in the Ru(III)/PQ$^{+\cdot}$ geminate radical (ion) pair occurs with a time constant of 25 ps [123].) Because the T $\rightarrow$ S transition is fast in type 1 dyads, back ET is consequently very rapid because the spin restriction to the reaction is effectively eliminated owing to the fast spin relaxation at the d^5 metal. The situation is very different for the charge separated state in the type 2 dyads. Thus, as shown in Scheme 20 the charge separated state consists essentially of a covalently linked organic radical anion and organic radical cation. Owing to this fact, the unpaired spins reside in orbitals having little contribution from the metal, and consequently there is little spin-orbit coupling. Because spin-orbit coupling is weak, the S $\rightarrow$ T transition (**44** $\rightarrow$ **45**) is slow and accordingly the rate of back ET is significantly retarded.

At this point we turn to consider the driving force dependence of ET in the Ir$_2$ dyads studied by Gray and co-workers [106,107]. Figure 12 illustrates a plot of $\ln k_{ET}$ vs. ΔG_{ET} for Ir$_2$ dyads **34a–d**. The rate data shown in this plot is taken from three distinct reaction types for the four Ir$_2$ dyads (refer to Scheme 16): photo-induced forward ET from the excited singlet state of the Ir$_2$ chromophore (**36** $\rightarrow$ **38**), forward ET from the triplet excited state of the Ir$_2$ chromophore (**37** $\rightarrow$ **38**), and back ET from the charge separated state (**38** $\rightarrow$ **35**). The remarkable feature is that the entire set of rate data correlates with a single parabolic log (rate)–free energy correlation calculated by using the semiclassical Marcus expression [Eq. (4), $\lambda = 1.06$ eV, $k° = 1.5 \times 10^{11}$ s^{-1}]. The rate data for the Ir$_2$ dyads contain four examples which occur in the inverted region (e.g., $|\Delta G_{ET}| > \lambda$). As expected based on the theory, the rates of these highly exothermic reactions increase as the driving force decreases. However, an unusual feature is that the inverted region rates are fitted well by the semiclassical model [Eq. (4)]. This is surprising, since for highly exothermic inverted region reactions nuclear tunneling is expected to occur for high-frequency vibrational modes. The net effect of nuclear tunneling is to weaken the driving force dependence of the reaction in the inverted region.

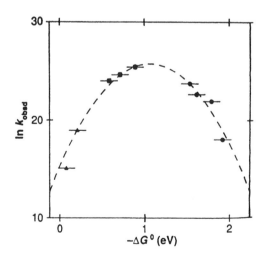

FIGURE 12 Plot of $\ln k_{ET}$ vs. ΔG_{ET} for Ir$_2$ dyads **34a–d**. (Reprinted with permission from Ref. 106.)

(Compare the plots of quantum and semiclassical theory that are shown in Fig. 2. The relatively faster rates and weaker driving force dependence in the inverted region for the correlation calculated by using the quantum expression is caused by nuclear tunneling in high-frequency vibrational modes that are coupled to ET.)

Gray suggests that the unusually steep inverted region driving force dependence observed for the Ir_2 dyads arises because low-frequency vibrational modes are the dominant modes coupled to the back ET reaction [124]. The strong coupling of low-frequency modes to ET may arise because of the change in the Ir-Ir bond order that occurs upon single electron oxidation of the Ir_2 chromophore (see above). The unusual driving force dependence observed in the Ir_2 dyads has been further substantiated by recent temperature dependence studies, which reveal that there is a substantial activation energy for back ET in the inverted region [124]. The observation of a strong temperature dependence indicates that nuclear tunneling is unimportant and that back ET is classical (i.e., back ET is coupled predominantly to low-frequency vibrational modes).

VIII. DISTANCE DEPENDENCE OF PHOTOINDUCED ELECTRON TRANSFER IN METAL-ORGANIC DYADS

Relatively little work has been carried out concerning the distance dependence of "long-range" (nonadiabatic) ET in metal-organic dyads. This stems mainly from the difficulty associated with chemical synthesis of spacer assemblies that are suitably functionalized to coordinate the metal center and to hold the electron donor or acceptor group at a defined distance from the metal. Moreover, analysis of distance dependence of ET data is complicated by the fact that the overall distance dependence of k_{ET} arises from at least two factors that vary with donor–acceptor separation: (1) the donor–acceptor electronic coupling (H_{DA}) decreases with increasing separation distance [Eq. (7)]; (2) the outer-sphere reorganization energy (λ_s) increases with increasing separation distance [Eq. (8)]. In principle it is possible to separately determine the distance dependence of these two terms; however, in practice this separation is at best, problematic. Therefore, for the purposes of this review we simply survey the metal-organic systems that have been examined and report the overall distance dependence of k_{ET} as a β value which is extracted from a plot of $\ln k_{ET}$ as a function of donor–acceptor separation [Eq. (6)].

Schanze and Sauer were the first to report a detailed study of long-range photoinduced ET in a series of metal-organic dyads [94]. These authors synthesized dyads **50a–e** (Scheme 22 and Table 4) that comprise a $Ru(diimine)_3^{2+}$ chromophore covalently linked to a p-benzoquinone acceptor using a series of oligo-L-proline peptide spacers. Oligo-proline peptide spacers are used because previous studies suggest that these peptides are conformationally restricted and

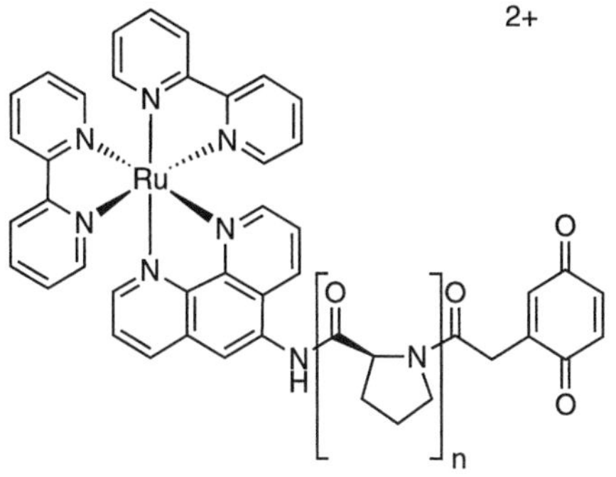

50

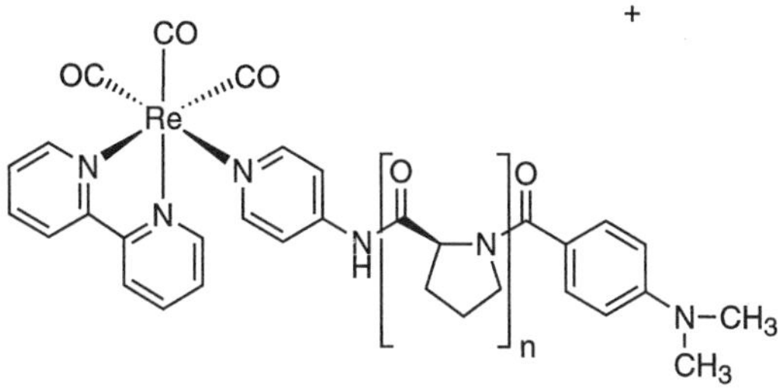

51

SCHEME 22

hold the peptide end groups at a well-defined separation distance [125]. Rates of forward ET from the MLCT excited state of the Ru(II) chromophore to the quinone acceptor are determined by luminescence decay techniques and the observed ET rates vary by approximately four orders of magnitude for the dyads with 0 to 4 proline spacers (See Table 4). Although the kinetic studies are complicated owing to the fact that the proline spacers adopt a range of conformations in the aprotic solvent used in the study (CH_2Cl_2), the authors estimate that $\beta \geqslant 0.75$ Å^{-1} [see Eq. (6)] for **50a–e**.

Several years later Schanze and Cabana reported a study of the distance dependence of photoinduced ET in a second series of oligo–proline bridged dyads (**51a–c**, Scheme 22 and Table 4) [126,127]. In this study the authors were able to study photoinduced ET in CH_3OH, a solvent that is demonstrated to stabilize the extended conformation of the proline spacers [126]. Rate constants for forward ET in **51a–c** are determined by using luminescence decay techniques, and an overall distance dependence of $\beta = 1.0$ Å^{-1} is obtained from the rate data (Table 4). By

TABLE 4   Distance Dependence of Electron Transfer in Metal-Organic Dyads[a]

Compound	n	Solvent	$r_{DA}/\text{Å}$[b]	$\Delta G_{ET}/eV$	k_{ET}/s^{-1}	Ref.
50a	0	CH_2Cl_2	10	-0.35	1.0×10^{9}[d]	94
50b	1	CH_2Cl_2	13.1	-0.35	5.6×10^{8}[d]	94
50c	2	CH_2Cl_2	16.2	-0.35	1.7×10^{7}[d]	94
50d	3	CH_2Cl_2	19.3	-0.35	2.1×10^{6}[d]	94
50e	4	CH_2Cl_2	22.4	-0.35	3.6×10^{5}[d]	94
51a	0	CH_3OH	9.0	-0.23	9.8×10^{7}	127
51b	1	CH_3OH	12.1	-0.23	5.3×10^{6}	127
51c	2	CH_3OH	15.2	-0.23	3.7×10^{5}	128
52a	2	CH_3CN	8.5	-0.36	5.9×10^{8}	128
52b	3	CH_3CN	9.7	-0.36	2.4×10^{7}	128
52c	4	CH_3CN	11.0	-0.36	5.4×10^{7}	128
52d	5	CH_3CN	12.2	-0.36	7.6×10^{6}	128
52e	6	CH_3CN	13.5	-0.36	8.7×10^{6}	128
52f	7	CH_3CN	14.7	-0.36	6.2×10^{6}	128
53a	1	CH_3CN	10.3[c]	$-0.42 \ (-1.65)$[e]	$5.9 \times 10^{10} \ (3.9 \times 10^{10})$[e]	129
53b	2	CH_3CN	11.5[c]	$-0.42 \ (-1.65)$[e]	$7.9 \times 10^{9} \ (1.6 \times 10^{10})$[e]	129
53c	3	CH_3CN	12.8[c]	$-0.42 \ (-1.65)$[e]	$2.0 \times 10^{9} \ (6.5 \times 10^{9})$[e]	129
53d	4	CH_3CN	14.1[c]	$-0.42 \ (-1.65)$[e]	$6.5 \times 10^{8} \ (3.2 \times 10^{9})$[e]	129
53e	5	CH_3CN	15.3[c]	-0.42	6.5×10^{7}	129
53f	7	CH_3CN	17.9[c]	-0.42	2.3×10^{6}	129
53g	8	CH_3CN	19.2[c]	-0.42	1.8×10^{5}	129
54a	3	DCE[f]	9.7	-0.12	6.1×10^{6}	105
54b	4	DCE[f]	11.0	-0.12	5.4×10^{6}	105
54c	5	DCE[f]	12.2	-0.12	4.4×10^{6}	105
54d	6	DCE[f]	13.5	-0.12	2.9×10^{6}	105
54e	7	DCE[f]	14.7	-0.12	2.3×10^{6}	105
54f	8	DCE[f]	15.9	-0.12	2.0×10^{6}	105

[a]All rates are for photoinduced forward ET unless otherwise noted.
[b]Approximate center-to-center separation distance unless noted otherwise. See original reference for details on distance approximation.
[c]Center-to-center separation distance approximated from edge-to-edge distances given in original reference.
[d]Rate estimated from multiexponential decay kinetics; see original reference for details.
[e]Data in parenthesis for back ET reaction.
[f]1,2-dichloroethane solvent.

examining the temperature dependence of k_{FET} for the proline-bridged oligomers, Schanze and Cabana demonstrate that the primary term that decreases with separation distance in **51a–c** is the donor–acceptor electronic coupling (H_{DA}).

Schmehl, Elliott, and co-workers reported an extensive study of the spacer dependence of photoinduced ET in several series of metal-organic dyads consisting of a Ru(diimine)$_3^{2+}$ chromophore covalently linked to a diquat electron acceptor [128]. In the present review we list only results for the series of poly-methylene linked dyads **52a–f** (Scheme 23 and Table 4); the reader is directed to the original

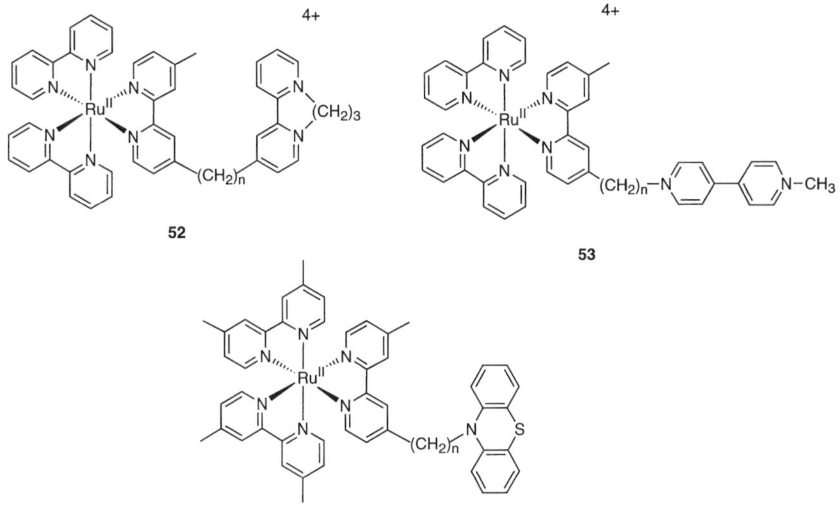

SCHEME 23

paper for rate data on compounds with *cis*- and *trans*-1,4-cyclohexyl- and 1,4-phenyl spacers [128]. Forward ET rates in **52a–f** were determined by using luminescence decay techniques (Table 4). Interestingly, the ET rate does not decrease monotonically as the length of the polymethylene chain increases in **52a–f**. By contrast, an odd–even variation in the ET rate is observed, the amplitude of which "damps" out as the length of the spacer increases (see Fig. 13 for a plot of the rate data). Although theory indicates that the through-bond component of electronic coupling may oscillate in an odd–even manner [54–59], the authors fall short of suggesting that their results provide an example of this effect. Rather, they suggest that it is likely that ET occurs by through space (solvent) pathways, and that the odd–even oscillation results from the effect of chain length on the ability of the donor and acceptor sites to come into close proximity to facilitate through space ET [128].

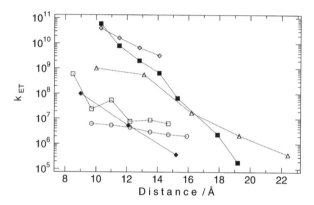

FIGURE 13   Plot of log k_{ET} as a function of approximate separation distance for metal-organic dyads; data from Table 4. Figure legend: ($\Diamond$) **53a–d** (back ET); (■) **53a–g** (forward ET); ($\triangle$) **50a–e**; (□) **52a–f**; (♦) **51a–c**; (○) **54a–f**.

Mallouk and co-workers reported an extensive study of the spacer and driving force dependence of the rate of ET in a series of dyads consisting of a Ru(diimine)$_3^{2+}$ chromophore covalently linked to a paraquat acceptor (**53a–g**, Scheme 23 and Table 4) [129]. Photoinduced ET in **53a–g** is very rapid, and therefore picosecond transient absorption methods are used to determine forward ET rates (Table 4). In some of the dyads (**53a–d**) the authors are able to monitor the disappearance of the charge separated state and consequently to determine the distance dependence of the rate of back ET as a function of spacer length. Several interesting conclusions are drawn from the data on **53a–g**. First, k_{FET} decreases strongly and monotonically as the length of the poly-methylene spacer increases, in sharp contrast to the "odd–even" effect observed by Schmehl and Elliott on **52a–f** [128]. The decay of k_{FET} is exponential and a plot yields an effective value of $\beta_{FET} = 1.38$ Å^{-1}. The observed spacer dependence of k_{FET} implies that: (1) the poly-methylene spacers adopt primarily extended conformations in **53a–g**; (2) the ability of the poly-methylene spacer to effectively separate the donor and acceptor sites is different for the ruthenium–paraquat and ruthenium–diquat dyads (e.g., **53** vs. **52**). A second interesting observation on **53a–d** is that the overall distance dependence of k_{BET} is much weaker than for k_{FET}. Indeed, a natural logarithmic plot of k_{BET} vs. distance yields $\beta_{BET} = 0.66$ Å^{-1}, which is less than 50% of β_{FET}. The outer-sphere reorganization energy (λ_s) in **53a** and **53b** is estimated by using a combination of theory and experimental data (fitting of rate vs. driving force dependence data for dyads **20a–h**). The authors then extrapolate to estimate the distance dependence of λ_s for the series of dyads **53a–g**. With this information in hand it is then possible to separate the distance dependence of λ_s and the electronic coupling matrix element (H_{AD}). This analysis reveals that the distance dependence of H_{AD} is approximately the same for forward and back ET. In other words, the large difference in β_{FET} and β_{BET} (determined from the slopes of the plots of ln k_{ET} vs. distance) is due to the distance dependence of λ_s. Interestingly, this experimental observation was predicted by a theoretical study that was published nearly a decade prior to the study of metal-organic dyads **53** [130].

Finally, Elliott and co-workers examined the spacer dependence of forward ET in the series of dyads **54a–f** (Scheme 23 and Table 4) which consist of a Ru(diimine)$_3^{2+}$ chromophore linked to a phenothiazine electron donor with poly-methylene spacers of varying length [105]. In this series the rates of forward ET were determined by monitoring the decay of the luminescence from the MLCT excited state. The driving force for forward ET is rather low in **54a–f** (see Table 4) and this fact, coupled with the large donor–acceptor separation distances, makes forward ET relatively slow (Table 4). Interestingly, in dyads **54a–f** k_{FET} depends only weakly on the length of the poly-methylene spacer. This suggests that ET occurs predominantly via a through space (solvent) pathway and that the observed ET rates are controlled mainly by the dynamics of conformational isomerization of the poly-methylene spacer.

A comparison of the overall distance dependence of k_{ET} for all the metal-organic dyads is presented in Fig. 13. The points for each series are connected with lines to make it easier to visualize the trends in the data sets. While this comparison only reveals qualitative trends, several features are worthy of note. First, as noted above, the overall distance dependence of the ruthenium–diquat and the ruthenium–paraquat series (**52** and **53**, respectively) is quite different, even though (1) the same poly-methylene spacer system is used in both systems, and (2) the charges on the donor and acceptor sites are the same in both systems. One possible explanation for the difference in the behavior of the poly-methylene spacer in **52** and **53** is that the dynamics of forward ET are significantly different in the two systems. Specifically, for a given spacer length, k_{FET} is approximately 10 times larger in **53** than in **52** (this difference arises because forward ET is more exothermic in **53**). Since ET is generally slower in dyads **52**, the poly-methylene chain has more time to explore different conformations, allowing through space (solvent) ET pathways to predominate, and introducing spacer conformational dynamics into the observed ET rates. Note that this effect is even more exaggerated in dyads **54**, which also contain a poly-methylene spacer. In this system k_{FET} is almost independent of the spacer length.

A second point is that k_{ET} decays strongly with distance in almost all of the metal-organic dyads which contain conformationally rigid spacers. Moreover, the overall distance dependence (e.g., k_{ET} vs. r) observed for the metal-organic dyads is commensurate with what is observed for organic donor–acceptor systems [47–49] and for natural and nonnatural ET proteins [37–39]. The performance of the metal-organic dyads in long-range ET reactions is satisfying and provides evidence that these systems provide an attractive means to study nonadiabatic ET processes.

IX. CONCLUSION

This chapter provides a comprehensive overview of photoinduced ET in metal-organic dyads. The focus is on systems in which intramolecular ET occurs between a metal center and an organic donor (or acceptor) that are held in proximity by an organic spacer. Emphasis is placed on systems in which the thermodynamic driving force for ET is well defined.

It is clear that much has been learned about fundamental aspects of ET through studies of these covalently linked metal-organic dyads. The survey of photoinduced forward ET rate data reveals that in this process metal-organic dyads are "well behaved," i.e., their performance is in accord with modern theories of ET reactions. By contrast, the survey of charge recombination rate data reveals that the kinetics for back ET in the type 2 dyads is unusually slow compared with other systems. This unusual behavior is attributed to effects of electron spin-multiplicity on the charge recombination process.

Current research efforts in many of the laboratories which are responsible for the work described herein is focused on more complex supramolecular assemblies including metal-organic based triads and tetrads [29,31,32,131,132]. Other groups are taking the small-molecule dyads and incorporating them into complicated molecular assemblies and materials such as zeolites, SiO_2, TiO_2, DNA, and organic polymers [133–137]. These studies are taking the knowledge base that has been created through the fundamental studies described herein and applying it to the construction of useful materials and devices for future technologies [138].

X. ACKNOWLEDGMENT

We wish to acknowledge the National Science Foundation for supporting our work on electron transfer in metal-organic dyads (Grant No. CHE-9401620).

REFERENCES

1. For reviews on electron transfer reactions see Refs. 2–9 and for reviews on photo-induced electron transfer see Refs. 10–15.
2. Zwolinski, B. J.; Marcus, R. A.; Eyring, H. *Chem. Rev.* 1955, *55*, 157.
3. Marcus, R. A. *Annu. Rev. Phys. Chem.* 1964, *15*, 155.
4. Newton, M. D.; Sutin, N. *Annu. Rev. Phys. Chem.* 1984, *35*, 437.
5. DeVault, D. *Quantum-Mechanical Tunnelling in Biological Systems*, 2nd ed.; Cambridge University Press: New York, 1984.
6. *Tunnelling in Biological Systems*; Chance, B.; DeVault, D. C.; Frauenfelder, H.; Marcus, R. A.; Schrieffer, J. R.; Sutin, N. eds.; Academic Press: New York, 1979.

7. *Electron Transfer in Biology and the Solid State*; Johnson, M. K.; King, R. B.; Kurtz, D. M., Jr.; Kutal, C.; Norton, M. L.; Scott, R. A., eds.; *A.C.S. Advances in Chemistry Series No. 226*; ACS Press: Washington, DC (1990).

8. *Electron Transfer in Inorganic, Organic and Biological Systems*; Bolton, R. R.; Mataga, N.; McLendon, G., eds.; *A.C.S. Advances in Chemistry Series No. 228*; ACS Press: Washington, DC (1991).

9. Barbara, P. F.; Meyer, T. J.; Ratner, M. A. *J. Phys. Chem.* 1996, *100*, 13148.

10. *Photoinduced Electron Transfer, Parts A-D*; Fox, M. A.; Chanon, M. D. (eds.); Elsevier: Amsterdam, 1988.

11. *Photoinduced Electron Transfer I*; Mattay, J. ed.; *Topics in Current Chemistry No. 156*; Springer-Verlag: Berlin, 1990.

12. *Photoinduced Electron Transfer II*; Mattay, J. ed.; *Topics in Current Chemistry No. 158*; Springer-Verlag: Berlin, 1990.

13. *Photoinduced Electron Transfer III*; Matay, J. ed.; *Topics in Current Chemistry No. 159*; Springer-Verlag: Berlin, 1991.

14. *Photoinduced Electron Transfer IV*; Mattay, J. ed.; *Topics in Current Chemistry No. 163*; Springer-Verlag: Berlin, 1992.

15. *Photoinduced Electron Transfer V*; Mattay, J., ed.; *Topics in Current Chemistry No. 168*; Springer-Verlag: Berlin, 1993.

16. Knibbe, H.; Rehm, D.; Weller, A. *Ber. Bunsenges. Phys. Chem.* 1968, *72*, 257.

17. Rehm, D.; Weller, A. *Isr. J. Chem.* 1970, *8*, 259.

18. Gafney, H. D.; Adamson, A. W. *J. Am. Chem. Soc.* 1972, *94*, 8238.

19. Bock, C. R.; Meyer, T. J.; Whitten, D. G. *J. Am. Chem. Soc.* 1974, *96*, 4710.

20. Bock, C. R.; Meyer, T. J.; Whitten, D. G. *J. Am. Chem. Soc.* 1975, *97*, 2909.

21. Young, R. C.; Meyer, T. J.; Whitten, D. G. *J. Am. Chem. Soc.* 1975, *97*, 4781.

22. Anderson, C. P.; Salmon, D. J.; Meyer, T. J.; Young, R. C. *J. Am. Chem. Soc.* 1977, *99*, 1980.

23. Bock, C. R.; Connor, J. A.; Gutierrez, A. R.; Meyer, T. J.; Whitten, D. G.; Sullivan, B. P.; Nagle, J. K. *J. Am. Chem. Soc.* 1979, *101*, 4815.

24. Meyer, T. J. *Acc. Chem. Res.* 1989, *22*, 163.

25. Juris, A.; Balzani, V.; Barigelletti, F.; Campagna, S.; Belser, P.; von Zelewsky, A. *Coord. Chem. Rev.* 1988, *84*, 85.

26. Westmoreland, T. D.; Schanze, K. S.; Neveux, P. E., Jr.; Danielson, E.; Sullivan, B. P.; Chen, P.; Meyer, T. J. *Inorg. Chem.* 1985, *24*, 2596.

27. Balzani, V.; Scandola, F. *Supramolecular Photochemistry*; Ellis-Horwood: New York, 1991.

28. Cooley, S. F.; Larson, S. L.; Elliott, C. M.; Kelley, D. F. *J. Phys. Chem.* 1991, *95*, 10694.

29. Larson, S. L.; Cooley, L. F.; Elliott, C. M.; Kelley, D. F. *J. Am. Chem. Soc.* 1992, *114*, 9504.

30. Larson, S. L.; Elliott, C. M.; Kelley, D. F. *J. Phys. Chem.* 1995, *99*, 6530.

31. Mecklenburg, S. L.; Peek, B. M.; Schoonover, J. R.; McCafferty, D. G.; Wall, C. G.; Erickson, B. W.; Meyer, T. J. *J. Am. Chem. Soc.* 1993, *115*, 5479.

32. Mecklenburg, S. L.; McCafferty, D. G.; Schoonover, J. R.; Peek, B. M.; Erickson, B. W.; Meyer, T. J. *Inorg. Chem.* 1994, *33*, 2974.

33. Balzani, V.; Juris, A.; Venturi, M.; Campagna, S.; Serroni, S. *Chem. Rev.* 1996, *96*, 759.

34. Harriman, A.; Ziessel, R. *Chem. Commun.* 1996, 1707.
35. Beer, P. D. *Chem. Commun.* 1996, 689.
36. Barigelletti, F.; Flamigni, L.; Collin, J.-P.; Sauvage, J.-P. *Chem. Commun.* 1997, 333.
37. Isied, S. S.; Ogawa, M. Y.; Wishart, J. F. *Chem. Rev.* 1992, *92*, 381.
38. Ogawa, M. Y.; Moreira, I.; Wishart, J. F.; Isied, S. S. *Chem. Phys.* 1993, *176*, 589.
39. Bowler, B. E.; Raphael, A. L.; Gray, H. B. *Prog. Inorg. Chem.* 1990, *38*, 259.
40. Durham, B.; Millet, F. *J. Chem. Ed.* 1997, *74*, 636.
41. Netzel, T. L. *J. Chem. Ed.* 1997, *74*, 646.
42. Wasielewski, M. R. *Chem. Rev.* 1992, *92*, 435.
43. Gust, D.; Moore, T. A. *Adv. Photochem.* 1991, *16*, 1.
44. Oevering, H.; Paddon-Row, M. N.; Heppener, M.; Oliver, A. M.; Cotsaris, E.; Verhoeven, J. W.; Hush, N. S. *J. Am. Chem. Soc.* 1987, *109*, 3258.
45. Sutin, N. *Prog. Inorg. Chem.* 1983, *30*, 441.
46. Ulstrup, J.; Jortner, J. *J. Chem. Phys.* 1975, *63*, 4358.
47. Miller, J. R.; Beitz, J. V. *J. Chem. Phys.* 1981, *74*, 6746.
48. Miller, J. R.; Beitz, J. V.; Huddleston, R. K. *J. Am. Chem. Soc.* 1984, *106*, 5057.
49. Closs, G. L.; Miller, J. R. *Science* 1988, *240*, 440.
50. Beratan, D. N.; Hopfield, J. J. *J. Am. Chem. Soc.* 1984, *106*, 1584.
51. Beratan, D. N. *J. Am. Chem. Soc.* 1986, *108*, 4321.
52. Larsson, S. *J. Am. Chem. Soc.* 1981, *103*, 4034.
53. Beratan, D. N.; Betts, J. N.; Onuchic, J. N. *Science* 1991, *252*, 1285.
54. Miller, J. R.; Paulson, B. P.; Bal, R.; Closs, G. L. *J. Phys. Chem.* 1995, *99*, 6923.
55. Paulson, B.; Pramod, K.; Eaton, P.; Closs, G.; Miller, J. R. *J. Phys. Chem.* 1993, *97*, 13042.
56. Curtiss, L. A.; Naleway, C. A.; Miller, J. R. *Chem. Phys.* 1993, *176*, 387.
57. Curtiss, L. A.; Naleway, C. A.; Miller, J. R. *J. Phys. Chem.* 1995, *99*, 1182.
58. Paulson, B. P.; Curtiss, L. A.; Bal, B.; Closs, G. L.; Miller, J. R. *J. Am. Chem. Soc.* 1996, *118*, 378.
59. Sengupta, B.; Curtiss, L. A.; Miller, J. R. *J. Chem. Phys.* 1996, *104*, 9888.
60. Isied, S. S.; Vassilian, A.; Wishart, J. F.; Creutz, C.; Schwartz, H. A.; Sutin, N. *J. Am. Chem. Soc.* 1988, *110*, 635.
61. Wang, Y.; Schanze, K. S. *Inorg. Chem.* 1994, *33*, 1354.
62. Paris, J. P.; Brandt, W. W. *J. Am. Chem. Soc.* 1959, *81*, 5001.
63. Klassen, D. M.; Crosby, G. A. *J. Chem. Phys.* 1968, *48*, 1853.
64. Demas, J. N.; Crosby, G. A. *J. Mol. Spectrosc.* 1968, *26*, 72.
65. Demas, J. N.; Adamson, A. W. *J. Am. Chem. Soc.* 1971, *93*, 1800.
66. Demas, J. N.; Adamson, A. W. *J. Am. Chem. Soc.* 1973, *95*, 5158.
67. Demas, J. N.; Crosby, G. A. *J. Am. Chem. Soc.* 1971, *93*, 2841.
68. Crosby, G. A.; Hipps, K. W.; Elfring, W. H., Jr. *J. Am. Chem. Soc.* 1974, *96*, 629.
69. Hager, G. D.; Crosby, G. A. *J. Am. Chem. Soc.* 1975, *97*, 7031.
70. Hager, G. D.; Watts, R. J.; Crosby, G. A. *J. Am. Chem. Soc.* 1975, *97*, 7037.
71. Hipps, K. W.; Crosby, G. A. *J. Am. Chem. Soc.* 1975, *97*, 7042.
72. *Energy Resources Through Photochemistry and Catalysis*; Grätzel, M., ed.; Academic Press: New York, 1983.
73. Hagfeldt, A.; Grätzel, M. *Chem. Rev.* 1995, *95*, 49.
74. Wrighton, M.; Morse, D. L. *J. Am. Chem. Soc.* 1974, *96*, 998.

75. Giordano, P. J.; Wrighton, M. S. *J. Am. Chem. Soc.* 1979, *101*, 2888.

76. Wrighton, M. S.; Geoffroy, G. L. *Organometallic Photochemistry*, Academic Press: New York, 1979, ch. 2.

77. Luong, J. C.; Nadjo, L.; Wrighton, M. S. *J. Am. Chem. Soc.* 1978, *100*, 5790.

78. Schanze, K. S.; MacQueen, D. B.; Perkins, T. A.; Cabana, L. A. *Coord. Chem. Rev.* 1993, *122*, 63.

79. Della Ciana, L.; Hamachi, I.; Meyer, T. J. *J. Org. Chem.* 1989, *54*, 1731.

80. Sullivan, B. P.; Abruña, H.; Finklea, H. O.; Salmon, D. J.; Nagle, J. K.; Meyer, T. J.; Sprintschnik, H. *Chem. Phys. Lett.* 1978, *58*, 389.

81. Root, M. J.; Deutsch, E.; Sullivan, J. C.; Meisel, D. *Chem. Phys. Lett.* 1983, *101*, 353.

82. Westmoreland, T. D.; Le Bozec, H.; Murray, R. W.; Meyer, T. J. *J. Am. Chem. Soc.* 1983, *105*, 3932.

83. Chen, P.; Danielson, E.; Meyer, T. J. *J. Phys. Chem.* 1988, *92*, 3708.

84. Chen, P.; Curry, M.; Meyer, T. J. *Inorg. Chem.* 1989, *28*, 2271.

85. Schoonover, J. R.; Chen, P.; Bates, W. D.; Dyer, R. B.; Meyer, T. J. *Inorg. Chem.* 1993, *32*, 1167.

86. Meyer, T. J. *Prog. Inorg. Chem.* 1983, *30*, 389.

87. Perkins, T. A.; Pourreau, D. B.; Netzel, T. L.; Schanze, K. S. *J. Phys. Chem.* 1989, *93*, 4511.

88. Perkins, T. A.; Humer, W.; Netzel, T. L.; Schanze, K. S. *J. Phys. Chem.* 1990, *94*, 2229.

89. Cooley, L. F.; Headford, C. E. L.; Elliott, C. M.; Kelley, D. F. *J. Am. Chem. Soc.* 1988, *110*, 6673.

90. Elliott, C. M.; Freitag, R. A.; Blaney, D. D. *J. Am. Chem. Soc.* 1985, *107*, 4647.

91. Yonemoto, E. H.; Riley, R. L.; Kim, Y. I.; Atherton, S. J.; Schmehl, R. H.; Mallouk, T. E. *J. Am. Chem. Soc.* 1992, *114*, 8081.

92. Ohno, T.; Yoshimura, A.; Prasad, R. R.; Hoffman, M. Z. *J. Phys. Chem.* 1991, *95*, 4723.

93. Kosower, E. M.; Cotter, J. C. *J. Am. Chem. Soc.* 1964, *86*, 5525.

94. Schanze, K. S.; Sauer, K. *J. Am. Chem. Soc.* 1988, *110*, 1180.

95. Worl, L. A.; Duesing, R.; Chen, P.; Della Ciana, L.; Meyer, T. J. *J. Chem. Soc. Dalton Trans.* 1991, 849.

96. Chen, P.; Westmoreland, T. D.; Danielson, E.; Schanze, K. S.; Anthon, D.; Neveux, P. E., Jr.; Meyer, T. J. *Inorg. Chem.* 1987, *26*, 1116.

97. Chen, P.; Duesing, R.; Tapolsky, G.; Meyer, T. J. *J. Am. Chem. Soc.* 1989, *111*, 8305.

98. Chen, P.; Duesing, R.; Graff, D. K.; Meyer, T. J. *J. Phys. Chem.* 1991, *95*, 5850.

99. Chen, P.; Mecklenburg, S. L.; Meyer, T. J. *J. Phys. Chem.* 1993, *97*, 13126.

100. MacQueen, D. B.; Perkins, T. A.; Schanze, K. S. *Mol. Cryst. Liq. Cryst.* 1991, *194*, 113.

101. MacQueen, D. B.; Schanze, K. S. *J. Am. Chem. Soc.* 1991, *114*, 1897.

102. Perkins, T. A.; Hauser, B. T.; Eyler, J. R.; Schanze, K. S. *J. Phys. Chem.* 1990, *94*, 8745.

103. Wang, Y.; Hauser, B. T.; Rooney, M. M.; Schanze, K. S. *J. Am. Chem. Soc.* 1993, *115*, 5675.

104. Lucia, L. A.; Wang, Y.; Nafisi, K.; Netzel, T. L.; Schanze, K. S. *J. Phys. Chem.* 1995, *99*, 11801.
105. Larson, S. L.; Elliott, C. M.; Kelley, D. F. *Inorg. Chem.* 1996, *35*, 2070.
106. Fox, L. S.; Kozik, M.; Winkler, J. R.; Gray, H. B. *Science* 1990, *247*, 1069.
107. Farid, R. S.; Chang, I.-J.; Winkler, J. R.; Gray, H. B. *J. Phys. Chem.* 1994, *98*, 5176.
108. Roundhill, D. M.; Gray, H. B.; Che, C.-M. *Acc. Chem. Res.* 1989, *22*, 55.
109. Marshall, J. L.; Hopkins, M. D.; Miskowski, V. M.; Gray, H. B. *Inorg. Chem.* 1992, *31*, 5034.
110. Ruthkosky, M.; Kelly, C. A.; Castellano, F. N.; Meyer, G. J., submitted.
111. Buckner, M. T.; Matthews, T. G.; Lytle, F. E.; McMillin, D. R. *J. Am. Chem. Soc.* 1979, *101*, 5846.
112. Rader, R. A.; McMillin, D. R.; Buckner, M. T.; Matthews, T. G.; Casadonte, D. J.; Lengel, R. K.; Whittaker, S. B.; Darmon, L. M.; Lytle, F. E. *J. Am. Chem. Soc.* 1981, *103*, 5906.
113. Casadonte, D. J.; McMillin, D. R. *J. Am. Chem. Soc.* 1987, *109*, 331.
114. McMillin, D. R.; Kirchoff, J. R.; Goodwin, K. V. *Coord. Chem. Rev.* 1985, *64*, 83.
115. Palmer, C. E. A.; McMillin, D. R.; Kirmaier, C.; Holten, D. *Inorg. Chem.* 1987, *26*, 3167.
116. Wasielewski, M. R.; Niemczyk, M. P.; Svec, W. A.; Pewitt, E. B. *J. Am. Chem. Soc.* 1985, *107*, 1080.
117. Lucia, L. A.; Schanze, K. S. *Inorg. Chim. Acta* 1994, *225*, 9078.
118. Damrauer, N. H.; Cerullo, G.; Yeh, A.; Boussie, T. R.; Shank, C. V.; McCusker, J. K. *Science* 1997, *275*, 54.
119. Striplin, D. R.; Crosby, G. A. *Chem. Phys. Lett.* 1994, *221*, 426.
120. Olmsted, J., III; Meyer, T. J. *J. Phys. Chem.* 1987, *91*, 1649.
121. Ohno, T.; Yoshimura, A.; Shioyama, H.; Mataga, N. *J. Phys. Chem.* 1987, *91*, 4365.
122. Steiner, U. E.; Wolff, H.-J.; Ulrich, T.; Ohno, T. *J. Phys. Chem.* 1989, *93*, 5147.
123. Wolff, H.-J, Bürβner, D.; Steiner, U. E. *Pure and Appl. Chem.* 1995, *67*, 167.
124. Gray, H. B., personal communication.
125. Stryer, L.; Haughland, R. P. *Proc. Nat. Acad. Sci. U.S.A.* 1967, *58*, 719.
126. Cabana, L. A.; Schanze, K. S. *A. C. S. Adv. Chem.* 1990, *226*, 101.
127. Schanze, K. S.; Cabana, L. A. *J. Phys. Chem.* 1990, *94*, 2740.
128. Ryu, C. K.; Wang, R.; Schmehl, R. H.; Ferrere, S.; Ludwikow, M.; Merkert, J. W.; Headford, C. E. L.; Elliott, C. M. *J. Am. Chem. Soc.* 1992, *114*, 430.
129. Yonemoto, E. H.; Sauper, G. B.; Schmehl, R. H.; Hubig, S. M.; Riley, R. L.; Iverson, B. L.; Mallouk, T. E. *J. Am. Chem. Soc.* 1994, *116*, 4786.
130. Brunschwig, B.; Ehrenson, S.; Sutin, N. *J. Am. Chem. Soc.* 1984, *106*, 6858.
131. McCafferty, D. G.; Friesen, D. A.; Danielson, E.; Wall, C. G.; Saderholm, M. J.; Erickson, B. W.; Meyer, T. J. *Proc. Nat. Acad. Sci. U.S.A.* 1996, *93*, 8200.
132. Coe, B. J.; Friesen, D. A.; Thompson, D. W.; Meyer, T. J. *Inorg. Chem.* 1996, *35*, 4575.
133. Yonemoto, E. H.; Kim, Y. I.; Schmehl, R. H.; Wallin, J. O.; Shoulders, B. A.; Richardson, B. R.; Haw, J. F.; Mallouk, T. E. *J. Am. Chem. Soc.* 1994, *116*, 10557.
134. Pfennig, B. W.; Chen, P. Y.; Meyer, T. J. *Inorg. Chem.* 1996, *35*, 2898.

135. Argazzi, R.; Bignozzi, C. A.; Heimer, T. A.; Castellano, F. N.; Meyer, G. J. *J. Am. Chem. Soc.* 1995, *117*, 11815.

136. Thornton, N. B.; Schanze, K. S. *New. J. Chim.* 1996, *20*, 791.

137. Ley, K. D.; Whittle, C. E.; Bartberger, M. D.; Schanze, K. S. *J. Am. Chem. Soc.* 1997, *119*, 3423.

138. Schanze, K. S. and Schmehl, R. H. *J. Chem. Ed.* 1997, *74*, 633, and references therein.

4

Photochemistry and Photophysics of Liquid Crystalline Polymers

David Creed

Center for Macromolecular Photochemistry and Photophysics, University of Southern Mississippi, Hattiesburg, Mississippi

I. INTRODUCTION

Liquid crystalline (LC) materials are among the simplest of the so-called supramolecularly organized materials. They are known in both "small molecule" and polymeric forms. Appropriately designed molecules or macromolecules can spontaneously pack into partially organized LC phases over certain ranges of temperature (thermotropic materials) or in suitable solvents (lyotropic materials). LC materials are intermediate between three-dimensionally ordered crystals and isotropic liquids. Their chemistry and photochemistry is of fundamental interest because their constituent molecules are not only ordered, to a degree dependent on the phase type, but retain some ability to translate. Ignoring defects, crystalline materials are "perfectly" three-dimensionally ordered but molecular motion is restricted to small movements about an equilibrium position. This restricts the possibilities for interesting chemistry. The molecules in isotropic liquids and amorphous solids have, in general, no specific ordering with respect to one another although, on the molecular level, species such as dimers or even higher order aggregates can exist.

The photochemistry of LC polymers is not only interesting for fundamental reasons (vide supra) but because they can perhaps be formed into useful materials—films, fibers, rods, etc., with specifically tailored mechanical and/or optical properties. Their photochemical reactions may be used to modify these properties in an easily controlled, "switchable" manner. There is already a considerable body of knowledge on the possible practical applications of a few photochemical reactions of LC polymers. Such possible applications are discussed elsewhere [1–6] and are only briefly touched upon in this chapter. Furthermore, this chapter does not include an extensive compilation of information about the many types of LC polymers, the many methods used to synthesize and process these materials, their detailed properties, and the theoretical basis of their formation and properties. The reader is referred elsewhere [7,8] to capable reviews of these topics. However, a brief introductory review of the main types of LC polymers and their properties that are especially relevant to their photochemistry is given in Section II.

The photochemistry of small molecule liquid crystals has been thoroughly reviewed elsewhere [9]. Saeva [10] published a brief preliminary review in 1974 of the photochemistry of LC polymers. Krongauz [4] has reviewed photochromic LC polymers, those with spiropyran and azo dye chromophores and their possible applications. Shibaev [2] has recently reviewed SCLC polymers, mostly those containing azo dye chromophores for possible optical storage of information. Kreuzer et al. [3] have reviewed cyclic, oligomeric LC siloxanes as possible optical recording materials. Finally Bowry-Devereaux [5] has reviewed some aspects of the photopolymerization of LC monomers and of the photochemistry of LC polymers with particular reference to information storage applications. This chapter is an attempt to review all the literature to June 1997 on fundamental aspects of the photophysics and photochemistry of LC polymers. The material is organized by chromophore type in Section III. A brief discussion is also given in Section III of a few examples of the use of fluorescence as a photophysical probe of the microstructure of LC polymers. This chapter does not include discussion of photopolymerization of LC monomers [5,11], an interesting research topic but one in which the only photochemistry involved is in the generation of initiating radicals.

II. LIQUID CRYSTALLINE POLYMERS

LC polymers can either be *thermotropic*, their LC phases or mesophases are formed by heating a solid or cooling a liquid or solid, or *lyotropic*, the mesophases are formed by association of molecules in solution. The photochemistry and photophysics of molecules in lyotropic phases such as detergent micelles and lipid bilayers has been a very active research area for about 30 years. This extensive literature [12] is not reviewed in this chapter which is necessarily restricted to thermotropic LC polymers.

A. Mesogens, Spacers, and Molecular Architecture

All LC materials have structural segments known as *mesogens* or *mesogenic groups*. Most commonly, these are segments of the repeat unit which are rigid and rodlike, with a high length to width ratio. Less commonly, the mesogens are disc-like and can afford so-called discotic liquid crystals. Some photochemically interesting elongated, rodlike mesogens are shown in Fig. 1. There are so-called *main-chain liquid crystalline polymers* (MCLC) that consist entirely of connected, rigid mesogenic groups, but most of the MCLC polymers whose photochemistry has been studied have rigid mesogens connected through flexible *spacer groups* or *spacers* such as $(CH_2)_n$, $(CH_2CH_2O)_n$, and $CH_2OSi(CH_3)_2CH_2O$-$CH_2Si(CH_3)_2CH_2O$. In most *side-chain liquid crystalline polymers* (SCLC), the mesogens are connected to main chains such as poly(alkyl acrylate), poly(alkyl methacrylate), polysiloxane, poly(ethylene oxide), etc., by flexible spacers. The latter serve two principal functions. They increase the solubility and lower the

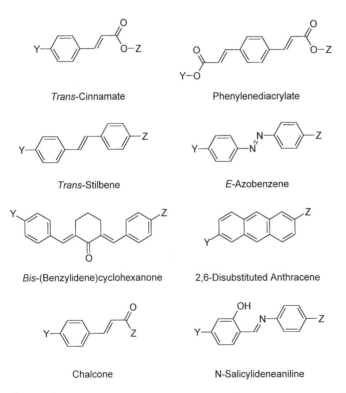

FIGURE 1 Photochemically reactive mesogenic groups ("mesogens").

transition temperatures of otherwise intractable rigid rod MCLC polymers, thereby making the materials easier to process. In the case of SCLC polymers, they not only affect the polymer solubility and therefore processibility, but the spacer group between the mesogen and the main chain is often essential to the observation of LC behavior. The spacer is believed to decouple the molecular motions of the main chain from the mesogen, allowing the latter to pack in a regular fashion in the LC mesophase and the former to maintain a random-coil, statistical arrangement [13]. The existence of the LC phases then results from a compromise between the competing tendencies of the mesogens to arrange aniso-tropically and the chain(s) to adopt a disordered, random conformation. It should be noted that chromophores that, typically, do not give rise to small molecule LC compounds, or to LC polymers, when they are the *only* potential mesogenic group in the polymer, can often be incorporated to some extent into LC copolymers that have one of the more typical "rodlike" mesogens as a major component of the copolymer structure without complete loss of the LC properties [4].

There are a number of possible *molecular architectures*; different arrange-ments of mesogens and flexible spacers in LC polymers. Two of the most common arrangements are shown in Fig. 2 along with a hybrid structure having mesogens in both the main and side chains. Many other structural variations are illustrated elsewhere [8,14]. As noted above, the mesogen can be disc-shaped and so-called discotic LC polymers are well known. However, although there have been some elegant photophysical experiments by Markovitsi and others using small molecule discotic LC materials [15], there does not seem to have been any work yet described on the photochemistry and photophysics of discotic LC polymers, so these materials will not be further discussed.

B. Mesophases and Their Characterization

Thermotropic LC materials can form a large number of different phases or *mesophases* upon heating and/or cooling. A detailed discussion of all these mesophases is outside the scope of this review, but the arrangements of rodlike mesogens in the phases that are most frequently encountered in LC polymer research are crudely illustrated in Fig. 3. The least ordered, is the *nematic* or *N* mesophase. This phase occurs immediately before *isotropization* or *clearing*, melting to the (nonbirefringent) isotropic phase (*I*), upon heating the polymer. In the *N* mesophase, the mesogens point in the same general direction (called the *director*), there is directional order, but there is no additional positional ordering such as the formation of layers. The *cholesteric* or Ch mesophase is an *N* phase formed by chiral molecules in which layers of nematically ordered mesogens are twisted with respect to one another. In the *smectic* or *S* mesophases, the mesogens not only point in the same general direction but form layers. They have positional as well as directional order. The different types of *S* phases differ in the manner in

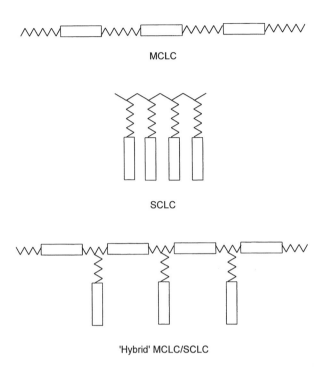

MCLC

SCLC

'Hybrid' MCLC/SCLC

FIGURE 2   Some "molecular architectures" of liquid crystalline polymers incorporating "rodlike" mesogens and flexible "spacers."

which the mesogens are packed within layers and/or what angle the director makes with respect to the layer plane (i.e., whether the mesogens adopt an arrangement like a pack of cards or are tilted with respect to one another). When the director is at 90° to the layer plane, the layer thickness is that of the mesogen. For any other angle, the layer thickness is less than the length of the mesogen. The smectic A (S_A) is the least organized S phase. It is layered with the director at 90° to the layer plane, but the mesogens are packed irregularly within the layers. The smectic C (S_C) is an interesting phase in which the director is not at right angles to the layer plane and the packing of the mesogens is irregular. The smectic C* (S_{C*}) is a S_C type of phase formed by chiral molecules. The mesogens in the layers are twisted with respect to the mesogens in the layers above and below. The smectic B (S_B) is similar to S_A except that the packing of mesogens within the layers is regular. In neither S_A, S_C, $S_C{}^*$, nor S_B are the mesogens restricted in their ability to rotate about their long axes. In several even more organized S phases, closer to being crystalline, the mesogens are also restricted in their ability to rotate about their

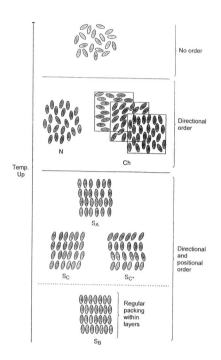

FIGURE 3 Mesogen arrangements in phases commonly formed by liquid crystalline polymers.

long axes. In general, both MCLC and SCLC polymers form N phases, but S phases are more commonly observed with SCLC polymers.

LC polymers cast into thin films are normally in the *homogeneous* or *planar* alignment in the mesophases. In this alignment the mesogens are, on average, parallel to the surface boundary (e.g., the surface of a glass plate), typically at right angles to the axis of visual observation. This arrangement leads to the observation of birefringence in films viewed using polarized optical microscopy (POM). In so-called *homeotropic* LC phases, the mesogens are oriented perpendicular to a surface boundary, typically in line with the axis of visual observation of, say, a thin film cast on a glass plate. Using POM, homeotropic phases appear non-birefringent (dark through crossed polarizers), but mechanical disturbance usually reveals the more normal birefringence of the homogeneous phase, and this enables the observer to distinguish this phase from isotropic phases which are also nonbirefringent. Of course, solid MCLC and SCLC polymers, like other polymers, can also be partially crystalline (K) or amorphous (A) following precipitation or casting from a solvent prior to heating.

Three other important points regarding the properties of mesophases should be mentioned. First, thermodynamically stable *enantiotropic* mesophases are those phases which form upon both heating the polymer above the glass transition temperature, T_g, or by cooling the polymer from the isotropic melt. In the absence of thermal degradation these mesophases can be formed reproducibly upon heating or cooling. Some mesophases are thermodynamically unstable. These *monotropic* mesophases form only upon cooling the polymer from a phase existing at a higher temperature. They are much less well studied than enantiotropic phases because their formation is harder to control. Second, rapidly cooling of an LC phase of a polymer to below T_g can result in a metastable glassy phase such as the *glassy nematic* (NG) phase, in which the ordering of the original mesophase is "locked in." If cooling is from the isotropic melt an amorphous glassy phase may be formed. The formation of nematic or amorphous glasses is rarely observed for small molecule LC materials which usually crystallize rather than form glasses. The existence of enantiotropic of monotropic mesophases is, however, common to both polymeric and small molecule LC materials. Finally, LC mesophases of all types derived from both small molecules and polymers are only ordered on the microscopic level, in so-called *microdomains* containing variable numbers of molecules, typically ca. 10^4. Macroscopic ordering (throughout the bulk of the material), the production of a *monodomain*, of LC polymers can be achieved, but it requires the orientation of the mesogens by application of electric or magnetic fields to a fluid film, by casting the polymer on a rubbed surface, by shearing the fluid phase (between glass plates, for example), or by stretching a film or fiber.

The existence of different phases in any polymer is best detected using *differential scanning calorimetry* (DSC). Assignment of mesophase type in LC polymers is usually based on *polarized optical microscopy* (POM) and x-ray scattering measurements. In ideal cases (and in the hands of an experienced observer!), mesophase types can be assigned by the different textures seen using POM. However, the characteristic textures, almost always seen for the different mesophases of small molecule LC materials, are not always clearly seen by POM of LC polymers. Ultimate proof of phase type usually requires variable temperature x-ray scattering measurements. This technique can afford both intermesogen and, in the case of S phases, interlayer spacings.

A number of characteristic temperatures are important in LC polymer work. The *glass transition temperature*, T_g, is that temperature below which segmental motion of the main chain of the polymer does not occur, although motions (e.g., rotation) of side-groups may occur. The *isotropization* or *clearing temperature*, T_i, is the temperature at which the polymer enters the isotropic melt from one of its mesophases and the birefringence of the mesophase disappears. Temperatures are often quoted more specifically defining where phase transitions occur. For example, T_{NI} would be the temperature where the nematic phase enters the isotropic melt. In this case, of course, T_{NI} is the same as T_i.

III. PHOTOCHEMISTRY AND PHOTOPHYSICS

The photochemistry of small molecule LC materials has been an active area of research for many years and has been reviewed recently [9]. The photochemistry of LC polymers, per se, has received much less attention although two brief reviews have appeared [5,10], and there has been a considerable effort to apply some simple photochemical transformations such as *trans-cis* photoisomerization, to the development of practical devices [1–6]. This section is divided into three parts. In Part A, chromophore aggregation, which seems to be important in almost all the cases in which careful UV-Vis and/or fluorescence studies of films of pure LC polymers have been made, is explicitly discussed. Part B is devoted to a thorough review, organized by chromophore type, of the photochemistry and related photophysics of LC polymers. No attempt has been made to extensively cross-reference the work on LC polymers to the hundreds of papers and reviews on analogous non-LC compounds. However, when it seemed particularly appropriate or interesting, experiments related to optical applications of the photochemistry of LC polymers are briefly described. In Part C, a few experiments are described in which a "classical" photophysical method, fluorescence spectroscopy, is used to probe the microstructures of some LC polymers.

A. Chromophore Aggregation

A common statement of the "First Law of Photochemistry" is that in order for a photochemical reaction to take place, light must be absorbed. Although there are alternative ways of making excited states, light absorption by a chromophore remains by far the most important. Light absorption by "pure" polymers, as in a cast film, for example, has an obvious characteristic: the effective chromophore concentration in a pure polymer is usually very high. This is of obvious importance to photochemistry but is sometimes ignored or glossed over. Assuming unit density (1 g ml^{-1}) for a homopolymer and a repeat unit molar mass of 200 g mol^{-1}, the effective chromophore concentration is 5 mol dm^{-3}, orders of magnitude higher than most photochemists would use in the typical solution phase photochemical or photophysical study. Processes involving chromophore association in both the ground and excited states ought to be extremely important in pure polymers, including LC polymers. Even in dilute solution the chromophore in a polymer may be able to interact with another chromophore in an intrachain process in either the ground or an excited state.

A more specific type of chromophore aggregation should be considered in crystalline or partly crystalline polymers and in the mesophases of an LC polymer. In the latter the mesogen *is* the chromophore. Our understanding of the origin of mesophases is that they arise, in large part, from the tendency of the rigid rod (or disclike) mesogens to pack most efficiently. *It is therefore reasonable to expect that chromophore/mesogen association will be a very important factor in the photochemistry of pure LC materials!* Such associations should, in many cases, lead to perturbations of the absorption spectra of the type that have been frequently discussed for crystalline materials. The simple theory of chromophore interactions has been elegantly expounded by Kasha [17]. The perturbations expected for different orientations of the transition moments of the chromophores in a ground state dimer are illustrated in Fig. 4. Comparison with the (idealized) mesogen arrangements shown in Fig. 3 is instructive. The two most extreme aggregate geometries (or, strictly speaking, arrangements of the transition dipoles) are the "card pack" H-aggregation (sometimes referred to as "K-aggregation") and "end-to-end" J-aggregation. The card pack type of stacking occurs most obviously in the S_A and S_B types of phase. The end-to-end type of stacking may occur in both S and N phases. Chromophore "tilting" surely occurs in the S_C phase. Stacking of the transition moments (chromophores if the transition moment is parallel to the long axis of the molecule) into the card pack type of arrangement (H- or K-aggregation) is expected to lead to an "allowed" transition shifted to higher energy and a "forbidden" transition shifted to lower energy. The converse is true for end to end interaction of the transition moments in so-called J-aggregates. Tilting the transition moments from $0°$ to one another (H-aggregate) through to $180°$ (J-aggregate) leads to the allowed transition shifting from the blue to the red. At $54°$, the transition is predicted to be at the same energy as that of the isolated chromophore.

Some crude photochemically relevant generalizations can be made based on

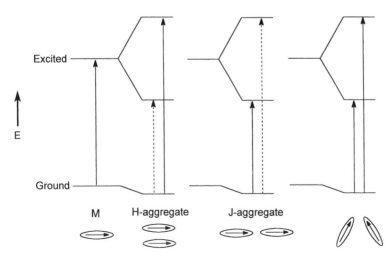

FIGURE 4 Exciton band structures in dimers with several orientations of transition dipoles. Dotted arrows indicate dipole forbidden transitions. (Adapted from Kasha [17]).

the model depicted in Fig. 4. Since the lowest energy singlet–singlet transition is forbidden in *H*-aggregates, the lowest energy excited singlet aggregate that presumably forms following light absorption at any wavelength (Kasha–Vavilow law) should be longer lived and enhanced intersystem crossing to the triplet state may occur in *H*- (*K*-) aggregates. In general, *J*-aggregates should have shorter lived singlet states than *H*-aggregates because transitions to the ground state from the lowest singlet state are allowed in *J*-aggregates. Intersystem crossing, that competes with other singlet decay processes, should be more probable from *H*-aggregates [17]. Finally, since the spin allowed transitions of large molecules are never strictly forbidden, it can be expected that *all aggregates will show some enhancement of absorption to the red of the lowest energy, longest wavelength "allowed" transition of the "isolated" chromophore*. This could be a critical effect where it is undesirable to have red enhanced absorption in, for example, nonlinear optical materials or structural materials subjected to sunlight. Finally, despite the general process of aggregation of chromophores/mesogens, one would expect that a certain fraction of chromophores that behave as if they were isolated, at least during their excited state lifetimes, will exist in pure polymers. This will be a consequence of the inevitable heterogeneity of the microstructure in any phase. In summary, red-enhanced absorption and chromophore heterogeneity should be common in pure LC polymers and may have unexpected photochemical consequences. Chromophore heterogeneity also makes it difficult to use UV-Vis or fluorescence spectroscopy to measure relative or absolute quantum yields of reaction. Therefore, changes in IR spectra have been more commonly used to obtain relative reactivity information. In the latter measurements, scattering of excitation light, which can obviously be a problem for comparisons of the relative reactivities of crystalline or LC materials with amorphous or isotropic materials (e.g., actinometer solutions), is rarely discussed!

B. Mesogenic and Other Chromophores

Cinnamates

Cinnamates occupy an important place in the history of photochemistry. Schmidt and his co-workers [18] used the solid state photochemistry of cinnamic acid and its derivatives to develop the idea of "topochemical control" of photochemistry in the crystalline state. Minsk [19] developed poly(vinyl cinnamate) as the first polymer for photoimaging. The cinnamate chromophore is still commonly incorporated in photopolymers of all types, including LC polymers, to enable them to be photochemically cross-linked [20], and a number of reports of the photochemistry of such MCLC and SCLC polymers are summarized below.

MCLC Polymers.   Creed et al. [21–27] have studied MCLC polymers such as **1** in which an aryl cinnamate chromophore is connected by polymethylene

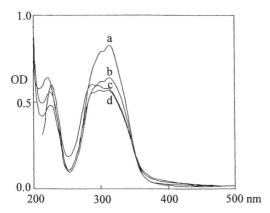

1, x = 9 or 10

and bis-siloxane flexible spacers. Photochemical cross-linking of such polymers could enhance their mechanical properties or be useful in photoimaging or information storage [27]. Upon heating "as cast" films of this type of polymer a nematic (N) mesophase is observed. Upon further heating into the isotropic melt (I) and cooling to room temperature the N organization is retained in a glassy nematic phase (NG) below the T_g. The UV-Vis spectra of these polymers are highly dependent upon phase type [21]. In the I phase, the spectrum resembles that observed in solution, although somewhat broadened. In the N and NG phases the UV-Vis spectrum is considerably perturbed with weak red-shifted absorption and strong blue-shifted absorption perhaps indicative of H-aggregate formation (Fig. 5). It should be pointed out that this effect might also be explained by different proportions of H- and J-aggregates, with the former predominant, but Occam's razor (assume that the simpler of two hypotheses is correct!) and the fact that irradiation into the red-shifted tail of the absorption [28] leads to preferential cycloaddition (unlikely for J-aggregates unless energy transfer to H-aggregates can occur) favors the simpler hypothesis. The extent of perturbation of the UV-Vis

FIGURE 5 UV-Vis spectra of high (HMW) and low (LMW) molecular weight thin films of polymer, **1**, at room temperature. (a) "As cast" HMW film with no thermal history. (b) HMW film cooled rapidly from the isotropic melt. (c) HMW film cooled slowly from the isotropic melt. (d) LMW film cooled rapidly from the isotropic melt. (Reprinted with permission from Creed et al. [25]. Copyright 1994 American Chemical Society.)

spectrum depends on the molecular weight of the polymer, its thermal history [25], and the thickness of the cast films [26]. The extent of perturbation is qualitatively greater for lower molecular weight samples. Slow annealing in the N phase leads to greater spectral perturbation [25], and thicker films display relatively smaller effects than thinner films [26]. Lower molecular weight samples presumably have lower viscosities at a given temperature. Higher molecular weight, more viscous samples require lengthier times of annealing for the mesogens to associate. Aggregation in very thin films ($<$ ca. 20 nm) is presumably more subject to effects of surface roughness.

Irradiation (313 nm) of films of polymers such as **1** leads to UV-Vis, IR, and solid-state ^{13}C nmr changes attributed to a combination of 2 + 2 photocycloaddition and photo-Fries rearrangement [21], the latter presumably leading to cinnamoylphenol moieties, **2**, in the irradiated polymer. Films irradiated at 313 nm in

2

any of the solid or fluid phases yellowed due to photo-Fries product and became insoluble due to cross-linking. Thermal degradation did not occur under any of the irradiation conditions. Photo-Fries rearrangement was observed to increase [21,25] in the order NG $<$ "as cast" $<$ I (least viscous). It is tempting to attribute this effect to increasing fluidity as the temperature is increased, but it was subsequently observed [23] that 366-nm excitation of NG films where the aggregate absorption is greater than that of "isolated" chromophores leads almost exclusively to photocycloaddition (Fig. 6). This wavelength dependence is not observed when **1** is irradiated in a "good" solvent such as dichloromethane, but can be observed when chromophore aggregation is induced by adding a nonsolvent to a solution of **1** in a good solvent [25]. The increase in photo-Fries rearrangement in the order NG $<$ "as cast" $<$ I is therefore likely to be due to the decreasing extent of absorption at 313 nm by aggregated chromophores in the order NG $>$ "as cast" $>$ I (lowest concentration of aggregated chromophores).

Irradiation of lower molecular weight samples in the fluid N phase at 313 or 366 nm led to an unusual result [21]. In the first few seconds of irradiation the perturbed spectrum of the N phase exhibited hyperchromism (an increase in absorbance) and its shape became similar to that of the spectrum of the isotropic melt. This effect is also observed upon triplet sensitization which, like 366-nm irradiation, suppresses photo-Fries rearrangement [28]. It has not yet been proved that this effect is accompanied by a phase change from N to I induced by photoproducts essentially acting as impurities in the mesophase. The effect could be at the microscopic level where formation of a cyclobutane dimer or other photoproduct could interrupt H-type aggregated chromophore stacks, or confor-

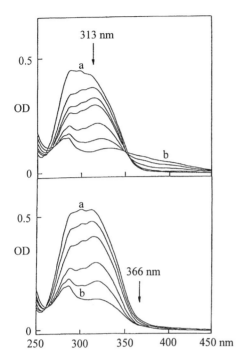

FIGURE 6 UV-Vis spectral changes upon 313- and 366-nm irradiation of glassy nematic films of HMW **1**. (a) Before irradiation. (b) After irradiation. Residual absorption at ca. 280 nm is attributed to cyclobutane photoproducts. Residual absorption beyond ca. 340 nm is mostly due to photo-Fries rearrangement products. (Reprinted with permission from Creed et al. [23]. Copyright 1990 American Chemical Society.)

mational changes associated with dimer formation could similarly disrupt several pairs of aggregated chromophores. A similar effect has been observed by Stumpe and his co-workers [29] in polymers containing the azobenzene chromophore.

The spectral shape, but not the photochemical behaviors of films of **1**, can be almost perfectly duplicated [25] using dispersions of the model compound 4-pentyloxyphenyl-4′-pentyloxycinnamate, in poly(methyl methacrylate). In fact the UV-Vis absorption spectra of **1** in various phases can be well simulated [25] assuming these spectra are due only to the sum of two chromophores, one aggregated with a λ_{max} of ca. 275 nm and the other isolated with a λ_{max} of ca. 313 nm.

The photo-Fries rearrangement of polymers such as **1** leads to yellowing, an undesirable behavior for many potential applications. The photochemistry of analogous MCLC polymers has been reported [30] in which the alkoxyphenyl ester groups of **1** have been replaced by 4,4′-*trans*-disubstituted cyclohexyl groups. These relatively rigid, symmetrically substituted, cycloalkyl groups enable LC character to be retained with retention of the ability to undergo 2 + 2 photocycloaddition but suppression of the photo-Fries rearrangement. Triplet sensitization [28] of polymers such as **1** also leads only to photocycloaddition and, possibly some *trans-cis* isomerization. Some preliminary results have been reported [26] on the effects of phase type on the process of triplet sensitization. Based on a simple stoichiometric analysis, a single ketocoumarin triplet sensitizer can sensitize the disappearance of about 80 cinnamate chromophores in "as cast" films of the polymer but over 300 chromophores in the more organized NG films [26]. Irradiation-induced hyperchromism masks this effect in the (fluid) N phase but the occurrence of hyperchromism upon triplet sensitization indicates that formation of cycloadducts and/or *cis*-isomers, both of which can be formed from the triplet states of *trans*-cinnamates, can be responsible for this unusual effect. It is tempting, based on the triplet sensitization results, to suggest that the wavelength dependence of the photochemistry of **1**, viz. cycloaddition and no photo–Fries rearrangement upon 366-nm irradiation [23], results from enhanced intersystem crossing (ISC) in H-aggregates [17]. However, direct measurements of ISC and/or triplet quenching measurements in the different mesophases are needed to prove this hypothesis.

Chang and co-workers [31] have studied poly(ether-esters) of general structure **3** that also contain several types of aryl cinnamate groups. Their studies were

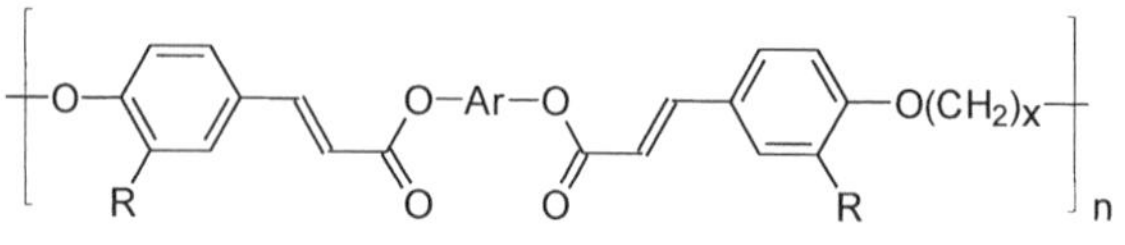

3,   R = H or OCH3,   x = 6 or 10

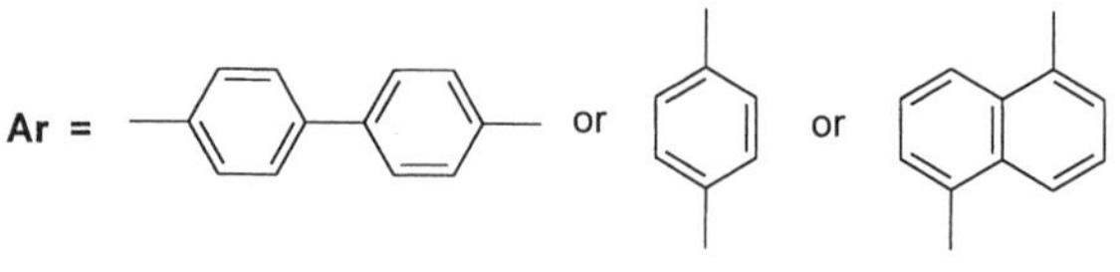

Ar =

motivated by interest in materials that are biaxially strengthened by photo-crosslinking through the cinnamate moiety. Some of these polymers have N mesophases, some S mesophases. The stabilities of mesophases with respect to the aryl ester group, Ar, were in the general order 4,4'-biphenyl > 1,4-phenyl > 1,5-naphthyl. The polymer with a 1,4-phenyl ester group and R=H did not cross-link either upon heating to the S phase or upon 350-nm irradiation (Rayonet lamps) at ambient temperature. However, it cross-linked upon irradiation above 50°C. Irradiated polymers become insoluble, the DSC transitions disappeared, and there was a considerable reduction in the 1630 cm^{-1} band in the IR spectrum due to the cinnamate double bond. The relative rates of photochemical reaction, as judged by the disappearance of the 1630 cm^{-1} band, increased approximately threefold between 150°C, just below T_m at 175°C, and 200°C, in the S mesophase. This is one of the few reports of how the relative reactivities of LC polymers change in a mesophase.

Kricheldorf and his co-workers [32–35] have studied a number of poly-(ester-imides) based on an unusual monomer, **4**, incorporating both cinnamic acid

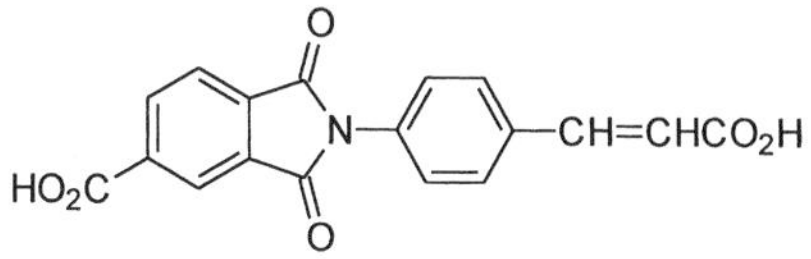

4

and phthalimide moieties, and chiral spacer groups. Some of these polymers have cholesteric (Ch) mesophases which when macroscopically oriented by, for example, shearing in the melt, could have so-called "Grandjean" textures. Such polymers would be colored, due to scattering of selected wavelengths. These colors would normally be transitory and would disappear once macroscopic orientation was lost upon removal of the orienting force, but the authors suggested that they could be fixed into the film by locking in the macroscopic orientation by photo-cross-linking. A number of polymers were studied and their structures in both the solid and fluid states were characterized. For example [33], both homopolymer **5** and copolymers **6** formed Ch mesophases but only the latter formed a (red)

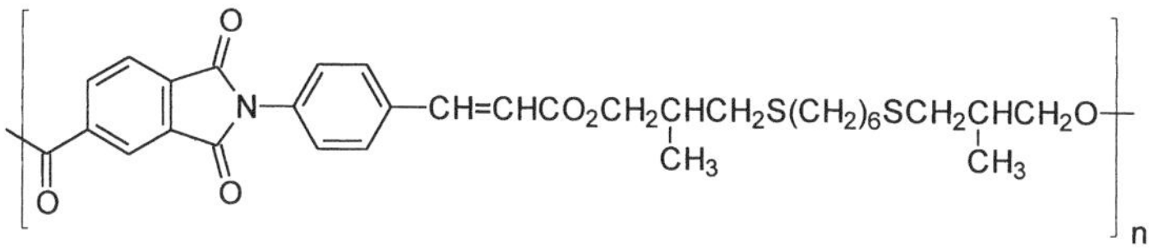

5

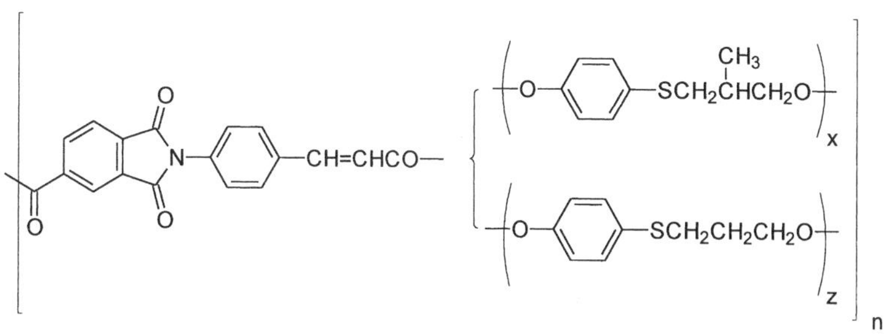

6a. x/z = 4
6b, z = 0

Grandjean texture upon shearing between glass plates. The Grandjean textures of some of the polymers changed color with temperature presumably as the pitch of the helical domains changed. No UV-Vis spectra of these polymers were published although the author's comment [34] on the apparent enhancement of absorption observed in thin films between 380–390 and 350 nm of related polymers such as **5**, **6a**, and **6b**. Irradiation ($\lambda \leq 360$ nm) of these materials led to "rapid" cross-linking relative to much slower thermal degradation and cross-linking opening up the possibility of fixing textures and colors at selected temperatures. More detailed photochemical results have been reported [32,35] for random copolymers such as **7** and for the analogous cholesteric polymer **8**. Irradiation

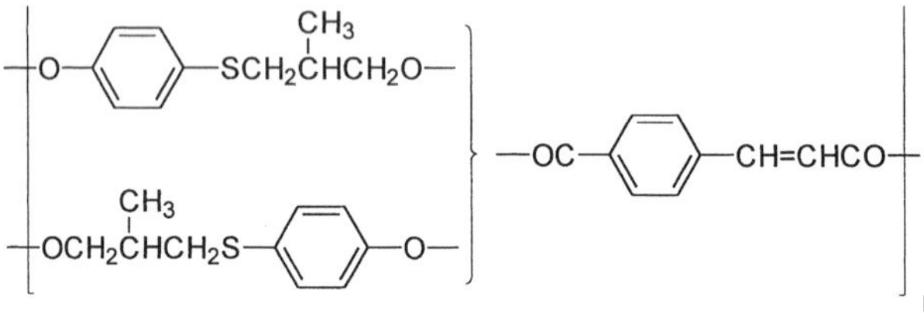

7

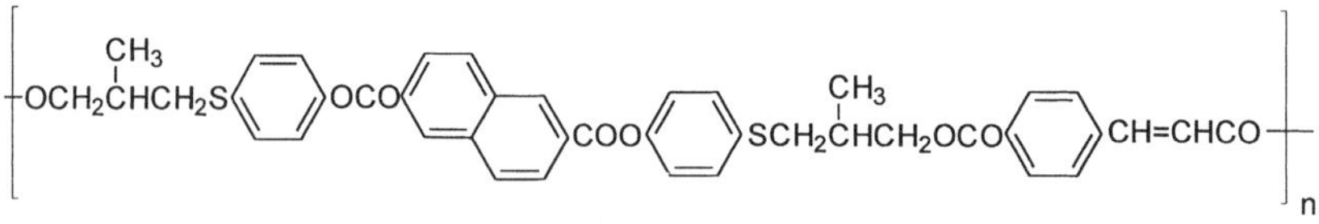

8

(313 nm) of these polymers in chloroform led to initial changes attributed to both 2 + 2 photocycloaddition and *trans-cis* photoisomerization. Absorption increases at 360 nm at longer times of irradiation were attributed to photo-Fries rearrangement. Interestingly, spin coated films of **6** annealed at 70°C but irradiated in the glassy state at room temperature showed no evidence of photo-Fries reaction. This type of rearrangement was observed for films of **8** but through the naphthalene-2,6-dicarboxylic acid moieties. Irradiated films became insoluble due to crosslinking *via* 2 + 2 photocycloaddition. The published UV-Vis spectrum [33] of the film of **6** seems to be broadened and to show enhanced absorption above 350 nm relative to the solution spectrum and indicates chromophore aggregation but this effect was not discussed. The perturbed UV-Vis spectrum of films of **7** was noted [34].

Navarro [36] has reported several soluble, low transition temperature MCLC polymers containing *ortho*- or *para*-linked units and cinnamates or phenylene-*bis*-acrylates in the main chain. One series of cinnamate-containing polymers is shown as **9**. Some of the polymers seemed to be partially crystalline in the solid state. All had N and/or S mesophases which were retained in the solid glasses upon cooling to room temperature. Some of the polymers underwent thermal degradation and cross-linking above 300°C. Irradiation ($\lambda_{max} = 300$ nm)

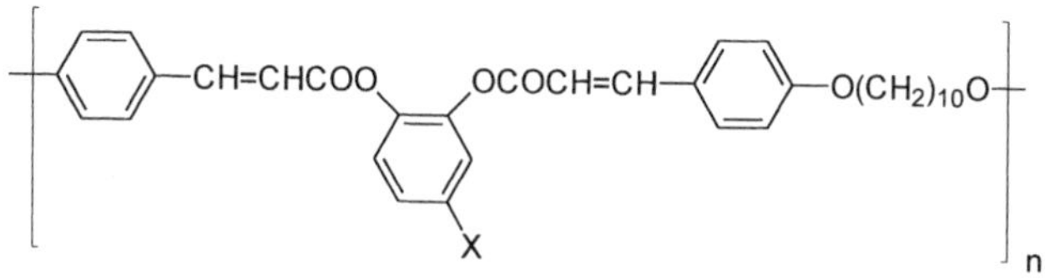

9, X = H, CH3, CHO, NO2

of thin films of the polymers resulted in the loss of the alkene stretching band in the IR at 1630 cm^{-1}. The irradiated films were resistant to solvents and presumably cross-linked.

SCLC Polymers. There has been a series of papers on the synthesis and photochemistry of SCLC polymers containing the 4-alkoxyphenyl-4′-alkoxycinnamate chromophore, the same chromophore as that of the MCLC polymer, **1**. Ritter and co-workers [37] were the first to examine the photochemistry of any SCLC cinnamate polymer. The DSC data reported for polymer **10** indicated the

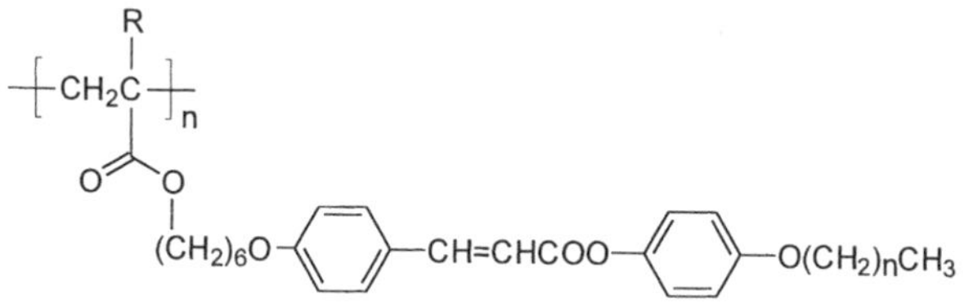

10, R = H, n = 2
14, R = CH3, n = 4

existence of two LC phases, one *N*, the other presumably *S*. UV-Irradiation (450 W, Hg lamp) of the polymer resulted in the loss of the DSC transitions and changes in the permeability of 1-butanol through polyamide composite membranes containing a film of the polymer. Unfortunately, no UV-Vis spectra or other information about this interesting material were reported. Keller [38] used the same chromophore but a polysiloxane main-chain to make a series of SCLC polymers, **11**, all of which had S_C mesophases. His idea was to make elastomers in which the cross-linking reaction is performed by UV irradiation of the polymer

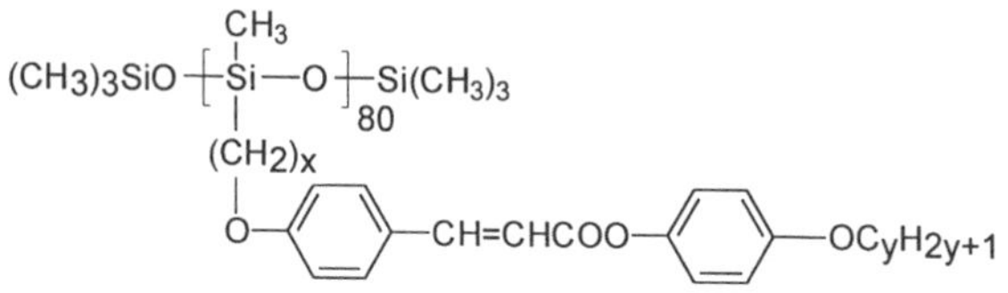

11, x = 5-10, y = 1-7

in the glassy state with the mesogens macroscopically ordered, and, thereby, to prepare organic thin films with controlled orientations of functional groups. Glassy films (heat treated then cooled to room temperature) of **11** ($x = 8$, $y = 6$), showed significantly perturbed UV-Vis spectra. There was no comment as to the nature of the perturbation. Upon irradiation (low-pressure Hg lamp) cross-linking occurred and the UV-Vis spectrum showed enhancement of absorption out to past 450 nm, presumably due to photo-Fries rearrangement.

Noonan and Caccamo [39] reported a detailed study of the synthesis and photochemistry of SCLC polymers such as **12** (also with the 4-alkoxyphenyl-4′-

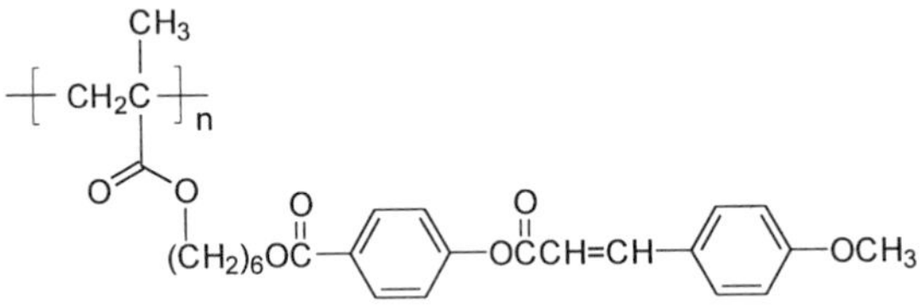

12

alkoxycinnamate chromophore) with the idea of determining whether synergistic effects in photochemical reactions could be achieved in organized, thermotropic LC polymers. Polymer **12** has a N mesophase over a wide temperature range. Its UV-Vis spectrum in dichloromethane showed a λ_{max} at about 315 nm compared to a value of about 290 nm, with a shoulder at 315 nm, in an "as cast" film. This effect was attributed to the existence of a (metastable) conformation in the cast film in which the phenyl ester group was not in plane with the cinnamate moiety. Films containing the triplet sensitizer, 3,3′-carbonylbis-(5,7-dipropyloxycou-marin) were irradiated above 405 nm for an unspecified time. The spectrum of the irradiated film resembled the solution spectrum prior to irradiation. This effect was attributed to photochemical changes (e.g., isomerization, conformational changes, and cyclobutane formation) disrupting the order of the side chains and allowing the residual mesogens to adopt their most thermodynamically stable orientations. The perturbed film spectrum could presumably be due to chromo-phore aggregation and the changes on irradiation to preferential consumption of aggregates as suggested for other LC polymers [21,23,29]. Noonan and Caccamo [39] noted that the complexity of the probable reactions precluded the use of UV-Vis monitoring to obtain relative quantum yields so they made photographic speed measurements of different molecular weight samples of **12** relative to a non-LC cinnamate containing copolymer, **13**. Corrected for the increased sensitivity to cross-linking which occurs with increasing molecular weight, SCLC polymer, **12**, triplet sensitized by the ketocoumarin, was approximately eight times as sensitive as non-LC polymer **13**. This interesting observation was attributed to the effect that the microscopic ordering of the polymer matrix had on the triplet energy transfer process. Unfortunately, these measurements seem to have only been

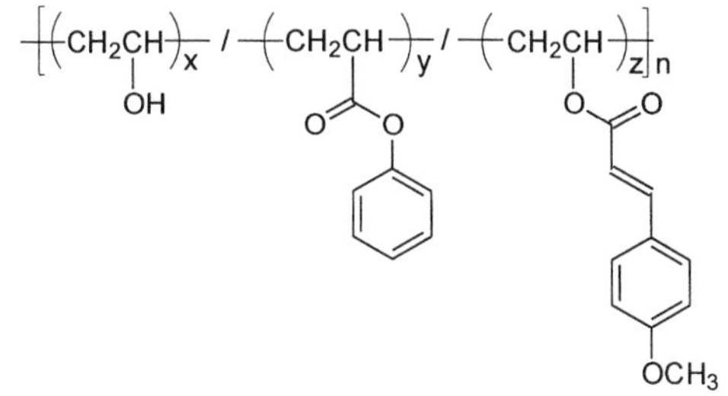

13

conducted on as cast films, not LC films, neither in the mesophase, nor in any glassy state that might exist below T_g.

Creed and co-workers [25,40,41] have reported preliminary results on the photochemistry of SCLC polymer **14**. This polymer is partially crystalline at room temperature and preliminary results suggest [40,41] it has three mesophases, N and two smectic. Its solution UV-Vis spectrum was virtually identical to that of a simple model compound, 4'-pentyloxyphenyl-4-pentyloxycinnamate. UV-Vis spectra of polymer films only resembled the solution spectrum when the films were heated to the isotropic melt, at or above 145°C. At lower temperatures the spectra were significantly perturbed, presumably by chromophore aggregation. The extent of perturbation was most dramatic in the as cast film or in films cooled to room temperature. Under these conditions the "normal" band maximum at about 313 nm was completely replaced by a new H-aggregate band at about 275 nm with extensive "tailing" of the spectrum toward the visible. Spectra of films of the mesophases were intermediate between these two extremes. Irradiation (313 nm) of as cast films led to cross-linking, presumably due to 2 + 2 photocycloaddition, and yellowing presumably due to photo-Fries rearrangement [41]. Irradiation at 70°C, in the lower temperature S phase (possibly S_B), led to [40] a large hyperchromism, much larger than that observed [21] for MCLC polymer, **1**, that was attributed to disruption of aggregates as photochemistry occurred. A similar, smaller effect was observed upon irradiation of another S phase (possibly S_A) at 105°C. Subsequent spectral changes in films irradiated at all temperatures indicated the occurrence of both 2 + 2 cycloaddition and photo-Fries rearrangement.

Chien and Cada [42] have prepared optically active and photoactive SCLC copolymers, **15**, with the 4-alkoxyphenyl-4'-alkoxycinnamate chromophore, with the intention of creating LC polysiloxane networks that could be used to prepare macroscopically oriented organic ferroelectric polymers for electro-optical devices. Optical activity was introduced into the polymer by the use of a chiral spacer. Those copolymers which were mesogenic exhibited properties characteristic of a S_{C*} phase. UV-irradiation of thin films of the polymers in their mesomorphic states at 90°C, led to a loss of the IR absorption at 1635 cm^{-1} that is due to the cinnamate double bond, and to cross-linking. Long-term irradiation led to

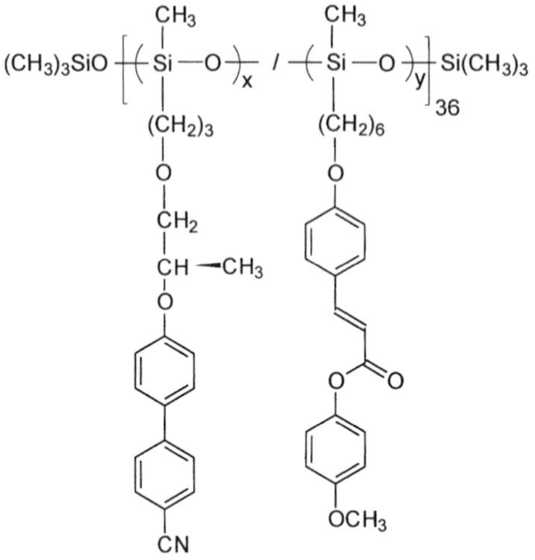

15

loss of up to 24% of the cinnamates and then reaction stopped, presumably because all chromophores in orientations accessible to reaction(s) were eventually consumed. Transition temperatures were increased upon irradiation indicating stabilization of the smectic order by cross-linking. Irradiation in the I phase (above the clearing temperature) depressed the transition temperatures, slightly suggesting that cross-linking under these conditions destabilizes the ordered state and suppresses reorientation of the mesogens. The authors also report yellowing of the polymers during irradiation attributed to photo-Fries rearrangement. Yellowing could be suppressed by long-wavelength ($\lambda \geqslant 313$ nm) irradiation.

Mitchell, Gilbert, and their co-workers [43–47] have explored the synthesis [43–45] and some aspects of the photochemistry [46–48] of SCLC polymers containing the aryl cinnamate chromophore and analogs of this chromophore. Some examples are the copolymer, **16**, homopolymers of general structure, **17**, homopolymer, **12**, and copolymer, **18**. An optimized synthesis of siloxane containing cinnamate polymers, analogous to **11**, that was reported by Keller [38], was also developed [45]. The copolymer **16**, when prepared with a 9:1 ratio of the 4-cyanophenylbenzoate and 1-naphthylpropenoate mesogens, had a N phase when prepared with a deliberately low degree of polymerization (ca. 19) and a S phase at greater (but unspecified) molar mass [46]. Films of this copolymer were annealed at 58°C in the N mesophase on a surface treated glass slide to develop a surface aligned monodomain and were then irradiated through a mask at 340 ± 20 nm. The frozen LC texture appeared, after cooling to room temperature, to be the same as it did prior to irradiation. However, the irradiated regions of the film underwent the transition to the isotropic melt at a lower temperature than the unirradiated regions. This property enabled an image to be stored on the irradiated

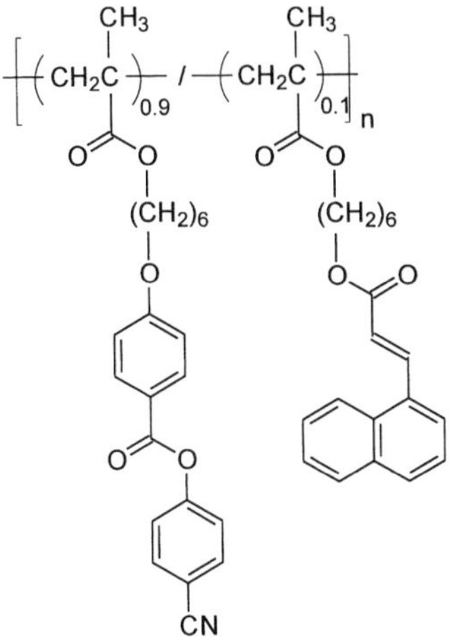

16

polymer if it was cooled rapidly to room temperature. The irradiated polymer was not cross-linked, presumably because the 1-naphthylpropenoate mesogen, which can undergo 2 + 2 photocycloaddition, was only 10% of the repeat units. The effect of irradiation on the transition temperature was attributed to E to Z photo-isomerization to the extent of 64%. Whitcombe, Gilbert, and Mitchell [47] also observed that copolymers of 4-methoxyphenyl-4′-(6-methacryloyloxyhexyloxy) cinnamate with methyl methacrylate showed evidence for only photo-Fries rearrangement on broad band irradiation in dichloromethane. More interestingly, the same authors reported [48] several aspects of the photochemistry of the copolymer, **18**, and, especially, the use of the photo-Fries rearrangement as a means of influencing the stability of a LC system. Irradiation (315 ± 8 nm) of **18**, containing 20% of the photoactive cinnamate mesogen, in dichloromethane led to spectral changes associated with photo-Fries rearrangement. Increasing degrees of rearrangement of the approximately linear aryl cinnamate mesogen to the non-linear hydroxychalcone rearrangement product (presumably not a mesogen) led to increasing depression of the clearing temperatures of films cast from the polymer irradiated in solution. Spin cast films of polymer **18** were also irradiated in the

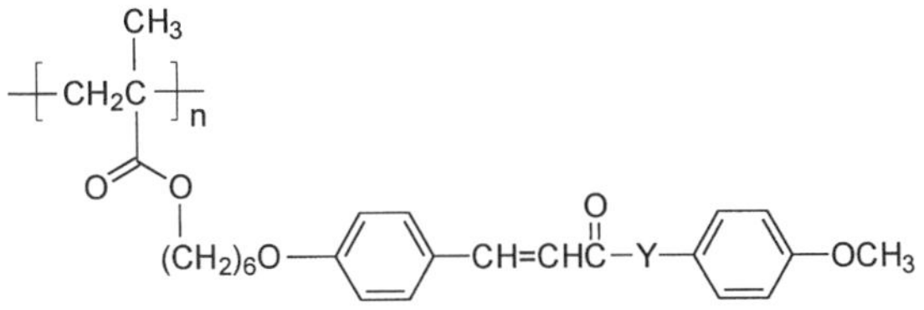

17, Y = O, or NH, or OCH2

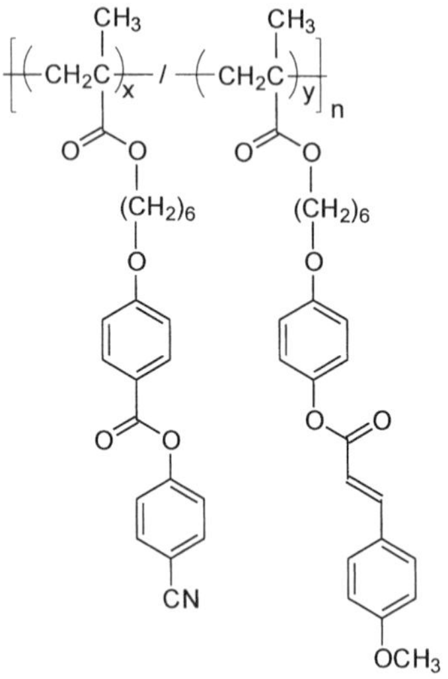

18

room temperature glassy state, the LC state, and the isotropic melt (*I*). Irradiation in the glassy and *I* states led to changes mainly associated with photo-Fries rearrangement. However, all the films became insoluble, indicating some cross-linking, presumably via cycloaddition. Irradiation in the LC state led to changes in the UV-Vis spectrum indicative of increased 2 + 2 photocycloaddition and very little rearrangement. The authors suggested that in the ordered LC state, a favorable alignment of side-chains increases the likelihood of (bimolecular) cycloaddition which then competes more effectively with (unimolecular) rearrangement.

Ikeda et al. [49] have incorporated the 4-methoxycinnamate chromophore through an ethylene glycol ester spacer group into an alternating copolymer of maleic anhydride and vinyl ethers with pendant 4′-methoxy-4-biphenyl moieties, also connected to the main chain with ethylene glycol spacer groups. The resulting polymer, **19**, which has a degree of substitution of the cinnamate moiety of 13%, shows an unspecified LC mesophase over a broad temperature range. Irradiation (310 nm) of spin cast films in the mesophase led to loss of the cinnamate IR band at 1650 cm^{-1} and insolubilization of the film. The irradiated sample retained some birefringence indicating that the order present in the LC mesophase was not completely disrupted.

Stumpe and co-workers [50] have investigated the photochemistry of SCLC poly(olefin sulfone)s such as **20** and a copolymer with the same main chain but with 95% of the cinnamate groups replaced with saturated tetradecyl groups. "As cast" or heat-treated thin films of the polymers were irradiated at 313 nm using both nonpolarized and linearly polarized light. Initial irradiation led to an initial increase in the absorbance of the films at 280 nm followed by a decrease. The

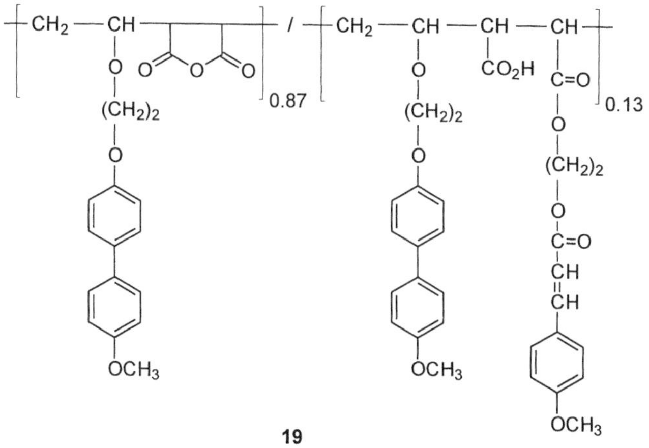

19

20

increase in absorbance was attributed to initial destruction of *H*-aggregates such that the number of "free" *E* isomers of the cinnamoyl chromophores generated by deaggregation was larger than the number lost by photocycloaddition and isomerization (the alkyl ester chromophore in this type of polymer cannot undergo photo-Fries rearrangement). The same hyperchromic effect has been observed for other LC polymers (vide supra) upon initial irradiation. Interestingly, the irradiation of **20** with linearly polarized light which, it was hoped, might lead to photoalign-

ment and thus anisotropically modified films, did not lead to the induction of optical anisotropy by angular dependent photoselection. This lack of photoselection could be caused by chromophore aggregation, energy transfer, or the primary order of the film. The copolymer has an insufficient number of chromophores for aggregation to occur, but only a small degree of photoinduced anisotropy was observed when its films were irradiated. Energy transfer and/or reorientation of the remaining cinnamate chromophores by rotational diffusion were suggested as possible explanations for the small extent of angular dependent photoselection observed in these interesting materials.

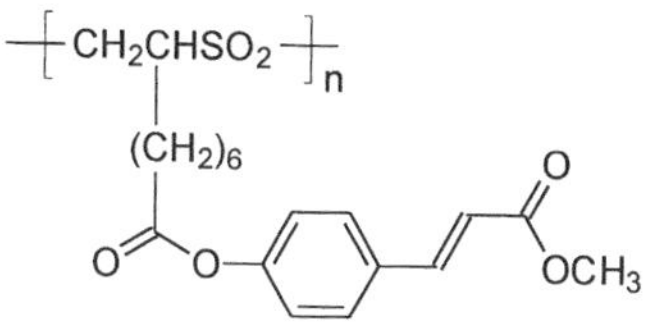

Kawatsuki and co-workers [51,52] have reported studies of the photochemistry of SCLC methacrylate and acrylate polymers, **21**, that have mesogenic groups based on the 4-alkyloxy-4'-cinnamoylbiphenyl chromophore. All polymers exhibited [51] at least one LC phase. Above the melting temperature, T_m, the

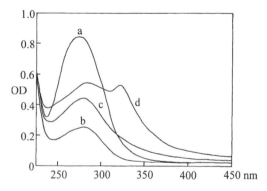

21a, R = H, m = 6
21b, R = CH₃, m = 6
21c, R = H, m = 2
21d, R = CH₃, m = 2

polymers with six carbon spacers, **21a** and **21b**, had both a S and a N mesophase whereas those with the two carbon spacer, **21c** and **21d**, had only an N mesophase. On cooling below T_m, the LC texture observed under the polarizing microscope was preserved. All the polymers showed a shift in λ_{max} in the UV-Vis spectrum from 282 to 283 nm in dichloromethane to ca. 272 to 274 nm in an "as cast" thin film, and, in the film, the absorption band was broader and tailed to the red of the main band. The blue shift was attributed to aggregates with a "parallel or head to head arrangements" of the chromophores, and the red shift to aggregates with a head-to-tail arrangements of the chromophores. A copolymer of 10 mol % of the methacrylate monomer used to make **21b** and **21d** and 90 mol % of methyl methacrylate, had almost the same UV-Vis spectrum in a thin film as it did in solution supporting the idea of chromophore aggregation being responsible for UV-Vis changes in thin films of the homopolymers. The UV-Vis spectra of thin films of SCLC polymer, **21b**, showed complex behavior upon heating from room temperature to the isotropic melt at 230°C (Fig. 7). The absorption maximum

FIGURE 7   UV-Vis spectra of a film of polymer, **21b**, at different temperatures. (a) Room temperature. (b) Heated to 130°C. (c) Heated to 145°C. (d) Cooled from I melt to room temperature. (Adapted from data of Kawatsuki and co-workers [51]).

around 275 nm decreased upon heating above T_m (70°C) by a factor of about 5. Upon heating above 130°C, the 275-nm band increased and new absorption appeared above 300 nm. As the temperature was further raised toward the isotropization temperature, the absorbance of the major band recovered to almost that of the film prior to heating but was shifted slightly to 278 nm, and the enhanced red-shifted "tail" above 300 nm disappeared. Upon cooling from the I phase, a new band maximum was observed at 330 nm (Fig. 7). The effects were reproducible upon repeated heating and cooling suggesting that they were not due to thermal degradation. The authors suggested that the aggregation was enhanced by the heating and cooling treatment and that, for the methacrylate polymers, the head-to-tail aggregation (formation of J-aggregates) was especially enhanced. The polymers were irradiated in both dichloromethane and as cast films using a high-pressure Hg lamp. Irradiated films became insoluble due to cross-linking. Spectral changes were consistent with 2 + 2 photocycloaddition in all cases with photo-Fries rearrangement being observed in solution. There was little evidence for this rearrangement in the irradiated films. Annealed thin films, presumably in a glassy LC state, showed complex UV-Vis changes with a small increase in absorbance upon initial irradiation, attributed to disruption of chromophore association, followed by continual loss of absorption with little indication of enhanced absorption of a photo-Fries type of product.

Kawatsuki et al. [52] also studied SCLC polymers such as **22** in which the

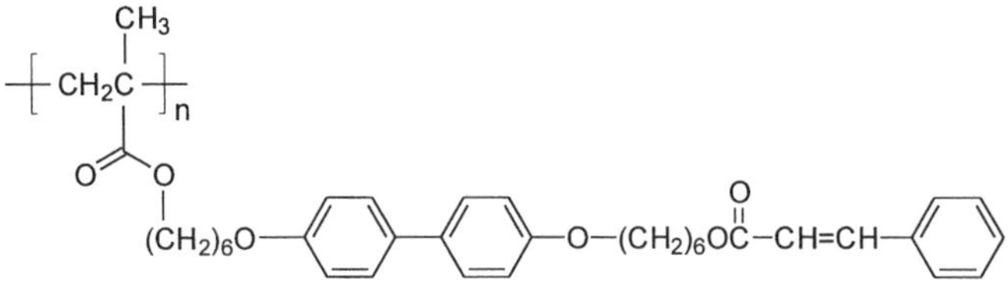

22

side-chain has a biphenyl group which is attached to the cinnamate mesogen via a flexible spacer rather than being part of the mesogen, as it is in **21**. Polymer **22** and analogs with different spacer lengths had N mesophases but their thin film UV-Vis spectra did not show evidence for chromophore aggregation. Irradiation above 290 nm led to loss of IR absorption due to the cinnamate double bond and insolubilization, presumably due to 2 + 2 photocycloaddition. Macroscopically oriented, transparent films of the polymers were obtained by pressing the molten polymer between polyamide coated prerubbed glass plates. Films of **22**, prepared in this manner, retained their anisotropic ordering after irradiation in the N phase at 95°C. In contrast, oriented thin films of the LC polymer **21b**, irradiated in the N phase at 165°C, showed a loss of the anisotropic alignment that had been obtained at the same temperature prior to irradiation. Polarized light microscopy showed disorder of the monodomains of **21b** in the irradiated regions. The anisotropic ordering of irradiated films of **21d** was retained upon heating to an extent dependent on the time of irradiation. Films of **22**, irradiated until 68% of the cinnamate chromophores had been lost, retained their anisotropic alignment upon heating to over 300°C!

Phenylenediacrylates

Tazuke, Ikeda, and their co-workers [53,54] were the first to use "phenylene-diacrylate" (PDA) as a mesogen and chromophore to explore the possibility of forming two-dimensionally reinforced polymer materials by photochemical cross-linking of a thermotropic LC polymer. Polyesters, **23**, of *para*-phenylene-diacrylic acid and several diols were synthesized and found to exhibit LC behavior upon annealing above T_g. Polymer **23a** exhibited [53] the texture of a N meso-

$$\left[\begin{array}{c} O \\ \parallel \\ CCH=CH \end{array} \right. \!\!\!-\!\!\! \left\langle \begin{array}{c} \\ \end{array} \right\rangle \!\!\!-\!\!\! CH=CH \begin{array}{c} O \\ \parallel \\ C \end{array} O(CH_2CH_2O)_x \left] \right._n$$

23a, x = 2; 23b, x = 3; 23c, x = 4

phase by polarized light microscopy. The fluorescence of the polymers and of a model compound, diethyl phenylenediacrylate, was studied in detail [54]. Polyester films showed both monomer and "excimer-like" emission (Fig. 8), with the latter predominating. The ratio of excimer to monomer fluorescence intensity

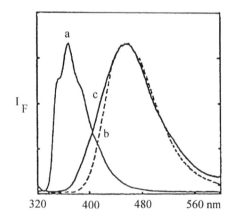

FIGURE 8 Normalized fluorescence spectra of diethyl phenylenediacrylate (Et$_2$PDA) and Polymer, **23a**. (a) Et$_2$PDA (1.0×10^{-5} mol dm^{-3}) in 1,2-dichloroethane. (b) Et$_2$PDA (2.0 mol dm^{-3}) in 1,2-dichloroethane. (c) Thin, air-saturated film of **23a**. (Reprinted with permission from Ikeda et al. [54]. Copyright 1990 American Chemical Society.)

(I_E/I_M) was temperature dependent. An abrupt drop in I_E/I_M was observed at the clearing temperature, consistent with the chromophores being aligned parallel to one another in a face-to-face arrangement favorable to formation of excimers, an arrangement presumably disrupted in the isotropic melt. Upon cooling from the I phase, an abrupt drop in I_E/I_M was not observed, consistent with the slow formation of LC domains that had been observed by light microscopy. No UV-Vis spectra were reported but the authors suggested that these excimers were probably excited aggregates, having the same emission characteristics of an excimer, on the basis of time-resolved fluorescence measurements of the monomer (isolated chromophore) emission at 370 nm and the excimer emission at 470 nm in films at 90°C. The excimer emission shows no rising component that would be consistent with excited isolated chromophores being quenched to form true excimers. The excimer-like emission was analyzed using double and triple exponential functions. Using either method, fluorescence lifetimes were longer for **23b** and **23c** than for **23a** in the LC phase but were very similar in the isotropic melt. This effect was attributed to the greater order of the LC state in **23b** and **23c** that was supported by the higher values for ΔS measured for the LC phases of **23b** and **23c** relative to **23a**. It was suggested that the mesogens in the LC state of **23a** have a greater mobility, thereby facilitating bimolecular quenching processes and leading to a higher rate of nonradiative decay.

The photochemistry of these polymers was also investigated [53]. Films of polymer **23b** were irradiated (313 nm) in air in the LC phase at 35°C. During irradiation, the films became insoluble, the C=C stretching band at 1635 cm^{-1} almost disappeared and the C=O stretching band shifted from 1710 to 1720 cm^{-1} and two cyclobutane carbon signals were seen at δ 0 and 50 ppm by solid state ^{13}C nmr of the irradiated samples. Irradiation also led to disappearance of the DSC transition corresponding to the N to I phase change of the un-cross-linked polymer, and an increase in T_g. Disappearance of the IR band at 1635 cm^{-1} was used to monitor the temperature dependence of the photochemical reactivity of **23a** (Fig. 9). The relative rate of reaction was much higher in the temperature range where **23a** is in its LC phase. Below T_g (30°C) where segmental motion is completely suppressed, the relative reactivity was extremely low. Above the isotropization temperature (87°C), the reactivity was somewhat reduced. The authors concluded that the LC phase seemed to have the optimum balance of chromophore mobility and organization for maximum photochemical reactivity. Molecular weight effects on the photochemistry of **23a** and **23b** were also investigated using samples fractionated by preparative gel permeation chromatography. The temperature effect on photoreactivity was larger in higher molecular weight (HMW) samples than in lower molecular weight (LMR) samples (Fig. 9). However, the reactivity was highest for an unfractionated sample. The authors suggested that in HMW, the chromophore orientation may be favored but the mobility is restricted. In LMW, the regular orientation may not be achieved but the mobility

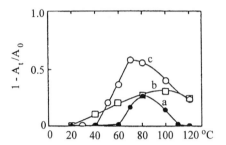

FIGURE 9 Temperature dependence of topochemical photodimerization of films of polymer, **23a**. (a) HMW polymer. (b) LMW polymer. (c) Unfractionated polymer. Relative reactivities were obtained by monitoring the change in absorbance $(A_0 - A_t)$ of the C=C stretching band at times t relative to $t = 0$ (A_0). (Reprinted with permission from Ikeda et al. [54]. Copyright 1990 American Chemical Society.)

may be favored. Both fractions may contribute in a complementary fashion in the unfractionated sample.

Gómez and Ostariz [55] also investigated PDA containing LC polymers with the idea of photochemically "fixing" LC phases and thereby improving their mechanical properties. MCLC copolymers, **24**, were obtained that contained

$$\left(\!\!\left[\!\!\left[\text{O}\!-\!\!\bigcirc\!\!-\!\overset{\overset{\text{O}}{\|}}{\text{C}}\right]_x\!\!\left[\text{O}\!-\!\!\bigcirc\!\!-\!\text{O}\right]_y\right]/\left[\overset{\overset{\text{O}}{\|}}{\text{C}}\text{CH=CH}\!-\!\!\bigcirc\!\!-\!\text{CH=CH}\overset{\overset{\text{O}}{\|}}{\text{C}}\right]_y\right)_n$$

24

p-PDA, *p*-hydroxybenzoic acid, and pyrocatechol moieties and which could be cross-linked both thermally and photochemically. The polymers had low crystallinity and entered *N* mesophases upon heating. A sample prepared by melt polymerization was thought to be partly cross-linked and formed highly transparent oriented films upon shearing. "As cast" films of an un-cross-linked sample, prepared by solution polymerization, were irradiated at room temperature with a low-pressure mercury lamp with emission center at 300 nm. Irradiated films became yellow, insoluble in all common organic solvents, and did not melt below 350°C. IR spectra showed the near disappearance of the C=C stretching band at 1630 cm^{-1}. These observations are indicative of cross-linking by 2 + 2 photocycloaddition and, possibly, photo-Fries rearrangement. Unfortunately no further photochemical studies were reported.

Chang and co-workers [56,57] also reported studies of thermotropic MCLC polymers containing the *para*-PDA chromophore. Their motivation was also to develop two-dimensionally reinforced polymer materials. Aromatic copoly-

(ester)s were prepared from p-PDA, with a mixture of methylhydroquinone and various hydroxy carboxylic acids, including 4-hydroxycinnamic acid. All these polymers showed a thermotropic N mesophase by optical microscopy. The polymers were quite thermally stable. Thermogravimetric analysis indicated that decomposition did not begin until above 300°C. The photochemistry of the polymers was studied by the disappearance of the 1630 cm^{-1} C=C stretching band in the IR spectrum and was strongly temperature dependent. Below T_g the polymers were virtually unreactive to 350-nm irradiation (Rayonet lamp). Above T_g photo-cross-linking rates increased greatly. In general, the lower the T_g, the higher the photochemical reactivity. It was suggested that above T_g, the segments containing the double bonds were sufficiently mobile to find the correct orientation for 2 + 2 photocycloaddition. Below T_g segmental mobility is frozen and the photoreactivity, therefore, was quite low. The N textures of the polymers were preserved in the final, photo-cross-linked materials at room temperature indicating that the LC order could be preserved in the final material.

Kricheldorf and Probst [58] have studied a series of photoreactive cholesteric copoly(ester-imide)s incorporating p-PDA, N-(4-carboxyphenyl)trimellitimide, **25**, and chiral diols such as **26**. Their intention was to develop chiral

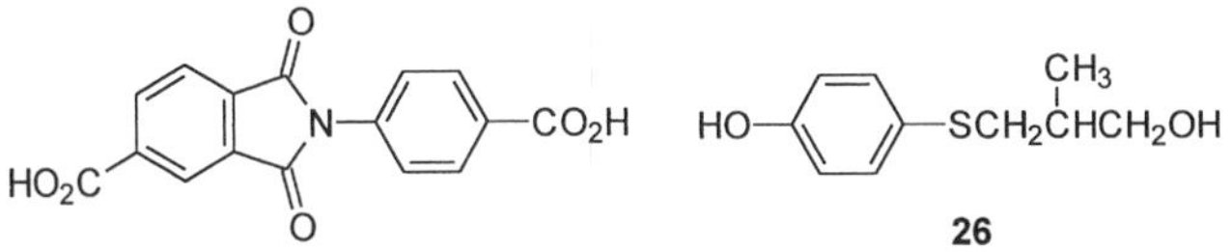

25   **26**

polymers which have a "Grandjean texture," which arises from macroscopic ordering of all domains, that could be "fixed" by photo-cross-linking of the material in the melt. Additionally, such materials might form S_{C*} phases with ferroelectric or piezoelectric properties. The polymers synthesized, such as **27** and

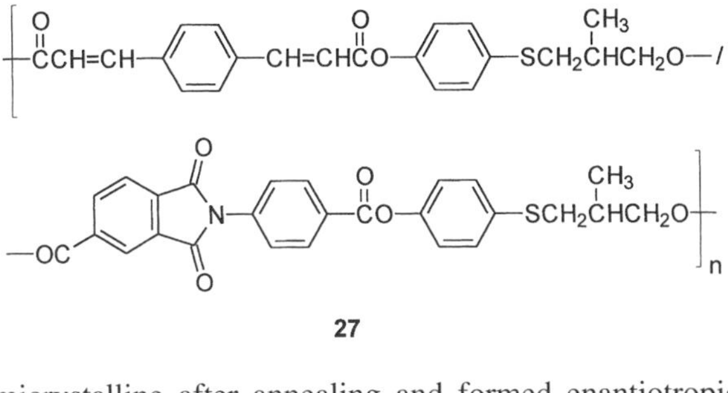

27

28, were semicrystalline after annealing and formed enantiotropic cholesteric melts above 180°C which showed Grandjean textures when the melts were sheared between glass plates. Some of the polymers supercooled down to 150°C

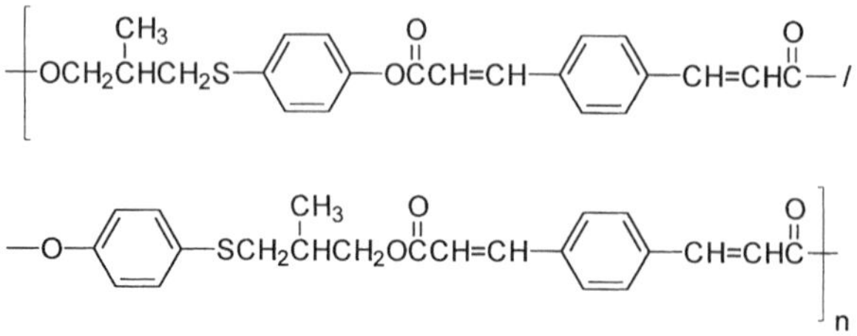

28

and their blue color in transmitted light was almost unchanged from 150 to 300°C. Irradiation ($\lambda \leq 360$ nm) enabled all the polymers to be cross-linked. The sensitivity increased as the proportion of PDA units increased. It was noted that a film of the random copolymer of general structure **28** had absorption that rose sharply at $\lambda \leq 400$ nm and that slow cross-linking occurred even in daylight for this polymer and the analogous alternating copolymer with the same functional groups. Since the PDA chromophore has virtually no absorption above about 380 nm, it is tempting to suggest that chromophore aggregates contribute to this effect.

Stilbenes

Stilbene and its derivatives have also been very important in the development of photochemistry and photophysics. The photochemical *trans-cis* isomerization of stilbene is possibly the most thoroughly studied photochemical reaction. The 2 + 2 photocycloaddition of stilbenes is also well known [59], although perhaps less thoroughly explored. Both these reactions are of interest when they occur in LC polymers. Isomerization changes the shape of a *trans*-stilbene moiety from that of a mesogen to that of a nonmesogen. Photocycloaddition is one of the most widely used photochemical cross-linking mechanisms. In comparison with the large number of papers (vide infra) on analogous azobenzene-containing polymers, there have been relatively few studies of the photochemistry of stilbene-containing polymers and even fewer of stilbene-containing LC polymers. In contrast, there has been considerable recent interest in "small molecule" stilbene aggregates and their excited state behavior [60]. This latter work is obviously of great relevance to work on stilbene polymers.

Saeva mentions, in an early review [10] on LC materials and aspects of their photochemistry and photophysics, the irradiation of an unspecified stilbene containing polymer and changes in its physical properties. Creed et al. [25,61,62] have reported several observations of the photophysics and photochemistry of a *trans*-stilbene 4,4'-dicarboxylate containing MCLC polyester, **29**, one of a series of such polymers with different spacers synthesized by Jackson, Morris, and co-workers [63]. This polymer is partly crystalline in the "as cast" state and has a N mesophase over a narrow temperature range (177–186°C). In solution, the structured UV-Vis absorption [62] and fluorescence [25,61] spectra are almost

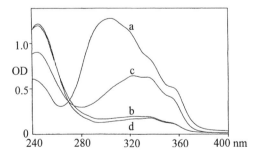

29

identical to those of a simple model compound such as diethyl *trans*-stilbene-4,4'-dicarboxylate. UV-Vis absorption spectra [62] of as cast and heated films were broadened by both blue- and red-shifted absorption relative to the solution spectrum. These broadened spectra (Fig. 10) strongly suggested the occurrence of extensive chromophore aggregation. It is not possible to tell whether two or more types of aggregates are present or whether both blue and red-extended absorption is from a single species. On heating, complex changes were observed in the UV-Vis absorption spectrum. Initial heating above T_g enhanced the blue-shifted absorption. Presumably, once chain motion can occur, the mesogens can begin to pack more efficiently. Fluorescence spectra [25,61] of as cast films were red-shifted, broad, and structureless relative to fluorescence in solution or from model compounds. Although these spectra were "excimer-like" and suggested parallel interaction of the chromophores, fluorescence excitation spectra [25] showed they arose, at least in part, from excitation of aggregates rather than only from excitation of unaggregated stilbene chromophores. Annealing the films and cooling to room temperature caused a loss of excimer-like emission and the appearance of much weaker structured emission [61] that was tentatively assigned to a different kind of aggregate, formed as crystallization of the polymer is enhanced. The fluorescence decay from as cast films had three components [61] of lifetimes

FIGURE 10 UV-Vis spectra of polymer, **29**. (a) Thin film at 65°C (above T_g) before irradiation. (b) Thin film after 313 nm irradiation. (c) Film irradiated at 313 nm followed by irradiation at 254 nm. (d) Film irradiated at 313 nm, then 254 nm, and again at 313 nm. (Reprinted with permission from Creed et al. [62]. Copyright 1993 Society of Photo-Optical Instrumentation Engineers.)

1.6, 6.8, and 18.2 ns, supporting the idea of three emitting species, unaggregated stilbene and two different aggregates. When as cast thin films were irradiated (313 nm) several effects were observed [62]. In the early stages of irradiation blue-shifted aggregate absorption disappeared preferentially and the residual absorption, although somewhat blue-shifted, had the shape and vibrational structure of the isolated chromophores (e.g., in solution). The excimer-like fluorescence disappeared [61]. These observations suggested that the aggregates are preferentially consumed. Since cyclobutane dimers seem to be formed (vide infra), it seems likely that these are *H*-aggregates. Films became cross-linked and insoluble, suggesting cross-linking by 2 + 2 photocycloaddition. Irradiation at 254 nm, where cyclobutane dimers absorb, of the films photo-cross-linked at 65°C by 313-nm irradiation, led to restoration of film solubility, and partial regeneration of the UV-absorption above 280 nm, presumably by cleavage of cyclobutane dimers (Fig. 10). However, the regenerated absorption was somewhat different from the original absorption, leading to the suggestion [25] that some of the dimers in the irradiated film had cleaved at 254 nm to give the *cis*-stilbene chromophore. The "maximum overlap" cyclobutane dimer of *trans*-stilbene can cleave, in a concerted fashion, to give *either* two *trans*- or two *cis*-stilbenes, via paths *A* and *B*, respectively (Fig. 11). The other stereospecifically formed dimer of *trans*-stilbene can *only* cleave, by a concerted mechanism, to two *trans*-stilbenes. The film with 254 nm regenerated absorption could be irradiated again at 313 nm with loss of absorption above 280 nm.

Bis(benzylidene)cycloalkanones

Ganghadara and Kishore [64–66] have published an interesting series of papers on the photochemistry of MCLC [64,65] and SCLC [66] polymers containing various bis(benzylidene)cycloalkanones as mesogens/chromophores and a tetra-ethylene glycol-derived flexible spacer to lower the transition temperatures. The MCLC polymers [64], **30**, had *N* mesophases, identified by POM. Polymer **30a**

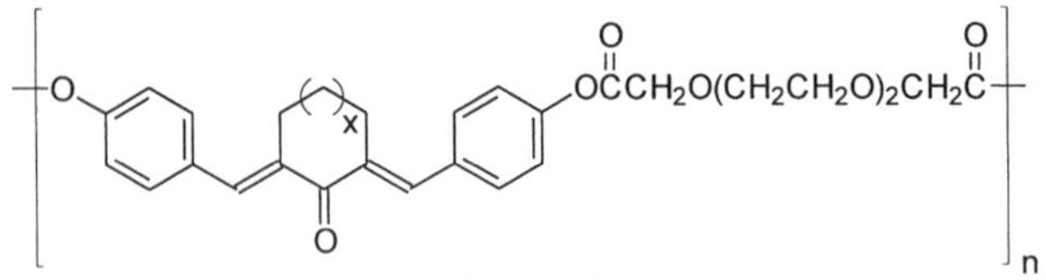

30a, x = 0
30b, x = 1
30c, x = 2

crystallizes upon cooling back to room temperature, but **30b** and **30c** formed glassy LC states that retained the LC texture and presumably the N organization, upon cooling down below their glass transition temperatures. Films of the polymers showed complex changes to their UV-Vis spectra upon heating from room

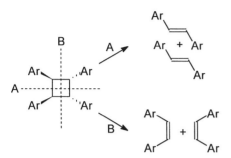

FIGURE 11   Different pathways for concerted "Retro 2 + 2 Photocycloaddition" (photo-cleavage) of a cyclobutane dimer.

temperature to the I melt. The changes in absorbance at the peak maximum at 322 nm (A_{322}) and the first differential of the absorbance with respect to temperature (dA_{322}/dT) were correlated with the transitions seen by DSC (Fig. 12). A drop of A_{322} between 40 and 70°C (B to C in Fig. 12) matched the T_g of **30b**, and a second drop between 100 and 120°C (E to F in Fig. 12) started at a crystal to crystal transition and continued until the polymer was completely in the N phase. The increase from 140 to 180°C (G to H in Fig. 12) corresponded to the polymer going from the N mesophase to the I melt. In general, as A_{322} dropped, the spectrum

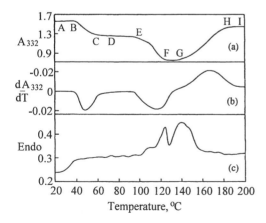

FIGURE 12   Temperature (T) effects on UV-Vis and DSC behavior of polymer, **30b**. (a) Absorbance at 322 nm (A_{322}) as a function of temperature (T). (b) Differential of A_{322} with respect to T, dA_{322}/dT, as a function of T. (c) Enthalpy changes as a function of T. (Reprinted with permission from Gangadhara and Kishore [64]. Copyright 1993 American Chemical Society.)

broadened and an enhancement of a red-shifted tail was observed. These two effects were attributed to chromophore aggregation. It was proposed that in the films cast at room temperature from chloroform, the chromophores are randomly oriented but, upon heating through the T_g, the polymer chains became mobile and the chromophores began to aggregate. This process, accompanied by a drop in A_{322} and changes in the spectrum (vide supra) continued through the crystal-to-crystal transition and into the N mesophase. Above the isotropization temperature, the chromophores become completely randomly oriented with a corresponding increase in A_{322}. Cooling effects were similar but less dramatic probably because of incomplete aggregation due to supercooling. Annealing the films for three hours at 90°C produced hypochromism and enhancement of a red-shifted tail comparable to that observed on the heating cycle. Films of **30** were irradiated (medium-pressure Hg lamp) and changes monitored by both UV-Vis and IR spectroscopy. Initial irradiation of a N film of **30b** at 130°C led to an initial increase in A_{322} and a small blue shift of the peak maximum. This effect was also observed at 75°C and was attributed to initial formation of EZ and/or ZZ geometrical isomers disrupting parallel stacking of chromophores. Further irradiation led to a decrease of A_{322} and eventual appearance of a 287-nm maximum assigned to 2 + 2 photodimers, a conclusion supported by studies on the model compound bis(benzylidene)cyclopentanone. Upon long-term irradiation, films became insoluble, T_g of the polymer shifted to higher temperature, and other DSC transitions seen in the unirradiated polymer vanished. Irradiation in the N phase led to a partial loss of the N texture seen by POM but irradiation in the nematic glass led to cross-linking without loss of the N texture. The decrease in the IR stretching band due to C=C at 1602 cm^{-1} relative to an unchanged C$-$H asymmetric band at 1295 cm^{-1} was used to measure the structure and phase (temperature) dependence of cross-linking. The reactivity of the polymers followed the order of decreasing ring size, **30c** < **30b** < **30a**. It was suggested that the geometry for cycloaddition became less favorable as ring size increased. The temperature dependence of the reactivity of **30b** was studied for a fixed 0.5 min of irradiation (Fig. 13) using IR method (vide supra). The reactivity increase above 40°C was attributed to increased aggregation above T_g. A drastic fall in reactivity above 140°C was attributed to the less ordered nature of the isotropic melt.

Ganghadara and Kishore [65] have extended their work on polymers **30a–30c** to include additional polymers with different spacer groups and higher molecular weight polymers prepared by interfacial rather than solution polymerization methods. MO calculations on model compounds indicated that increased ring size led to a less linear and less rigid mesogen which lowered the transition temperatures for a fixed spacer length. Some more detailed studies of the temperature dependence of irradiation of this type of polymer were described but no additional conclusions were made regarding the photochemistry and/or photophysics. The same authors [66] have also incorporated the bis(benzylidene)cyclo-

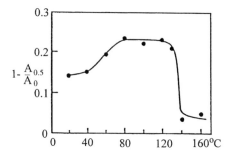

FIGURE 13 Temperature dependence of the photochemical reactivity of polymer, **30b**. The change in absorbance after 0.5 min of irradiation (A_0–$A_{0.5}$) of the C=C stretching band relative to the initial absorbance (A_0) was used to monitor reactivity. (Reprinted with permission from Gangadhara and Kishore [64]. Copyright 1993 American Chemical Society.)

hexanone chromophore into nematic SCLC homopolymers, **31**, and copolymers, such as **32** with a cholesteryl co-monomer, and studied their thermal properties and photochemical behavior. None of the polymers crystallized but all exhibited LC properties. Homopolymers, **31** went into N mesophases above T_g. Circular

31, x = 2, 4, 6, 8

dichroism (CD) was induced in the bis(benzylidene)cyclohexanone chromophore in the copolymers **32** by the cholesteric co-mesogen, and the induced CD effect was used to demonstrate the existence of cholesteric mesophases in these copolymers.

Unfortunately, temperature dependent UV-Vis spectra of polymers, **31**, were not reported. Films of the polymers were irradiated (medium-pressure Hg lamp) and the reactions monitored by UV-Vis and FTIR spectroscopy. Homopolymer, **31** ($x = 6$), was irradiated at 120°C in the N phase. A gradual decrease in the main absorption band of the chromophore at 352 nm was observed and the film eventually became completely insoluble. Interestingly, no initial increase in absorbance was observed, as had been seen for a small molecule LC model and for MCLC polymers [64] **30** (vide supra) and previously attributed to disruption of

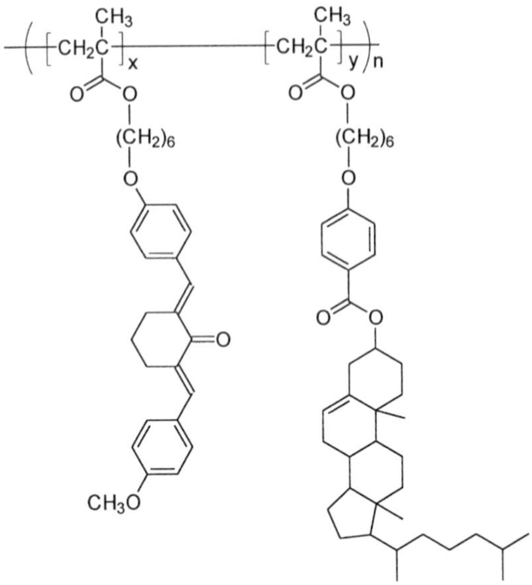

32

aggregation by geometrical isomerization. This lack of hyperchromism was attributed to the high-melt viscosity of **31** ($x = 6$). During irradiation the main absorption band decreased and new absorption attributed to photodimers appeared. The 1600 cm^{-1} band, due to C=C stretching, in the FTIR spectrum decreased and the polymer films became partly insoluble, indicating a maximum extent of cross-linking of 65% after 15 min irradiation. The temperature dependence of photo-cross-linking was studied using the disappearance of the C=C stretching band in the FTIR spectrum of **31** ($x = 6$). The rate of cross-linking increases above T_g (50°C) and peaks at about 80°C in the N mesophase. The rate is four to five times lower below T_g and at and above T_i. Cholesteric copolymer **32** was annealed in its mesophase, cooled into the LC glass and irradiated. Induced CD spectra diminished as irradiation proceeded. However, the chloroform insoluble fraction of the film retained induced CD absorption indicating some helicoidal structure had been retained even after photo-cross-linking.

Spiropyrans

Spiropyrans have been incorporated as photochromic moieties into many polymers [4,67]. Krongauz and his co-workers [68–72] have extended this work to include several types of SCLC polymers. Small molecule LC materials containing a spiropyran moiety have not yet been reported, presumably because typical spiropyrans are not sufficiently elongated. The covalent attachment of spiropyrans to rodlike mesogenic groups does afford "quasi liquid crystals"; compounds with unusual metastable mesophases [4]. However, these compounds do not show LC and photochromic behavior simultaneously. In contrast, spiropyran chromophores can be covalently incorporated as groups attached to the side-chain to a considerable extent in SCLC copolymers containing variable fractions of "conventional" mesogens, without loss of either the LC behavior of the polymer or the photochromism of the spiropyran (vide infra). This strategy affords some intriguing photochromic materials with possible applications in photoimaging and/or information storage.

Cabrera et al. [68] synthesized SCLC copolysiloxanes, **33**, containing a typical mesogen, 4-methoxyphenyl-(4'-alkoxybenzoate), and a spiropyran. In-

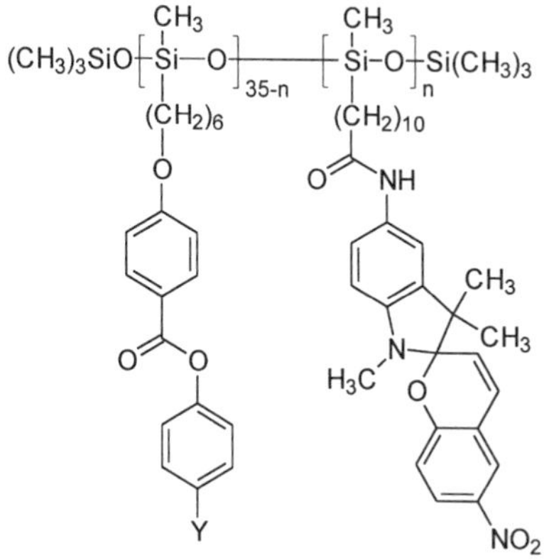

33, Y = OCH₃ or CN

creasing the proportion of the spiropyran chromophore up to 11 mol % narrowed the stability range of the mesophase mainly by lowering the isotropization temperature from that of the homopolymer of the benzoate ester alone. Increasing the mol % of the spiropyran had a much smaller effect on T_g, which remained below room temperature. Copolymer **33** ($R = OCH_3$) had fairly complex photochromic and thermochromic behavior (Fig. 14). The copolymer with 11 mol % spiropyran was pink and strongly birefringent upon casting at 25°C, within the mesophase stability range (ca. 15 to 30°C). Irradiation at 25°C ($\lambda > 500$ nm) gave a pale yellow color that, upon further UV irradiation ($\lambda = 365$ nm), gave a deep red color. The yellow color was attributed to the spiropyran chromophore ($\lambda_{max} \approx 370$ nm) and the deep red color to *aggregated* merocyanines ($\lambda_{max} = 550$ nm). Irradiation of a yellow film at -20°C, below the T_g, gave a blue color attributed to *isolated* merocyanine groups ($\lambda_{max} = 580$ nm) that could be converted back to the yellow spiropyran by visible light irradiation. The blue film was stable in the dark at -20°C but turned red above -10°C. The red film was thermally stable below T_g but turned yellow on irradiation with visible light above -10°C. The red color faded over several days at room temperature but in less than a minute above the clearing temperature. It was noted that the aggregated mero-

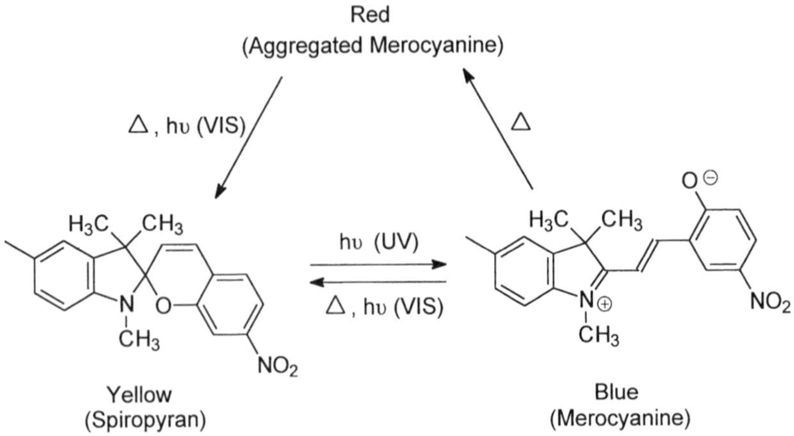

FIGURE 14 Spiropyran-merocyanine interconversions proposed for polymer, **33**. Adapted from Cabrera et al. [68].)

cyanines cause a physical cross-linking of the macromolecules, as was observed earlier [73] for non-LC spiropyran-containing polymers. Cabrera et al. [69,70] have extended work with copolysiloxanes, **33**, to polyacrylates, **34**, in which the

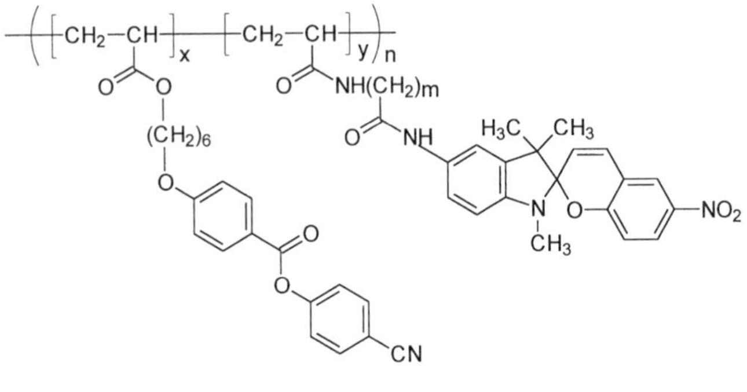

34

spiropyran-containing monomer is co-polymerized with a monomer containing the 4-cyanophenyl-(4-alkoxybenzoate) mesogen. Detailed thermal characterization of these materials was described [70]. Order parameter measurements on films aligned in an electrostatic field suggested that photochromic and mesogenic groups may be partly located in separate sites. These polymers show a N mesophase for copolymers with up to about 45 mol % of the spiropyran. Their photochromism was very similar to that of copolysiloxnes, **33**. Their isotropic films showed a very unusual transient translucence when they were mechanically disturbed (e.g., by touching or squeezing). It was shown that this effect was not due to formation of a homeotropic orientation of the molecules that could be mechanically perturbed but was probably due to partial restoration of LC order in the I film induced by the mechanical perturbation.

Analogous SCLC copolysiloxanes and copolyacrylates containing a spiro-naphthoxazine side-group have also been reported by the same group [71]. The spirooxazine chromophore is less prone to photochemical fatigue than the spiro-pyran chromophore. It was pointed out that this would be an advantage in application of these or related polymers as photo- or electro-sensitive materials. Spirooxazine copolymers of type **35** typically formed N mesophases which gave

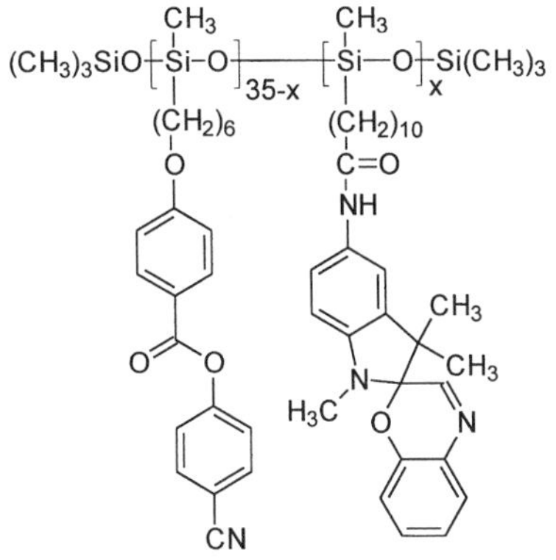

35

NG phases below T_g. The main difference between these spirooxazine polymers and the spiropyran polymers such as **33** (vide supra) was that the former showed no spectral evidence for aggregation of the merocyanines formed upon irradiation. The spirooxazine-containing polyacrylates, whatever the proportion of the spiro-compound, showed similar behavior upon irradiation from -20 to $+25°C$ in the mesophase, amorphous phase, and in tetrahydrofuran. Upon UV irradiation, the structureless spirooxazine absorption tailing in the visible region between 400 and 600 nm was replaced by a structured absorption with a maximum at ca. 630 nm and a shoulder at ca. 580 nm. These bands were attributed to two merocyanine isomers in thermal equilibrium. The complex thermal decay of these mero-cyanines was much slower for the acrylate polymers with increasing proportions of spirooxazine, an effect attributed to steric hindrance, by surrounding bulky photochromic groups, of thermal ring closure.

Azo Compounds

A large number of SCLC polymers with azoaryl photochromic chromophores have been studied [74] mainly with a view to applications in optical recording [2].

Cyclic, oligomeric LC siloxanes with azoaryl chromophores have also been studied and the results recently reviewed [3]. The elongated *E*- or *trans*-azobenzene moiety with one or two *para*-substituents can act as a mesogen, either alone or in copolymers with other types of mesogenic groups such as 4-substituted phenyl benzoates, biphenyls, etc. Following *E*- to *Z*-photoisomerization, the length to width ratio of an azobenzene is reduced [2,75] and the *Z*-isomer no longer behaves as a mesogen (Fig. 15). Many azoaryl compounds undergo high quantum efficiency *E*- to *Z*-photoisomerization, probably mainly through an *inversion through nitrogen* mechanism, rather than a rotation of an aryl group about the N=N bond mechanism [74,76,77]. The inversion mechanism requires a much smaller sweep volume that the rotation mechanism and therefore can occur in a polymer below T_g where segmental motion of the main chain cannot occur. The *Z*-isomer typically undergoes thermal reversion to the more thermodynamically stable *E*-isomer over a period of from minutes to hours, depending on both the structure of the azoaryl compound, the temperature, and other conditions [74]. Thermal back isomerization also occurs in polymers below T_g. Increased electron donor-acceptor character seems to facilitate this process [78]. The *Z*-isomer can also be converted photochemically back to the *E*-isomer. Under constant illumination, a photostationary state will be established whose composition depends on the absorption spectra of the azoaryl species, the medium, the temperature, etc. These readily reversible geometrical isomerizations and the concomitant change of the molecule from a mesogen to a nonmesogen and vice versa have made azoaryl groups attractive as chromophores for various types of materials, both LC (vide infra) and non-LC [79], for so-called "photooptical" information recording. An alternative method, "thermooptical recording" or "thermorecording," involves laser heating or contact heating of the LC material to induce a phase change [2].

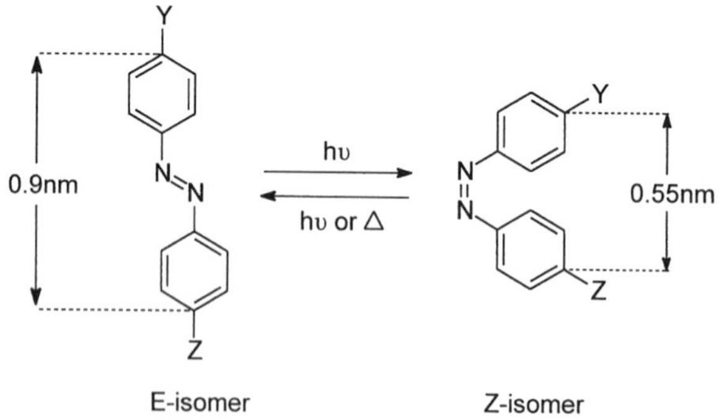

FIGURE 15 Geometry changes during isomerization of *E*- to *Z*-azobenzene. (Adapted from Irie [75].)

No photochemistry is involved in the latter method and it will not be considered further. The highlights of this work on azoaryl-containing LC polymers for photooptical recording will be briefly surveyed but the reader is directed to recent reviews [2,3,79], especially the excellent recent review by Shibaev [2] that covers the literature, including important and extensive contributions in the Russian language, thoroughly up until about 1994 and gives many more details about potential applications than would be appropriate for a chapter focusing on photochemistry and photophysics.

Eich, Wendorff, and their co-workers [80–83] were the first to use SCLC polymers for photooptical information recording. For example, the polymer, **36**,

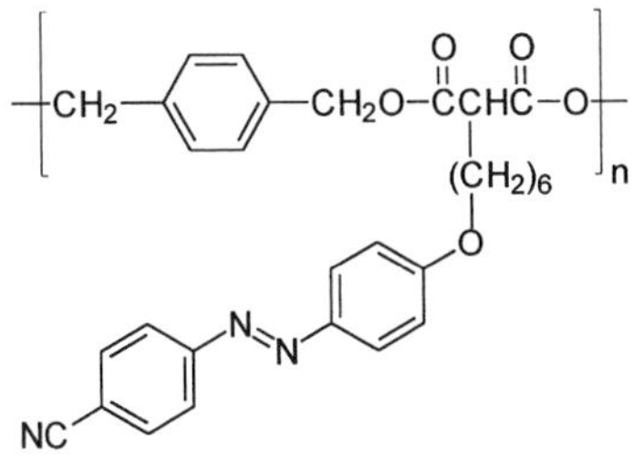

36

had both S_A and N mesophases. A clear, homeotropically aligned "monodomain" (macroscopically organized) film was generated by application of an electric field to a film of the polymer, held above its T_g between two conductive glass plates. The film was irradiated at room temperature with linearly polarized green light (514.5 nm) to cause E-Z photoisomerization in the irradiated region. This isomerization caused a reorientation of the side-groups with respect to the local molecular environment and, therefore, a change in refractive index in the irradiated area, that was measured at 632.8 nm (He–Ne laser), a wavelength that was not absorbed by the chromophore. A change in refractive index, $\Delta n_{nd} = 7 \times 10^{-3}$ was achieved. The storage image could be recognized by POM proving that optically induced molecular reorientation had been achieved in the irradiated area. The reorientation persisted after the light source was turned off and even after sufficient time had elapsed for thermal back isomerization of the azoaryl moieties to occur. It was suggested that the geometrical change generating the Z-pendent groups and their optically induced orientation together influenced the local LC orientation giving rise to a weak reorientation in the irradiated region. This effect was used to generate holograms and detailed studies were later published on this phenomenon [82,83]. In a later publication [84] it was shown that the polarized exciting light turned the optical axis of the LC polymer perpendicular to the plane of polarization. Heating the films above the isotropization (clearing) temperature erased the stored information. Spiess and his co-workers [85,86] used time-dependent infrared spectroscopy and forced Rayleigh scattering, in which a grating is induced in the sample from interference of two laser beams and its time evolution monitored by a subsequent diffraction experiment, to study the complex dynamics of photo-isomerization of azobenzene in non-LC copolymers and SCLC copolymers such as **37**. Two relaxation processes were observed in the glassy states of these

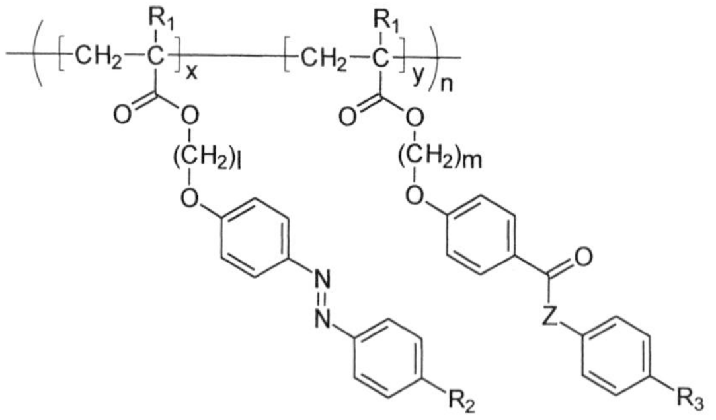

37, R_1 = H, I = m = 6, Z = O, R_2 = R_3 = CN
38, R_1 = CH$_3$, I = m = 6, Z = NH, R_2 = H, R_3 = OCH$_3$
40, R_1 = CH$_3$, I = 2, m = 6, Z = NH, R_2 = H, R_3 = OC$_4$H$_9$
48, R_1 = H, I = m = 3, Z = O, R_2 = R_3 = OCH$_3$

polymers. The faster process had a wide range of relaxation times and was attributed to reorientational relaxation of both the dye and the glassy matrix. The slower, exponential process was assigned to thermal back isomerization of the Z-isomer to the E-isomer.

Stumpe et al. [87] investigated an alkoxyazobenzene-containing SCLC copolymer, **38**, with 14 mol % of the azobenzene chromophore, that had both *S* and *N* mesophases. They prepared homeotropic, glassy LC films of **38** by anneal-ing and rapid cooling of the polymer in a polyimide coated glass cell. Conoscopi-cal examination revealed that the director of the mesogenic groups was oriented perpendicular to the substrate surface of the cell. Irradiation with both unpolarized and linearly polarized light (several wavelengths) led ti *E-Z* photoisomerization and macroscopic reorientation of the homeotropically oriented film to a homoge-neous orientation (director parallel to the substrate surface). A reorientation of both the (minority) photochemically active azoaryl mesogens and the (majority) photochemically inactive mesogens had occurred triggered by the *E-Z* photo-isomerization and rotational diffusion of the azoaryl mesogens. Irradiation with linearly polarized light also led to angular dependent photoselection and a local change of the orientation distribution of the azoaryl moieties (photoinduced dichroism). Both the induced dichroism and birefringence persisted after thermal *E-Z* isomerization but could be eliminated by regeneration of the original homeo-tropically aligned film upon heating to the isotropic state and cooling the isotropic melt. The angle of reorientation of the optical axis of the LC polymer film upon irradiation with linearly polarized light depended on the activating light intensity and exposure time. The optical contrast increased in similar fashion. A stationary state was achieved when the optical axes of the azo chromophores and those of the other mesogenic side groups were oriented perpendicular to the plane of polarization of the excitation light. The induced optical anisotropy was reversed when the sample was irradiated with light that was linearly polarized perpendicular to the initial, activating light. This reversible optical switching of photoinduced birefringence could be repeated over a number of irradiation cycles.

Ivanov et al. [88] studied three very similar polymers of general structure **39**, with different ratios of mesogens, that had either S_A or N mesophases, and **40**

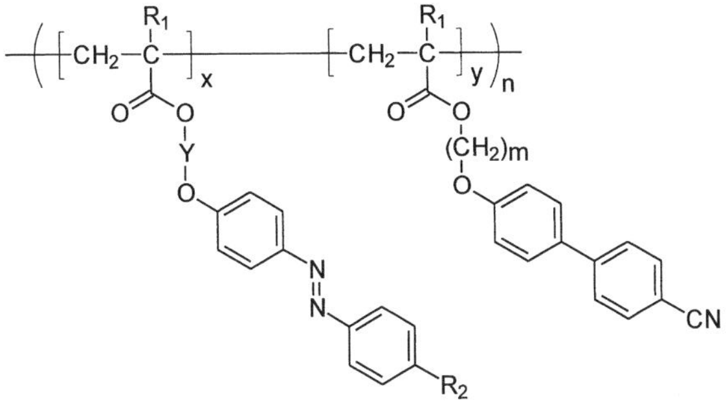

39, $R_1 = H$, $Y = (CH_2)_6$, $m = 4$ or 5, $R_2 = CN$
41, $R_1 = CH_3$, $Y = (CH_2)_2O(CH_2)_2$, $m = 2$, $R_2 = CN$
42, $R_1 = H$, $Y = (CH_2)_2O(CH_2)_2$, $m = 2$, $R_2 = CN$
43, $R_1 = H$, $Y = (CH_2)_6$, $m = 2$, $R_2 = CN$
44, $R_1 = H$, $Y = (CH_2)_2O(CH_2)_2$, $m = 6$, $R_2 = CN$
45, $R_1 = H$, $Y = (CH_2)_6$, $m = 3$, $R_2 = CN$
51, $R_1 = H$, $Y = (CH_2)_{11}$, $m = 11$, $R_2 = H$

(similar to **38**) with 20 mol % of the azobenzene chromophore. Laser-induced birefringence (Δn_{ind}) was about 10 times greater for homeotropically oriented films held above T_g than for films of the same compound below T_g. There was little difference between the behavior of S_A and N mesophases. All the polymers showed an optimum temperature for the maximum induction of birefringence. At and above the clearing temperature photoinduced birefringence was not observed. Lasker et al. [89] showed irradiation induced dichroism at both 365 nm (azobenzene) and 285 nm (benzanilide) for films of copolymers of type **40** and **41**, with variable fractions of mesogens. Infrared dichroism of the CN stretching bands showed of these polymers confirmed that both types of mesogenic group had been oriented perpendicular to the electric vector of the linearly polarized 488-nm irradiating light, presumably by a multistep cooperative process involving nearest neighbors. Lasker et al. [90] have also irradiated polymers of types **40**, **42**, **43**, and **44** in amorphous, optically isotropic films. Both photochromic and nonphoto-chromic side-groups were oriented perpendicular to the vector of the actinic light to an extent similar to the LC order. Changes in birefringence (Δn_{ind} = ca. 0.03) were much smaller than was observed (Δn_{ind} = ca. 0.08) for homeotropically aligned films [88] of the same materials. A uniformly planar aligned film of **40** (with 20 mol % azobenzene) was irradiated with linearly polarized light. In this experiment the degree of order, S, dropped from 0.4 to 0.16 during irradiation suggesting that the reorientation process disrupts the LC order. Four steps were proposed [90] to explain photochemical induction of optical anisotropy in amorphous films of LC polymers. First, absorption takes place that is proportional to the square of the angle between the electric vector of the exciting source and the transition moment of the chromophore (which coincides with the long axis of the azoaryl group). Second, there are repeated $E \rightleftharpoons Z$ (*trans* $\rightleftharpoons$ *cis*) photochemical and thermal isomerization steps with perturbations of the molecular environment. Third, repetition of the first two steps within the photoisomerization steady state results in a "successive reorientation of the photochromic moieties *via* photo-induced rotational diffusion steps" [90]. Ultimately the azoaryl moieties become oriented perpendicular to the electric vector of the exciting light and absorption can no longer take place. Finally, reorientation of nonphotochromic mesogene occurs.

Läsker, Stumpe and their co-workers [91–93] have also studied azoryl containing polymers on the "border" between SCLC and amorphous polymers. They pointed out that the preparation of oriented, nonscattering monodomain films of LC polymers is difficult. However, the values of irradiation-induced birefringence in easily made films of amorphous polymers are small. Perhaps polymers which were on the border between LC and non-LC character would combine the desirable properties of both types of materials. Two approaches were suggested. Amorphous films could be prepared by supercooling the isotropic melts of LC polymers, or inherently amorphous polymers containing rodlike side-groups could be used. Polarized light irradiation of both amorphous (supercooled isotropic) and initially aligned polymers such as **45** led to virtually the same induction of optical anisotropy. Comparison of the angular dependent UV absorption due only to the azobenzene chromophore with the IR absorption due to the stretching of the CN group on the biphenyl mesogen suggests that both mesogenic groups undergo reorientation even though only the azobenzene is excited and irradiations were carried out below T_g. Birefringence (Δn_{ind}) induced by linearly polarized light (488 nm) in amorphous films of SCLC polymer, **40a**, with 30 mol % of azobenzene, at 30°C was ca. 0.07, almost twice the value of 0.04 achieved for amorphous polymer, **40b**, with 40 mol % of azobenzene. As the temperature was raised, Δn_{ind} reached a maximum of about 0.11 above T_g (but below T_i) for **40a**, whereas Δn_{ind} declined for **40b** from its maximum value of about 0.04 at 30°C to zero at about T_g. Interestingly, films of **40** and related azoaryl-containing polymers can have their mesogenic groups completely aligned by photoselection during irradiation above T_g [91,93,94]. This is a new photochemical method of producing macroscopically oriented films ("monodomains"). "Classical" methods of aligning films, such as application of a magnetic field or annealing the polymer film cast on an oriented polyimide surface, failed for **40a**. Effects of the thermodynamic stability of the LC phase on the tendency of mesogens in several different SCLC polymer to be photo-oriented were also investigated [91,95–97]. In some polymers, irradiation caused the absorbance of the *E*-isomer to increase after an initial decrease. This as attributed to disruption of *H*-aggregates and generation of an increased number of nonaggregated *E*-isomers [29,95,97]. More generally, it was observed that in LC polymers with a higher enthalpic stability, photo-orientation by the linearly polarized light was restricted. In LC polymers with a lower enthalpic stability, photo-orientation was much more facile. It was suggested [91,93] that, in aligned SCLC polymer films, there is a competition between the self-organization of the side-groups and the photoselection process.

Recently is has been reported [98–104] that the layered structures of Langmuir–Blodgett (LB) films of several amphotropic (amphiphilic and thermotropic) SCLC copolymers (e.g., of type, **46** and **47**) can also be reoriented by UV-

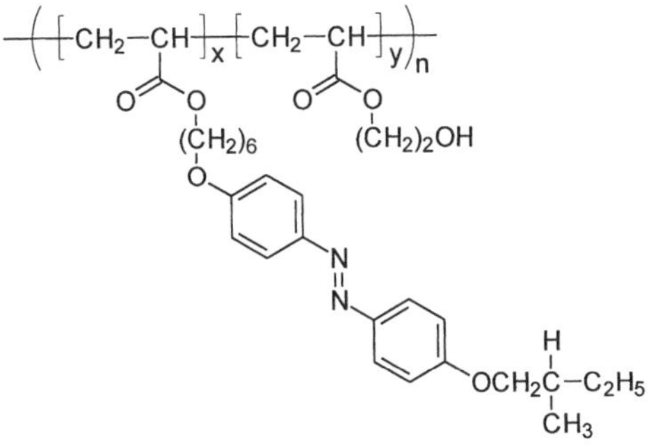

46

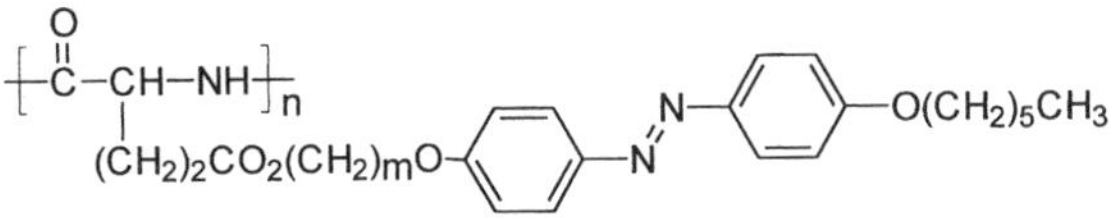

47, m = 2, 4, or 6

irradiation induced *E-Z* photoisomerization of the mesogenic azoaryl groups. Phase transitions from the constrained LB films into LC-like layered structures could be achieved both by heating and/or by excitation of the azoaryl mesogens. Thus LB films of polymer **46**, which also has a low-temperature *S* mesophase (between 56 and 86°C), were prepared and irradiated [101,103,104]. X-ray scattering and atomic force microscopy measurements indicated a vertically layered structure, characterized by a 4.6-nm double layer spacing, for the LB films. This structure was irreversibly destroyed on annealing or by UV irradiation (365 nm) at 63°C, affording a new lamellar structure, characterized by a 5.2-nm double layer distance, and very similar to the S structure found in the bulk phase of the polymer. Further UV irradiation resulted in production of predominantly Z-azobenzenes and an isotropic film, but further visible irradiation (436 nm) to regenerate E-azobenzenes resulted in regeneration of the S-type film. Optical anisotropy could be achieved using plane polarized light.

"Hairy rod" substituted amphotropic poly-L-glutamates of type **47** have also been studied in detail [29,103]. These polymers exhibited two mesophases, one S-like with a layered structure, the other a less-ordered Ch phase at higher temperatures. Multilayered LB assemblies exhibited "a bilayer structure of deformed hairy rods" in which the rodlike macromolecules were oriented in the dipping direction. The UV-Vis spectra of such layers had $\lambda_{max} = 329$ nm compared to 348 nm for the same chromophore in solution, indicating the formation of H-aggregates of the azobenzene chromophores. On annealing at 95°C, above the T_g, the "stressed" bilayer structure was lost and was replaced by a structure characteristic of the bulk S-like LC phase. The UV-Vis absorbance increased due to a reduction in H-aggregation. Irradiation (365 nm) of spin-coated, amorphous films (Fig. 16a) resulted in a uniform loss of the 348-nm band ($\pi-\pi^*$ transition) and an increase in the 440 nm band ($n - \pi^*$ transition) due to E-Z isomerization. However, irradiation of the LB films resulted in an initial decrease in the 329-nm band in the $\pi-\pi^*$ transition region and an increase at 440 nm, both attributed to

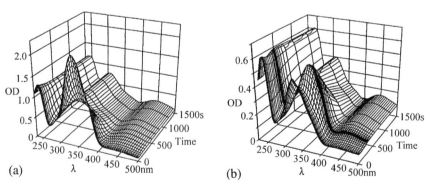

FIGURE 16 UV-Vis absorption spectra of films of **46** irradiated at 365 nm. (a) Spin-coated film. (b) Langmuir-Blodgett multilayer film. (Reprinted with permission from Stumpe et al. [29]. Copyright 1996 American Chemical Society.)

isomerization, followed by an increase (Fig. 16b) in absorbance and shift in the absorption maximum to the red to 345 nm (the "normal" $\pi-\pi^*$ transition), both attributed to loss of H-aggregates. Finally the absorption of unaggregated E-isomers at 345 nm diminished and was replaced by Z-isomer absorption at 310 and 440 nm. The orientational order of the chromophores also changed during irradiation. The films produced from the LB films by UV irradiation have a high Z/E isomer ratio and are isotropic. Subsequent visible light irradiation (488 nm) established a photostationary state with ca. 85% E-isomers and a layered S-type of structure similar to that produced by thermal treatment of the original LB multilayer films (vide supra). There are subtle differences in the behavior of these polymers, particularly in the orientational ordering of the chromophores, depending on the spacer lengths and different combinations of thermal and photochemical treatment. The reader is referred to the original papers [29,103] for further details of these fascinating physical and chemical manipulations of supramolecular ordering.

Ikeda, Tazuke, and their co-workers [106–108] have introduced and explored the idea of "photochemically triggered phase transitions" in which a photoreaction (e.g., photoisomerization) of chromophores perturbs a whole system, causing an isothermal phase transition, changing the physical properties of the system, and amplifying the effect of the photoreaction. This idea which, they point out, is analogous to the amplification of the initial signal that occurs in the visual process, was first applied to micellar, vesicular, and small-molecule LC systems [106], and then to SCLC polymers, sometimes with small molecule azobenzenes as dopants [106–109]. Polymer films were irradiated at 366 nm in the (birefringent) N mesophase and transition to the (clear) I melt that accompanied E-Z photoisomerization of the azoaryl chromophores, was monitored using the change in scattering of the 633-nm light from a He–Ne laser. Among the SCLC copolymers with azoaryl and phenyl benzoate side groups that were studied [107,108], **48** showed the highest rate of isothermal $N \rightarrow I$ phase transition. The rate of the phase transition was dependent on molecular weight. Lower molecular weight samples of **48** and several analogs had a higher $N \rightarrow I$ rate, and the rate increased with increasing temperature. The rate was found to correlate well with ΔS_{NI}, the entropy of the $N \rightarrow I$ transition. Less-ordered systems (low ΔS_{NI}) underwent faster $N \rightarrow I$ transition. Test patterns were generated by irradiation of films of **48** at 54°C in the N mesophase and then, to preserve them, they were cooled below T_g at ca. 30°C [105]. Similar results were obtained [108,110] using copolymers analogous to **48** but with cyanobiphenyl side groups replacing the phenyl benzoate side-groups of **48**. Using a photoresponsive "guest" molecule, 4-hydroxyazobenzene, which does not photoisomerize, Ikeda et al. [108] showed that phase transitions in these types of materials did not arise merely from heating caused by the irradiations (thermooptical recording mode). They also used [109,111] another small molecule dopant, 4-butyl-4′-methoxyazobenzene (BMAB) and pulsed laser excitation (355 nm) to show that if sufficient Z-isomer of BMAB were formed a single 10-ns laser pulse, then the relaxation of the matrix SCLC from N to I occurred in ca. 200 ms, comparable to the time scale for the same effect in low molecular weight LC systems. This was at temperatures, T, close to the N to I transition temperature, T_{NI} (the temperature where the birefringent phase was completely lost) and such that $T/T_{NI} = 0.998$. In a later study [112] the behavior of BMAB doped into an SCLC polymer was compared with that of a SCLC polymer with a covalently bound azobenzene chromophore. In these studies, a relaxation time, τ, was defined as the time required for the transmitted light from the He–Ne laser viewed through crossed polarizers, to drop to 50% of its initial value. For SCLC polymers doped with BMAB, τ decreased from ca. 100 to ca. 40 ms as T/T_{NI} decreased from 0.998 toward 1. There was almost no effect of the molecular weight of the polymer and a minimum single shot laser power at 355 nm of 0.22 J cm^{-2} was required to observe the transition. In copolymers such as **48**, containing covalently bound azobenzene chromophores, τ was as low as ca. 10 ms, significantly faster than for the polymer doped with BMAB.

More recently Ikeda and Tsutsumi [113] have used a SCLC polymer, **49**, in which *all* of the mesogens are photoactive, azoaryl groups. They irradiated (355 nm, Nd-YAG) surface aligned monodomain films of **49** in both the N phase at

49, R = OC$_2$H$_5$
50, R = CN

140°C and the glassy N phase at 23°C (below the T_g at 45°C). The N to I phase transition could be induced in 200 μs even below T_g by a single pulse of the laser. Ikeda and Tsutsumi point out [113] that, in LC materials, optical anisotropy arises from the anisotropically shaped mesogens being aligned along a common axis. However, unlike the situation in a crystal, the centers of gravity of the mesogens are randomly placed. When E-Z isomerization occurs, optical anisotropy disappears in a LC material because of the random orientation of the chromophores generated in the irradiated region and the absence of a regular alignment of the centers of gravity. The latter would persist, even after irradiation, in a true crystal. It is believed that this effect accounts for the microsecond response of the films of **49** used in this work. The irradiation-induced I phase was stable for at least 8 months, when the films were kept below T_g, despite the thermal back isomerization, in less than 24 hours, of the Z- to E- forms of the azoaryl mesogens. Increasing the temperature of the film above T_g resulted in restoration of the N phase. The analogous donor–acceptor substituted polymer, **50**, also showed [78] an N to I phase change in under 200 μs, but the thermal Z-E or *cis-trans* reversion occurred in 800 ms at 135°C, an order of magnitude faster than for **49**. Intra-molecular charge-transfer stabilization of the transition state that leads to thermal isomerization of Z-azoaryl moieties in **50** was proposed [78].

Yamamoto and co-workers [114] studied copolymers, **51**, containing both cyanobiphenyl and azobenzene mesogens. Irradiation (365 nm) of thin films of polymers having more than 50 mol % of azobenzene mesogens caused an isother-mal phase transition from the N to the I state. Cooling below T_g after irradiation preserved the image generated by irradiation. There was an optimum temperature range for LC formation for effective photorecording. Below T_g it was difficult to obtain high resolution. An excimer laser pulse (351 nm, ca. 20 ns) was used to measure the response time of the phase transition after irradiation. For the polymer with all azobenzene mesogens, the response time was 10 ms at 80°C, where T/T_{NI} = 0.970, and 600 ms at 70°C where T/T_{NI} = 0.85. These response times were considerably slower than the 0.2-ms response time reported by Ikeda and Tsu-tsumi [113] for polymer **49** (vide supra).

Han et al. [115] reported the use of novel SCLC polymers of type **52**, with two azobenzene mesogens in a side chain attached to a single tertiary carbon of the

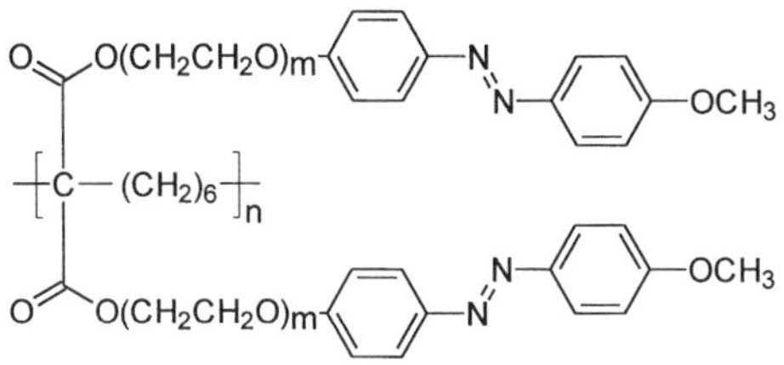

52

main chain, as materials for optical information storage. Thin films of these polymers had N mesophases. The polymers underwent E-Z photoisomerization upon UV-irradiation (365 nm) in both solution and in spin cast thin films. Thin films were annealed above T_g to develop a monodomain texture and irradiated through a photomask. The films were cooled below T_g to store the image devel-oped from the phase change from N to I induced by E-Z photoisomerization.

Angeloni, Chiellini, and their co-workers [116–118] reported the synthesis and characterization of both MCLC and SCLC polymers containing the 4,4'-bisalkoxyazobenzene chromophore. These interesting materials photoisomerized in solution but their photochemistry in an LC mesophase was not reported.

Oge and Zentel [119] have studied a ferroelectric SCLC copolymer, **53**, containing a photoisomerizable, chiral 4,4'-bisalkoxyazobenzene mesogen to-gether with a chiral 4-alkoxyphenyl-4'-alkoxybenzoate mesogen that is photo-chemically unreactive at wavelengths where the azobenzene moiety can be photo-excited. This polymer has both a ferroelectric S_{C*} phase and a S_A phase. Films

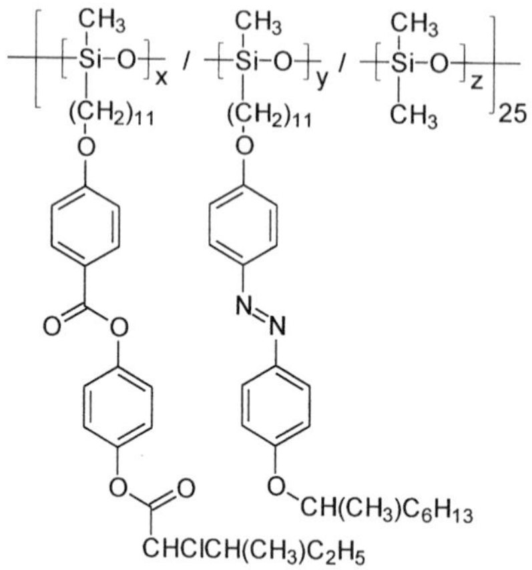

53

of **53** were contained in a rubbed polyimide coated glass cell and oriented by repeated heating and cooling in an electric field. UV irradiation (365 nm) led to a loss of polar order to the S_{C*} phase as measured by a drop in the spontaneous polarization (P_S) relative to the film kept in the dark. This effect was attributed to the phase being disrupted by *E-Z* photoisomerization. Visible light irradiation (>420 nm) to cause *Z-E* photoisomerization led to the restoration of P_S to near to its original value. Thermal *Z-E* re-isomerization was an order of magnitude faster in the LC phase than in solution.

Other Chromophores

Kricheldorf et al. [120] have synthesized a large number of MCLC polycarbonates containing 4,4'-disubstituted chalcone as a photochemically reactive mesogenic group. Achiral copolymers were prepared from 4,4'-dihydroxychalcone (DHC) alone or in combination with methylhydroquinone or 4,4'-dihydroxybiphenyl (DHB) with diphosgene to generate the carbonate diester linking group. Chiral polymers, for example, the alternating copolymer, **54**, and the random terpoly-

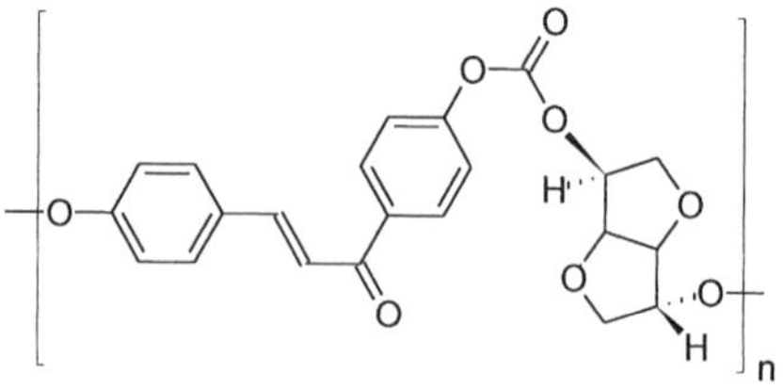

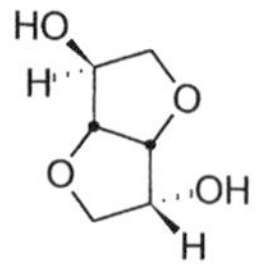

54   **56**

mers, **55**, were prepared using chiral isosorbide (1,4:3,6-dianhydro-D-sorbitol), **56**, as one of the co-monomers. A copolymer, **57a**, containing DHB was mono-

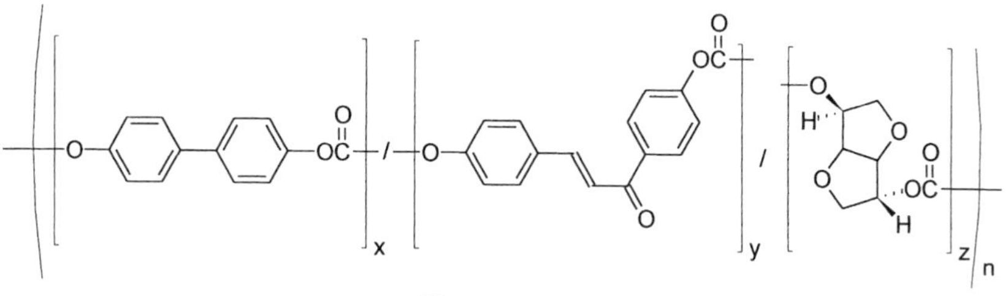

55a, x/y/z = 1/1/1
55b, x/y/z = 2/1/1

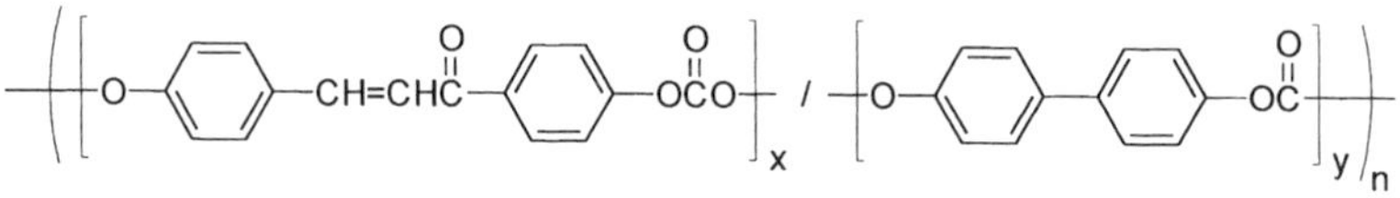

57a, x/y = 1/2
57b, x/y = 1/1

tropic. Copolymer **57b**, with a different composition, gave an enantiotropic N melt. Several ternary copolymers, **55**, exhibited enantiotropic cholesteric phases and Grandjean textures when sheared. Two of these copolycarbonates, **55a**, the only one of this series that was not LC, and **55b** (enantiotropic cholesteric) were irradiated (350 nm) in solution and as spin-coated films. In solution the absorbance dropped continually but recovered, in the dark, to the extent of 75–80%. This suggests reversible *trans-cis* isomerization is the major reaction. Irradiation of the "as cast" films led to loss of 323 nm absorption, no recovery of absorption when irradiation stopped, and films became insoluble due, presumably, to cross-linking by 2 + 2 photocycloaddition. Long-term irradiation of **55b** in the Ch mesophase at 250°C led to cross-linking and hardening but affected the Grandjean texture. The authors concluded that careful optimization of the conditions would be necessary to satisfactorily "fix" the texture. Kricheldorf et al. [121] have also reported the synthesis of "photosetting" N or Ch random polycarbonates also derived from 4,4′-dihydroxychalcone and isosorbide, **56**, but with more flexible hydroquinone-4-hydroxybenzoate replacing DHB, and copolymers of the chalcone and hydroquinone-4-hydroxybenzoate that do not contain isosorbide. The latter mostly have N mesophases. The terpolymers containing isosorbide units have Ch mesophases and show Grandjean textures. To date no photochemistry has been reported for these interesting chiral materials.

Stumpe and his co-workers [122–124] synthesized several SCLC methacrylate copolymers containing the photochromic N-salicylideneaniline (NSA) chromophore together with several other mesogenic and nonmesogenic groups. Copolymers, **58**, had from 0 to 100% of the NSA chromophore. Most of these polymers had an unspecified S mesophase. The NSA chromophore undergoes an excited state proton transfer (Fig. 17) to the Z-keto (often called *cis*-keto form)

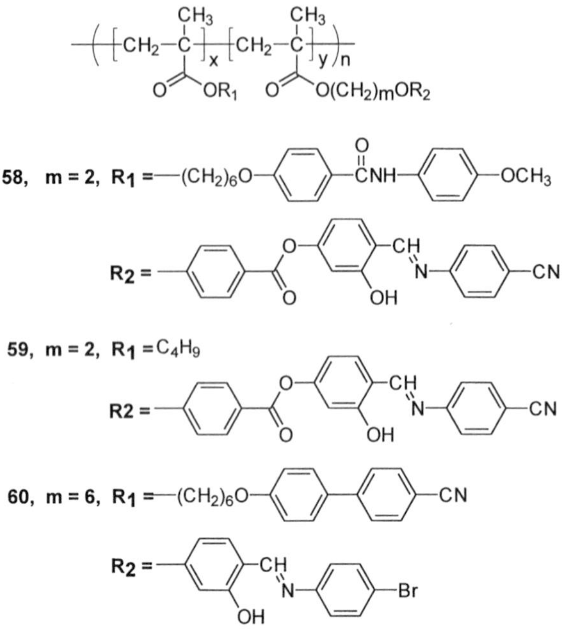

58, m = 2, R₁ = —(CH₂)₆O—

59, m = 2, R₁ = C₄H₉

60, m = 6, R₁ = —(CH₂)₆O—

which can then thermally isomerize to the more stable *E*-keto form (often called the *trans*-form). In solution the keto forms ultimately revert to the enol (benzenoid) tautomer. Irradiation (365 nm) of glassy films of the copolymers, **58**, led to phototautomerism to the *Z*-keto form (λ_{max} = 480 nm), followed by isomerization to the *E*-keto form (λ_{max} = 470 nm) which was stable for at least 12 months when

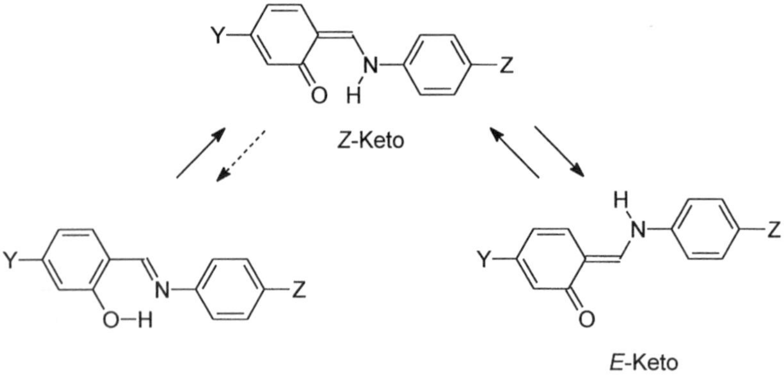

FIGURE 17 Enol (phenol)-keto tautomerism of the *N*-salicylideneaniline chromophore. Photochemical processes are indicated by solid arrows, thermal processes by dotted arrows.

the polymer was kept below T_g. In the S mesophase at a temperature 15K above T_g, the photoproduct reverted to the phenolic (benzenoid) form. In solution or in LC guest/host systems, the photoproduct could also be converted back to its original state by 470 nm irradiation. This process was suppressed in copolymer films, presumably below T_g. In copolymer, **59**, the E-keto form was only stable [123] in irradiated films at 77 K. At 298 K only traces of E-keto form were observed and another irreversible, unknown photoreaction depleted the enol absorption at 365 nm. A polymer, **60**, with a cyanobiphenyl group connected by a long (C_6) spacer to the methacrylate chain and with a bromo-substituted and two-ring, as opposed to a more rigid three-ring NSA group, as the photoreactive mesogen, was most resistant to the irreversible second photoreaction. It was suggested that the bromine substitution and higher flexibility of the spacer in **60** favored deactivation *via* photoautomerization thereby reducing the rate of the second, irreversible photoreaction at room temperature. A preliminary report has also been made [125] of the fluorescence properties of some of these NSA-containing SCLC copolymers using fluorescence microscopy of small regions (about 5 μm in diameter) of the polymer film. The polymers in solution, in spin coated films, and in the bulk solid, showed triple fluorescence attributed to the "normal" fluorescence of the benzanilide chromophore, proton transfer (PT) fluorescence, and twisted intramolecular charge transfer (TICT) fluorescence. The emission was much broader and blue shifted in the film, indicating a larger contribution of PT fluorescence caused by strong formation of intermolecular hydrogen bonds. The behavior of polymer films in the mesophase has not yet been reported.

The synthesis and some aspects of the photochemistry and photophysics of a 2,6-dialkoxy-substituted anthracene SCLC polymer, **61**, have been reported by

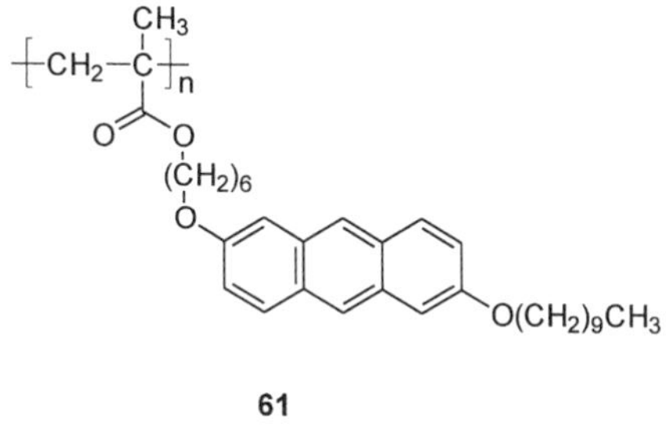

61

Creed et al. [125]. The 2,6-di-substitution pattern was chosen so that the anthracene moiety would be more likely to be elongated and to act as a mesogen. This polymer has an LC mesophase that is most probably N. Evidence was reported for chromophore association in this polymer even when it is highly diluted in a good solvent such as dichloromethane, but spectral perturbations due to aggregation are especially noticeable in films. Aggregation, as evidenced by a strongly blue-shifted band at 235 nm (the isolated chromophore absorbs at about 260 nm), progressively increases upon cooling from the mesophase at 155°C down to 51°C, well below the T_g at about 80°C. There is also weak absorption tailing to the red of the "normal" structured 2,6-dialkoxyanthracene absorption in the region between 300 and 450 nm. Similar aggregate absorption is observed when the polymer in a good solvent (CH_2Cl_2) is diluted with a nonsolvent such as methanol. Fluorescence spectra indicate the formation of excimers or excited aggregates at high-polymer concentrations in CH_2Cl_2, but films annealed above the isotropization temperature show broadened emission but no distinct excimer or aggregate maxima. Preliminary lifetime experiments indicate that films have at least three emitting species whose proportions and lifetimes depend on the thermal history of the film. UV-irradiated films (366 ± 10 nm) became partly insoluble. UV-Vis and fluorescence spectra of films irradiated under N_2 indicate the formation of two types of 4 + 4 cycloadducts, one from the usual 9,9′ to 10,10′ coupling and the other from 1,9′ to 4,10′ coupling [126] of the anthracene mesogens. Unknown photooxidation products formed in the presence of oxygen.

Ringsdorf and his co-workers [127] have prepared SCLC copolymers with acrylate, **62**, or methacrylate, **63**, main chains, phenyl benzoate mesogenic side-

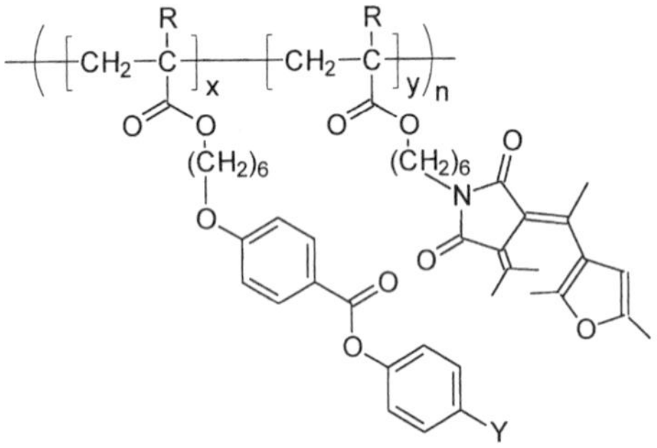

62, R = H, Y = CN
63, R = CH₃, Y = OCH₃

groups, and a photochromic $(E)/(Z)$-fulgimide side-group. The advantage of the fulgimide type of chromophore is that its photochromism is thermally irreversible up to 100°C. Fulgides, in general, have been developed to minimize fatigue in their photochromic behavior [128]. This resistance to thermal fatigue is in marked contrast to that of azoaryl compounds, spiropyrans, spirooxazines, and analogous materials (vide supra) in which the colored forms are not thermally stable, even at room temperature. Both acrylate and methacrylate based copolymers were LC, up to ca. 15 mol % of the fulgimide. The mesophases were N under most conditions although a S mesophase exists for the methacrylate copolymer at < ca. 2% of fulgimide. Increasing proportions of the fulgimide pendent group led to narrowing of the stability range of the mesophases. The fulgimide group in both solutions and films of the polymers can be ring closed upon 366 nm irradiation and the colored product ring opened again upon irradiation above 475 nm (Fig. 18). The absorption maximum of the colored form in the film was at 506 nm, bathochromically shifted from 496 nm in toluene. UV-irradiation led to a higher isotropization temperature for the mesophase, suggesting that the ring closed (colored) form distorts the packing of the mesogens less than do the fulgimide groups. A homeotropically oriented film of a copolyacrylate was prepared by heating the polymer to the *I* melt followed by slow cooling in an electric field. The absorption spectrum of the colored form in the film does not change as long as the film is kept in the dark and it is suggested that these materials could be useful in optical storage of information.

C. Photophysical Probes of LC Structure

Complex structures have frequently been probed using photophysical methods such as fluorescence or phosphorescence of intrinsic or added probes. There have been a few intriguing reports of the use of mesogen fluorescence as an intrinsic probe of LC polymer structures to obtain structural information that could not be obtained by other methods.

Ikeda et al. [54] studied the excimer-like fluorescence of phenylenediacrylate chromophores in MCLC polyesters of type **23**. These results are discussed in Section III.B of this chapter. A correlation was noted between the "excimer" fluorescence lifetime and the Δ*S* values for mesophase formation. The less-ordered, more mobile mesophases had shorter excimer lifetimes presumably because of more facile nonradiative decay. Kurihara, Ikeda, and Tazuke [129,130] also studied the emission properties of SCLC polymers of type **64**, containing the

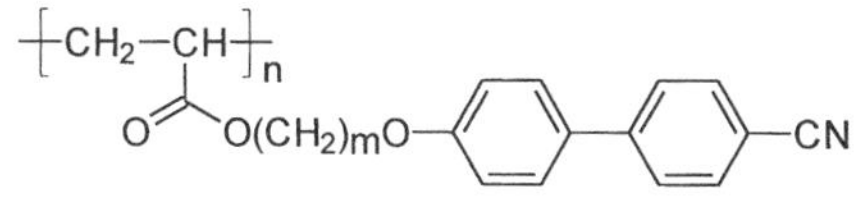

64a, m = 2
64b, m = 3
64c, m = 5
64d, m = 6
64e, m = 11

4-alkoxy-4′-cyanobiphenyl mesogen, in both solution [129] and in the mesophases [130]. In dilute solution in tetrahydrofuran, the intensity of "excimer-like" fluorescence at 420 nm increased in the order **64e** < **64a** ≈ **64c** ≈ **64d** < **64b**. The fluorescence of **64e**, in which the mesogen is essentially completely decoupled from the main chain by the long spacer, was virtually identical to that of a small molecule model compound, 4-pentyloxy-4′-cyanobiphenyl. Decay of both mono-

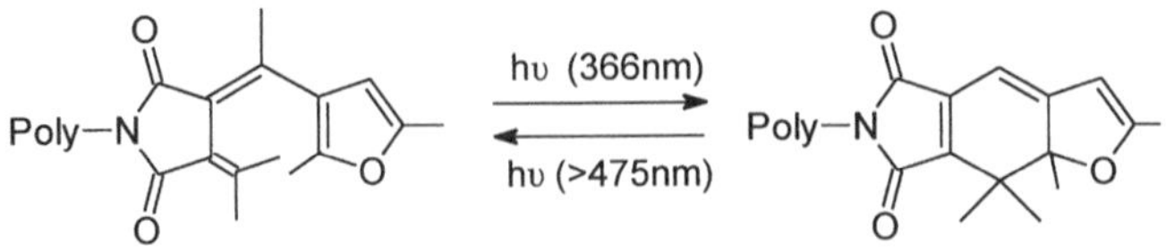

E-Fulgimide

FIGURE 18   Photochemical transformations of fulgimides.

mer like and excimer-like fluorescence was triple exponential in dilute solution (3×10^{-6} mol cm^{-3}) with no rise time of the excimer-like fluorescence. In more concentrated solutions of **64b**, time-resolved fluorescence spectra showed the formation of an even more red-shifted excimer-like emission that was tentatively assigned to a termolecular complex. In concentrated solution (1 mol dm^{-3}) a rising component of excimer emission was observed and was assigned to inter-molecular excimer formation. Fluorescence spectra of neat films of **64b** were also reported [130] in the I melt and N mesophase. Fluorescence intensities were higher in the N than in the I phase and the emission was slightly red-shifted in the N phase. The ratio (I_D/I_M) of the excimer-like to monomer fluorescence intensities at 420 and 350 nm, respectively, was obtained as a function of tempera-ture for all the polymers, **64**. An abrupt increase in I_D/I_M was observed upon cooling through T_i into the mesophase. It was suggested that the face to face organization of the mesogens in the mesophase favored excimer formation. In the case of **64e** which has a S_A rather than a N mesophase, no abrupt change in I_D/I_M was observed. Fluorescence decay curves in the N and I phases of **64b** were triple exponential at both 363 (monomer) and 450 nm (excimer). A rising component of 450 nm emission was observed. Time-resolved fluorescence spectra were re-ported. At $t = 0$, **64c** and **64d** both showed much more initial red-shifted emission and, therefore, a higher efficiency of excimer formation, than did the other polymers. It was suggested that the low efficiency of excimer formation in **64a** and **64b** in the neat film was due to the strong coupling of the fluorophore (mesogen) to the main chain. In solution, where the mesogens are highly mobile, **64b** had the highest excimer fluorescence. For **64e**, which has a S_A mesophase, the excimer formation efficiency was low despite the favorable packing for excimer formation. It was suggested that the high viscosity of the S_A phase suppresses reorientation to form the excimer geometry and thereby leads to a low efficiency of excimer formation.

Horie and his co-workers [131–135] have made extensive use of fluores-cence spectroscopy to probe the microstructures of MCLC polymers. The first such polymer to be studied [131,135] was the all aromatic, rigid rod polyester, **65**, with long, flexible side-chains attached to a pyromellitic ester mesogenic group.

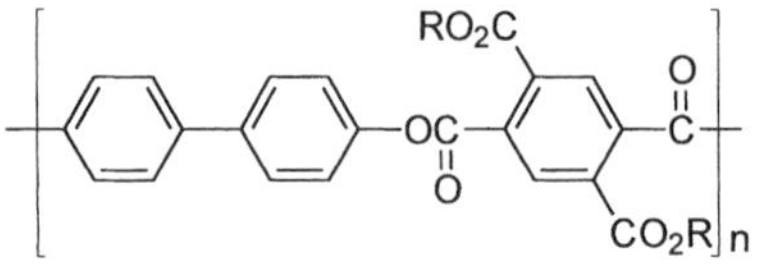

65, R = (CH₂)₁₆H

Polymer **65** formed two novel thermotropic mesophases with layered structure dependent on temperature. X-ray diffraction of the lower temperature phase suggested that the aromatic main chains are extended and packed into layers with positional order along their long axes. The chains are quite close (0.4 nm) within the layers and the layers are 2.17 nm apart with the "molten" alkyl groups occupying the space between them. The arrangement of biphenyl and pyromelli-tate groups within the layers could not be determined by x-ray methods. As the concentration of **65** was increased in chloroform from 10^{-4} to 10^{-1} mol (repeat unit) dm^{-3}, the structured emission with $\lambda_{max} = 355$ nm (excitation $\lambda_{max} = 287$ nm) was replaced with a structureless emission with $\lambda_{max} = 490$ nm with a new structureless excitation maximum at 410 nm. It was proposed that the new emission at high concentrations was due to a CT complex between the biphenyl moiety (electron donor) and the pyromellitic ester moiety (electron acceptor). The fluorescence and excitation spectra of a film of **65** heated to 100°C and cooled to room temperature to freeze in the LC structure were very similar to those of the 10^{-1} mol dm^{-3} solution. On the basis of this observation it was suggested that the packing of mesogens was such that the interaction of donor (biphenyl) and acceptor (pyromellitic ester) groups within a layer was maximized, i.e., in each adjacent pair of chains within a layer a biphenyl mesogen was next to a pyromel-litic ester mesogen on the neighboring chain.

Hashimoto et al. [132] studied the fluorescence of the MCLC polyesters, poly[(ethylene terephthalate)-co-(p-oxybenzoate)], **66**, and poly[(ethylene 2,6-naphthalene dicarboxylate)-co-(p-oxybenzoate)], **67**. The fluorescence of **66**, both at relatively high concentration (5×10^{-3} mol dm^{-3}) in hexafluoro-2-propanol,

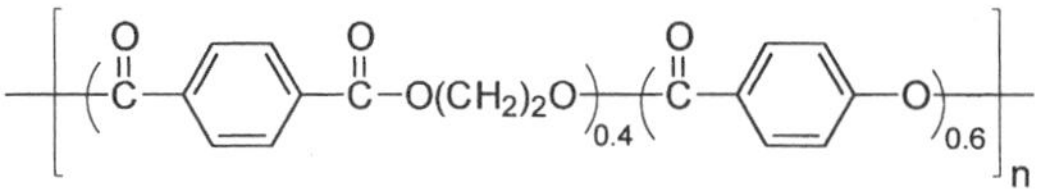

66

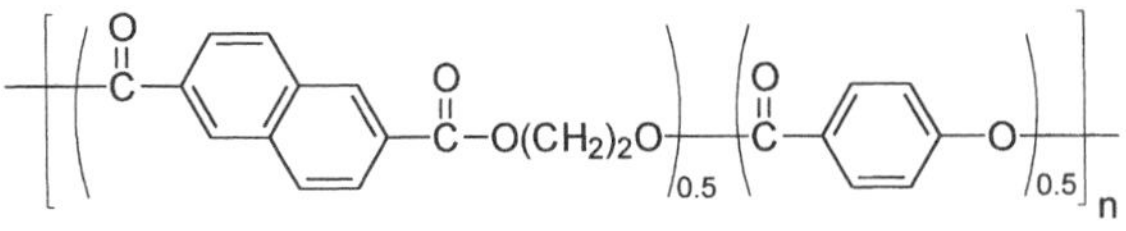

67

and from an "as cast" film, has a broad emission at $\lambda_{max} \approx 395$ nm, quite different from the dilute solution fluorescence with $\lambda_{max} = 325$ nm. The excitation spectrum of the film fluorescence had λ_{max} at 350 nm, greatly red-shifted from the absorption maximum at about 245 nm. The 395 nm fluorescence was therefore attributed to a ground state dimer. The fluorescence intensity (I_F) of the dimer increased and the fluorescence anisotropy ratio (r) decreased upon annealing at increasing temperatures, concomitant with development of the POM texture of the N phase. The lifetime of the emission was not changed upon heating suggesting the fluorescent species remained unchanged. The increase in excimer I_F was attributed to an increase in the concentrations of dimers upon development of the N mesophase. The drop in r was attributed to a change in dimer configuration upon heating. Similar behavior was seen for **67**. The dilute solution fluorescence was characteristic of the unassociated 2,6-naphthalenedicarboxylate chromophore, but in more concentrated solutions and films, a broad red-shifted excimer fluorescence was seen. Upon increasing the temperature of the film the fluorescence anisotropy ratio decreased from about zero for transparent films at room temperature to about -0.14 at about 200°C. It was suggested that the zero value of r implies that excimer fluorescence occurs *via* energy migration along the monomer units to a rather low concentration of excimer trap sites, as is observed for most polymers. However, the negative value of r at higher temperatures implies depolarization by energy migration is *not* occurring and fluorescence comes from a very high concentration of excimer trap sites possibly caused by the high degree of local molecular orientation in LC domains. More recently [133,135], the work on the fluorescence of **66**, with 60 mol % p-oxybenzoate mesogen, has been extended over a much larger temperature range (25 to 450°C). The polymer had a T_g at 115°C, a transition from the rubbery state to the N mesophase at 250°C, and from the N mesophase to the I melt at 350°C. The excimers of model compounds dimethylterephthalate and methyl methoxybenzoate had fluorescence λ_{max} at 381 and 352 nm, respectively, whereas a 1.0 mol dm^{-3} mixture of both emitted at 405 nm. The polymer fluorescence λ_{max} was at 400 nm in concentrated solution and 395 nm in a film [132]. The film emission was therefore assigned to an intermolecular CT complex. Between 115 (T_g) and 250°C (LC transition temperature) the fluorescence λ_{max} shifted from 394 to 430 nm, while λ_{max} for excitation remained at 340 nm. Below 115 and above 250°C there was little change in the λ_{max} of fluorescence. The fluorescence intensity (I_F) decreased from room temperature until about 287°C, increased somewhat until isotropization occurred (ca. 350°C) and then decreased drastically in the isotropic melt. Since the excitation λ_{max} remained constant while the fluorescence λ_{max} increased, it was proposed that a more stable excited state complex can form once T_g has been reached. This complex has a different electronic configuration. The drop in I_F above room temperature to 287°C (and an abrupt drop at T_g) was attributed to enhanced radiationless deactivation at higher temperatures. It was proposed [132] that the increase in I_F in the N phase between ca. 287 to 370°C was due to the increased concentration of complexes in the mesophase. Samples annealed at 305°C and cooled to 25°C had the same λ_{max} for fluorescence but much higher I_F than for nonannealed samples at 25°C.

Most recently Horie et al. [134,135] have used x-ray and fluorescence methods to investigate the microstructure of the 4,4'-dialkoxy-containing thermotropic polyester, **68**. This polyester is crystalline and enters a smectic H (S_H)

68

mesophase at 207°C and the I melt at 257°C. The S_H mesophase is analogous to S_C, with "tilted" mesogens (Fig. 3), but the mesogens are packed regularly within the layers in a "herringbone" arrangement and there is strongly hindered rotation about their long axes. The fluorescence of crystalline biphenyl and 4,4'-diacetoxybiphenyl had λ_{mx} at 360 and 362 nm, respectively. It is known that the biphenyl chromophores fully overlap in these model compounds. Fluorescence excitation spectra for these emissions differed from those for isolated biphenyl emission. Therefore, the ca. 360 nm fluorescence was assigned to that of a ground-state "complex." The long wavelength fluorescence (365–469 nm) of **68** at 26°C was excitation wavelength dependent and temperature dependent (Fig. 19) and was assigned to intermolecular ground state complexes of different geometries. For example, 320 nm excitation at 26°C led to the 365 nm fluorescence expected from a parallel arrangement of mesogens, analogous to the arrangement in the crystalline model compounds. On the other extreme, 400 nm excitation at 26°C led to ca. 470 nm fluorescence assigned to a tilted arrangement of mesogens leading to a stronger intermolecular CT interaction. Raising the temperature shifted the fluorescence (except for that at 340 nm assigned to isolated biphenyls) to the red (Fig. 19). The most dramatic shift was for the fluorescence assigned to the "parallel dimer" excited at 340 nm. Two "breaks" were seen at 175 and 253°C, the latter corresponding quite well with the S_H to I phase change at 257°C, but the former being somewhat below the K to S_H phase change at 207°C. Similarly, breaks were seen in I_F (which decreases with increasing temperature) at the same temperatures. It was suggested that local microstructure changes precede the actual (bulk) phase transition seen by DSC and wide-angle x-ray diffraction (WAXD) measurements. At temperatures well above the isotropization temperature, the fluorescence λ_{max} tended to ca. 429 nm for all excitation wavelengths between 320 and 360 nm suggesting only one kind of steady-state molecular interaction between mesogens in the isotropic melt. It was suggested [134] that the very long wavelength fluorescence between 453 and 469 nm (excited above

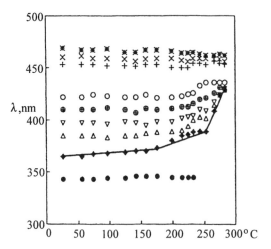

FIGURE 19 Temperature dependence of the fluorescence maxima of films of polymer, **68**, excited at 300 (●), 320 (◆), 330 (△), 340 (▽), 350 (⊕), 360 (○), 380 (+), 390 (×), and 400 (∗) nm during heating. (Reprinted with permission from Huang et al. [134]. Copyright 1996 American Chemical Society.)

380 nm) might be due to larger aggregates. This work [134,135] demonstrates in a rather striking manner how a photophysical technique can provide information about the microstructures in LC polymers additional to that provided by more conventional structure determination techniques such as WAXD.

IV. CONCLUSIONS

Many photochemists strive to understand *all* of the events, including photophysical events, beginning with light absorption by a chromophore and ending with the formation of a kinetically stable ("persistent") or thermodynamically stable product or products. In the light of this description of what many photochemists ideally strive to do, it is apparent that the study of the photochemistry of LC polymers in their different phases is in its infancy. In some cases the only photochemical observation that has been made is that a particular material cross-links upon irradiation. It is only fair to say that, for many materials scientists and engineers, the most interesting and important observations are the bulk mechanical, optical, or other properties of the material and does a specific physical or chemical treatment improve or degrade these desirable properties. However, it is my hope that this chapter will prove useful to all those working with LC polymers and might stimulate a larger interest within the community of photochemists in these fascinating and potentially important materials and in their photochemistry and photophysics.

ACKNOWLEDGMENTS

I thank the National Science Foundation, the State of Mississippi, and the University of Southern Mississippi for support during the writing of this chapter, and the colleagues who sent me reprints and preprints of their work.

REFERENCES

1. McArdle, C. B. In *Side Chain Liquid Crystal Polymers*; McArdle, C. B., ed., Chapman and Hall: New York, 1989, p. 357.
2. Shibaev, V. P.; Kostromin, S. G.; Ivanov, S. A. In *Polymers as Electrooptical and Photooptical Active Media*; Shibaev, V. P., ed., Springer-Verlag: Berlin, 1996, p. 37.
3. Kreuzer, F. H.; Bräuchle, Ch.; Miller, A.; Petri, A. In *Polymers as Electrooptical and Photooptical Active Media*; Shibaev, V. P., ed., Springer-Verlag: Berlin, 1996, p. 111.
4. Krongauz, V. In *Applied Photochromic Polymer Systems*; McArdle, C. B., ed., Chapman and Hall: New York, 1992 p. 121.
5. Bowry-Devereaux, C. In *Processes in Photoreactive Polymers*; Krongauz, V. V.; Trifunac, A. D., eds., Chapman and Hall: New York, 1995, p. 278.
6. McArdle, C. B. In *Applied Photochromic Polymer Systems*; McArdle, C. B., ed., Chapman and Hall: New York, 1992 p. 1.
7. McArdle, C. B., ed. *Side Chain Liquid Crystal Polymers*; Chapman and Hall: New York, 1989.
8. Percec, V. In *Handbook of Liquid Crystal Research*; Collings, P. J.; Patel, A. S., eds., Oxford University Press: New York, 1997, p. 259, and references therein.
9. Weiss, R. G. In *Photochemistry in Organized and Constrained Media*; Ramamurthy, V., ed., VCH Publishers: New York, 1991, p. 603.
10. Saeva, F. D. *Makromol. Chem. Suppl.* 1981, *5*, 58.
11. Hikmet, R. A. M.; Lub, J.; Maassen van der Brink, P. *Macromolecules* 1992, *25*, 4194, and references therein.
12. Spooner, S. P.; Whitten, D. G. In *Photochemistry in Organized and Constrained Media*; Ramamurthy, V., ed., VCH Publishers: New York, 1991, p. 691.
13. Finkelmann, H.; Happ, M.; Portugal, M.; Ringsdorf, H. *Makromol. Chem.* 1978, *179*, 2541. Finkelmann, H.; Ringsdorf, H.; Wendorff, J. H. *Makromol. Chem.* 1978, *179*, 273.
14. Percec, V.; Pugh, C. In *Side Chain Liquid Crystal Polymers*; McArdle, C. B., ed., Chapman and Hall: New York, 1989, p. 30.
15. Ecoffet, C.; Markovitsi, D.; Jallabert, C.; Strzelecka, H.; Veber, M. *Thin Solid Films* 1994, *83*, 242. Markovitsi, D.; Germain, A.; Millié, P.; Lécuyer, P.; Gallos, L.; Argyrakis, P.; Bengs, H.; Ringsdorf, H. *J. Phys. Chem.* 1995, *99*, 1005.
16. Demus, D.; Diele, S.; Grande, S.; Sackmann, H. *Adv. Liq. Cryst.* 1983, *5*, 1.
17. Kasha, M. *Radiation Research*, 1963, *20*, 55.

18. Cohen, M. D.; Schmidt, G. M. J. *J. Chem. Soc.* 1964, 1996. Cohen, M. D.; Schmidt, G. M. J.; Sonntag, F. I. *J. Chem. Soc.* 1964, 2000. Schmidt, G. M. J. *J. Chem. Soc.* 1964, 2014.

19. Minsk, L. M.; Van Deusen, W. P.; Robertson, E. M. *U. S. Patent* 1952, 2, 610, 120.

20. Reiser, A. *Photoreactive Polymers. The Science and Technology of Resists*; Wiley: New York, 1989, and references therein.

21. Creed, D.; Griffin, A. C.; Gross, J. R. D.; Hoyle, C. E.; Venkataram, K. *Mol. Cryst. Liq. Cryst.* 1988, *155*, 57.

22. Haddleton, D. M.; Venkataram, K.; Creed, D.; Griffin, A. C.; Hoyle, C. E. *Proc. North Am. Thermal Anal. Soc.*, *17th*, 1988, 2, 430.

23. Creed, D.; Griffin, A. C.; Hoyle, C. E.; Venkataram, K. *J. Am. Chem. Soc.* 1990, *112*, 4049.

24. Subramanian, P.; Creed, D.; Hoyle, C. E.; Venkataram, K. *Proc. SPIE—Int. Soc. Opt. Eng. (Photopolym. Device Phys. Chem. Appl.)* 1991, 2, 461.

25. Creed, D.; Cozad, R. A.; Griffin, A. C.; Hoyle, C. E.; Jin, L.; Subramanian, P.; Varma, S. S.; Venkataram, K. *ACS Symp. Ser.* 1994, *579*, 13.

26. Creed, D.; Cozad, R. A.; Griffin, A. C.; Hoyle, C. E., Jin, L. *Polym. Prepr. (Am. Chem. Soc. Div. Polym. Chem.)* 1996, *37*, 46.

27. Griffin, A. C.; Hoyle, C. E.; Gross, J. R. D.; Venkataram, K.; Creed, D.; McArdle, C. B. *Makromol. Chem., Rapid Commun.* 1988, *9*, 463.

28. Subramanian, P.; Creed, D.; Griffin, A. C.; Hoyle, C. E.; Venkataram, K. *J. Photochem. Photobiol. A: Chem.* 1991, *61*, 317.

29. Stumpe, J.; Fischer, Th.; Menzel, H. *Macromolecules* 1996, *29*, 2831.

30. Haddleton, D. M.; Creed, D.; Griffin, A. C.; Hoyle, C. E.; Venkataram, K. *Makromol. Chem., Rapid Commun.* 1989, *10*, 391.

31. Li, C.-H.; Lai, W.-W.; Hsu, K. Y.; Chang, T.-C. *J. Polym. Sci., Part A: Polym. Chem.* 1993, *31*, 27.

32. Stumpe, J.; Ziegler, A.; Berghahn, M.; Kricheldorf, H. R. *Macromolecules* 1995, *28*, 5306.

33. Kricheldorf, H. R.; Probst, N.; Gurau, M.; Berghahn, M. *Macromolecules* 1995, *28*, 6565.

34. Kricheldorf, H. R.; Probst, N.; Wutz, C. *Macromolecules* 1995, *28*, 7990.

35. Kricheldorf, H. R.; Hans, R.; Berghahn, M.; Probst, N.; Gurau, M. Schwarz, G. *React. Funct. Polym.* 1996, *30*, 173.

36. Navarro, F. *Macromolecules* 1991, *24*, 6622.

37. Koch, T.; Ritter, H.; Buchholz, N.; Knöchel, F. *Makromol. Chem.* 1989, *190*, 1369.

38. Keller, P. *Chem. Mater.* 1990, 2, 3.

39. Noonan, J. M.; Caccamo, A. F. *ACS Symp. Ser.* 1990, *435*, 144.

40. Singh, S.; Creed, D.; Hoyle, C. E. *Polym. Prepr. (Am. Chem. Soc. Div. Polym. Chem.)* 1993, *34*, 743.

41. Singh, S.; Creed, D.; Hoyle, C. E. *Proc SPIE Int. Soc. Opt. Eng.* 1993, *1774*, 2.

42. Chien, L.-C.; Cada, L. G. *Macromolecules* 1994, *27*, 3721.

43. Whitcombe, M. J.; Gilbert A.; Mitchell, G. R. *Br. Polym. J.* 1990, *23*, 77.

44. Whitcombe, M. J.; Gilbert, A.; Hirai, A.; Mitchell, G. R. *J. Polym. Sci., Part A: Polym Chem.* 1991, *29*, 251.

45. Barley, S. H.; Gilbert, A.; Mitchell, G. R. *Makromol. Chem.* 1991, *192*, 2810.

46. Legge, C. H.; Whitcombe, M. J., Gilbert, A.; Mitchell, G. R. *J. Mater. Chem.* 1991, *1*, 303.
47. Whitcombe, M. J., Gilbert, A.; Mitchell, G. R. *J. Polym. Sci., Part A: Polym. Chem.* 1992, *30*, 1681.
48. Whitcombe, M. J.; Gilbert, A.; Mitchell, G. R. *Polymer* 1993, *34*, 1347.
49. Ikeda, T.; Hasegawa, S.; Sasaki, T.; Miyamoto, T.; Lin, M.-P.; Tazuke, S. *Makromol. Chem.* 1991, *192*, 215.
50. Sapich, B.; Haferkorn, J.; Lasker, L.; Stumpe, J. *J. Inf. Rec.* 1996, *23*, 103.
51. Kawatsuki, N.; Sakashita, S.; Takatani, K.; Yamamoto, T.; Sangen, O. *Macromol. Chem. Phys.* 1996, *197*, 1919.
52. Kawatsuki, N.; Takatsuka, H.; Yamamoto, T.; Sangen, O. *Macromol. Rapid Commun.* 1996, *17*, 703.
53. Ikeda, T.; Itakura, H., Lee, C.; Winnik, F. M.; Tazuke, S. *Macromolecules* 1988, *21*, 3536.
54. Ikeda, T.; Lee, C. H.; Sasaki, T.; Lee, B.; Tazuke, S. *Macromolecules* 1990, *23*, 1691.
55. Gómez, F. N.; Ostariz, J. L. S. *J. Mater. Chem.* 1991, *1*, 895.
56. Li, C.-H.; Hsu, K.-Y.; Chang, T.-C. *J. Polym. Sci., Part A: Polym. Chem.* 1993, *31*, 1119.
57. Hsu, K.-Y.; Chang, T.-C.; Li, C.-H. *J. Polym. Sci., Part A: Polym Chem.* 1993, *31*, 971.
58. Kricheldorf, H. R.; Probst, N. *Macromol. Chem. Phys.* 1995, *196*, 3511.
59. Saltiel, J.; Sun, Y.-P.; In *Photochromism, Molecules and Systems*; Durr, H., Bouas-Laurent, H., eds., Elsevier: Amsterdam, 1990, p. 64. Waldeck, D. H. *Chem. Rev.* 1991, *91*, 415.
60. Whitten, D. G. *Acc. Chem. Res.* 1993, *26*, 502. Song, X.; Geiger, C.; Leinhos, U.; Perlstein, J.; Whitten, D. G. *J. Am. Chem. Soc.* 1994, *116*, 10340. Lewis, F. D.; Wu, T.; Burch, E. L.; Bassani, D. M.; Yang, J.-S.; Schneider, S.; Jager, W.; Letsinger, R. L. *J. Am. Chem. Soc.* 1995, *117*, 8785.
61. Creed, D.; Cozad, R. A.; Griffin, A. C.; Hoyle, C. E.; Jin, L. *Polym. Prepr. (Am. Chem. Soc. Div. Polym. Chem.)* 1996, *37*, 46.
62. Creed, D.; Cozad, R. A.; Hoyle, C. E. *Proc. SPIE Int. Soc. Opt. Eng.* 1993, *1774*, 69.
63. Jackson, Jr., W. J.; Morris, J. C. *J. Appl. Polym. Sci.: Appl. Polym. Symp.* 1985, *41*, 307.
64. Gangadhara; Kishore, K. *Macromolecules* 1993, *26*, 2995.
65. Gangadhara; Kishore, K. *Polymer* 1995, *36*, 1903.
66. Gangadhara; Kishore, K. *Macromolecules* 1995, *28*, 806.
67. Krongauz, V. In *Photochromism, Molecules and Systems*; Durr, H., Bouas-Laurent, H., eds., Elsevier: Amsterdam, 1990, p. 793.
68. Cabrera, I.; Krongauz, V.; Ringsdorf, H. *Angew. Chem. Int. Ed. Engl.* 1987, *26*, 1178.
69. Cabrera, I.; Krongauz, V.; Ringsdorf, H. *Mol. Cryst. Liq. Cryst.* 1988, *155*, 221.
70. Yitzchaik, S.; Cabrera, I.; Buchholtz, F.; Krongauz, V. *Macromolecules* 1990, *23*, 707.
71. Yitzchaik, S.; Ratner, J.; Buchholtz, F.; Krongauz, V. *Liquid Crystals* 1990, *8*, 677.
72. Krongauz, V. *Mol. Cryst. Liq. Cryst.* 1994, *246*, 339.
73. Krongauz, V. A.; Goldburt, E. S. *Macromolecules* 1981, *14*, 1382.
74. Xie, S.; Natansohn, A.; Rochon, P. *Chem. Mater.* 1993, *5*, 403.

75. Irie, M. In *Applied Photochromic Polymer Systems*; McArdle, C. B., ed.; Chapman and Hall: New York, 1992, p. 174.

76. Rau, H.; Lüddecke, E. *J. Am. Chem. Soc.* 1982, *104*, 1616.

77. Naito, T.; Horie, K.; Mita, I. *Macromolecules* 1991, *24*, 2907.

78. Tsutsumi, O.; Kanazawa, A.; Shiono, T.; Ikeda, T. *Macromol. Symp.* 1997, *116*, 117.

79. Natansohn, A.; Rochon, P.; Gosselin, J.; Xie, S. *Macromolecules* 1992, *25*, 2268.

80. Eich, M.; Wendorff, J. H.; Reck, B.; Ringsdorf, H. *Makromol. Chem., Rapid Commun.* 1987, *8*, 59.

81. Eich, M.; Wendorff, J. H. *Makromol. Chem., Rapid Commun.* 1987, *8*, 467.

82. Eich, M.; Wendorff, J. *J. Opt. Soc. Am.* B 1990, *7*, 1428.

83. Anderle, K.; Wendorff, J. H. *Mol. Cryst. Liq. Cryst.* 1994, *243*, 51.

84. Anderle, K.; Birenheide, R.; Eich, M.; Wendorff, J. H. *Makromol. Chem., Rapid Commun.* 1989, *10*, 477.

85. Wiesner, U.; Reynolds, N.; Boeffel, C.; Spiess, H. W. *Makromol. Chem., Rapid Commun.* 1991, *12*, 457.

86. Wiesner, U.; Antonietti, M.; Boeffel, C.; Spiess, H. W. *Makromol. Chem., Rapid Commun.* 1990, *191*, 2133.

87. Stumpe, J.; Müller, L.; Kreysig, D.; Hauck, G.; Koswig, H. D.; Ruhmann, R.; Rubner, J. *Makromol. Chem., Rapid Commun.* 1991, *12*, 81.

88. Ivanov, S.; Yakovlev, I.; Kostromin, S.; Shibaev, V.; Läsker, L.; Stumpe, J.; Kreysig, D. *Makromol. Chem., Rapid Commun.* 1991, *12*, 709.

89. Läsker, L.; Fischer, T.; Stumpe, J.; Kostromin, S.; Ivanov, S.; Shibaev, V.; Ruhmann, R. *Mol. Cryst. Liq. Cryst.* 1994, *246*, 347.

90. Läsker, L.; Fischer, T.; Stumpe, J.; Kostromin, S.; Ivanov, S.; Shibaev, V.; Ruhmann, R. *Mol. Cryst. Liq. Cryst.* 1994, *253*, 1.

91. Läsker, L.; Stumpe, J.; Fischer, T.; Rutloh, M.; Kostromin, S.; Ruhmann, R. *Mol. Cryst. Liq. Cryst.* 1995, *261*, 371.

92. Läsker, L.; Fischer, T.; Stumpe, J.; Ruhmann, R. *J. Inf. Rec.* 1994, *21*, 635.

93. Stumpe, J.; Läsker, L.; Fischer, T.; Rutloh, M.; Kostromin, S.; Ruhmann, R. *Thin Solid Films* 1996, *284–285*, 252–256.

94. Fischer, T.; Läsker, L.; Czapla, S.; Rubner, J.; Stumpe, J. *Mol. Cryst. Liq. Cryst.* in press.

95. Fischer, T.; Läsker, L.; Stumpe, J.; Kostromin, S. G. *J. Inf. Rec.* 1994, *21*, 639.

96. Stumpe, J.; Läsker, L.; Fischer, T.; Geue, T. *J. Inf. Rec.* 1994, *21*, 449.

97. Haferkorn, J.; Stumpe, J. *J. Inf. Rec.* 1995, *23*, 107.

98. Geue, T.; Stumpe, J.; Mobius, G.; Pietsch, U.; Schuster, A.; Ringsdorf, H.; Kaupp, G. *J. Inf. Rec.* 1994, *21*, 645.

99. Ziegler, A.; Geue, T.; Stumpe, J. *J. Inf. Rec.* 1996, *23*, 111.

100. Stumpe, J.; Fischer, T.; Geue, T.; Ziegler, A.; Menzel, H. *J. Inf. Rec.* 1996, *22*, 469.

101. Geue, T.; Pietsch, U.; Stumpe, J. *Thin Solid Films* 1996, *284–285*, 228.

102. Stumpe, J.; Fischer, T.; Ziegler, A.; Geue, T.; Menzel, H. *Mol. Cryst. Liq. Cryst.* in press.

103. Stumpe, J.; Geue, T.; Fischer, T.; Menzel, H. *Thin Solid Films* 1996, *284–285*, 606.

104. Geue, T.; Stumpe, J.; Pietsch, U.; Haak, M.; Kaupp, G. *Mol. Cryst. Liq. Cryst.* 1995, *262*, 157.

105. Ikeda, T.; Horiuchi, S., Karanjit, D. B.; Kurihara, S.; Tazuke, S. *Chem. Lett.* 1988, *1679.*
106. Ikeda, T.; Horiuchi, S., Karanjit, D. B., Kurihara, S.; Tazuke, S. *Macromolecules* 1990, *23*, 36.
107. Ikeda, T.; Horiuchi, S., Karanjit, D. B., Kurihara, S.; Tazuke, S. *Macromolecules* 1990, *23*, 42.
108. Ikeda, T.; Sasaki, T.; Ichimura, K. *J. Photopolym. Sci. Tech.* 1991, *4*, 191.
109. Ikeda, T.; Sasaki, T.; Kurihara, S. *Proc. Jap. Acad.* 1993, *69*, Ser. B, 7.
110. Ikeda, T.; Kurihara, S.; Karanjit, D. B.; Tazuke, S. *Macromolecules* 1990, *223*, 3938.
111. Ikeda, T.; Sasaki, T.; Kim, H.-B., *J. Phys. Chem.* 1991, *95*, 509.
112. Sasaki, T.; Ikeda, T.; Ichimura, K. *Macromolecules* 1992, *25*, 3807.
113. Ikeda, T.; Tsutsumi, O. *Science* 1995, *268*, 1873.
114. Haysahi, T.; Kawakami, H.; Doke, Y.; Tsuchida, A.; Onogi, Y.; Yamamoto, M. *Eur. Polym. J.* 1995, *31*, 23.
115. Han, Y.-K.; Kim, D.-Y.; Kim, Y. H. *Mol. Cryst. Liq. Cryst.* 1994, *254*, 445.
116. Angeloni, A. S.; Caretti, D.; Carlini, C.; Chiellini, E.; Galli, G.; Altomare, A.; Solaro, R.; Laus, M. *Liquid Crystals* 1989, *4*, 513.
117. Angeloni, A. S.; Campagnari, I.; Caretti, D.; Carlini, C.; Altomare, A.; Chiellini, E.; Galli, G.; Solaro, R.; Laus, M. *Gazz. Chim. Ital.* 1990, *120*, 171.
118. Chiellini, E.; Galli, G.; Altomare, A.; Solaro, R.; Angeloni, A. S.; Laus, M.; Carlini, C.; Caretti, D. *Mol. Cryst. Liq. Cryst.* 1992, *221*, 61.
119. Oge, T.; Zentel, R. *Macromol. Chem. Phys.* 1996, *197*, 1805.
120. Kricheldorf, H. R.; Sun, S.-J.; Sapich, B.; Stumpe, J. *Macromol. Chem. Phys.* 1997, *198*, 2197.
121. Kricheldorf, H. R.; Sun, S.-J.; Chen, C.-P.; Chang, T.-C. *J. Polym. Sci. Part A: Polym Chem.* 1997, *35*, 1611.
122. Kamphausen, C.; Stumpe, J. *J. Inf. Rec.* 1994, *21*, 1.
123. Kamphausen, C.; Ruhmann, R.; Stumpe, J. *J. Inf. Rec.* 1996, *23*, 97.
124. Pade, S.; Schmidt, H.; Stumpe, J.; Fischer, T. *J. Inf. Rec.* 1996, *23*, 117.
125. Creed, D.; Hoyle, C. E.; Griffin, A. C.; Liu, Y.; Pankasem, S. *ACS Symp. Ser.* 1995, *614*, 504.
126. Fages, F.; Desvergne, J.-P.; Frisch, I.; Bouas-Laurent, H. *J. Chem. Soc. (Chem. Commun.)* 1988, 1413.
127. Cabrera, I.; Dittrich, A.; Ringsdorf, H. *Angew. Chem. Int. Ed. Engl.* 1991, *30*, 76.
128. Heller, H. G.; Koh, K.; Elliot, C.; Whittall, J. *Mol. Cryst. Liq. Cryst.* 1994, *246*, 79.
129. Kurihara, S.; Ikeda, T.; Tazuke, S. *Macromolecules* 1991, *24*, 628.
130. Kurihara, S.; Ikeda, T.; Tazuke, S. *Macromolecules* 1993, *26*, 1590.
131. Sone, M.; Harkness, B. R.; Watanabe, J.; Yamashita, T.; Torii, T.; Horie, K. *Polym. J.* 1993, *25*, 997.
132. Hashimoto, H.; Hasegawa, M.; Horie, K.; Yamashita, T.; Ushiki, H.; Mita, I. *J. Polym. Sci. Part B: Polym. Phys.* 1993, *31*, 1187.
133. Huang, H. W.; Horie, K.; Yamashita, T. *J. Polym. Sci. Part B: Polym. Phys.* 1995, *33*, 1673.
134. Huang, H. W.; Horie, K.; Yamashita, T.; Machida, S.; Sone, M.; Tokita, M.; Watanabe, J.; Maeda, Y. *Macromolecules* 1996, *29*, 3485.
135. Horie, K.; Huang, H. *Macromol. Symp.* 1997, *116*, 105.

5

Photochemical Solid-to-Solid Reactions

Amy E. Keating
and Miguel A. Garcia-Garibay
University of California–Los Angeles, Los Angeles, California

I. INTRODUCTION

The study of photochemical reactions in organic crystals is a field rich in history [1] that has been the subject of intense activity only during the last 30 years. There are several excellent reviews on the subject to which the reader is referred for comprehensive coverage [2–11]. It has been shown that photochemical reactions in crystals tend to be quite different from those observed in solution or in the gas phase. The crystalline environment is capable of inducing high chemo-, regio-, and stereoselectivity with a degree of control that sometimes rivals that observed in enzymes. Some of the most remarkable examples of solid-state control reported in the literature include reactions that are only observed in the solid phase, the generation of optically pure compounds from *achiral* reagents [12–14], the preparation of single crystalline polymers [7,15], and the control of highly reactive intermediates [16–20]. As the number of reactions analyzed in crystals continues to increase and the potential of highly selective transformations where solvents can be omitted is increasingly appreciated, it has become desirable to search for reliable methods to carry out synthetic transformations in the crystalline solid state. However, the synthetic potential of solid-state photochemistry does not only depend on highly selective transformations but also requires generality and versatility with high chemical efficiency. In order to address how solid-phase reactions with these characteristics might be designed in the future, this article discusses factors that affect the progress of organic reactions in crystals. These include reaction conditions such as temperature and irradiation wavelength as well as sample properties such as particle size. In particular, we will emphasize how phase transformations may influence crystalline reactivity. After a short overview of solid-state photochemistry we will analyze selected literature examples to survey factors important to maintaining high selectivity and high conversion. The connection between molecular reactivity and macroscopic sample characteristics in the solid state has not been extensively studied, and a major conclusion from this work is that a more systematic characterization of such reactions is needed. We note that although there has been a great deal of recent work on two-component crystalline solids formed by hydrogen bonding, charge transfer, or ionic interactions, our analysis will concentrate on examples of one-component molecular crystals.

A. Reaction Control by Crystals

Most crystalline solids are thermally and photochemically stable. The solid state imposes severe restrictions on chemical reactivity in terms of (1) the number of components present in a crystalline phase, (2) the distance and orientation between prospective reactants, and (3) the limited motion that molecules may experience in a crystal. Early observations of photoreactive crystals suggested that the intrinsic photochemical properties of organic compounds become secondary and that their solid-state photochemistry is determined by the three-dimensional periodic environment where the reaction occurs. While this view places too much emphasis on structural factors, it applies with reasonable accuracy to a large number of cases investigated so far. Following a suggestion by Kohlshutter in 1918 [21] and relevant work by Schmidt et al. in the mid 1960s [22], reactions that occur under the influence and control of crystalline media are known as "topochemical reactions." The topochemical postulate suggests that reactions in crystals may only occur with a minimum of atomic and molecular motion. Bimolecular reactions rely on packing arrangements that place reaction partners within the proper distance and orientation while unimolecular rearrangements occur by mechanisms with least atomic and molecular motion.

1. Topochemical Control

The dimerization of olefins is among the earliest systematically documented reactions that occur in organic solids, with examples dating back to 1877 [23,24]. More recently, the potential of correlating information from x-ray structural analysis and photochemical reactivity to obtain insight into the reaction mechanism, and to predict or rationalize the observed products, was first recognized and exploited by Schmidt et al. [22].

In a systematic study of the $2\pi + 2\pi$ photodimerization of cinnamic acids it was shown that reactions in crystals occur with molecularity, selectivity, and efficiencies that are quite different from those observed in solution. Cinnamic acids and their derivatives were shown to crystallize in three distinct packing arrangements known as α-, β-, and γ-forms. The packing arrangement of a given crystal form determines whether or not a $2\pi + 2\pi$ photodimerization may occur in the solid state and which of the possible products will form (Scheme 1). Al-

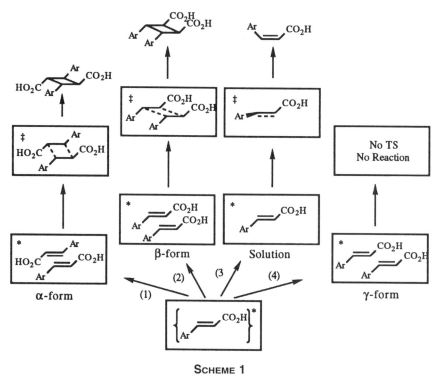

SCHEME 1

though photochemical excitation in solution results in isomerization of the double bond to reach a photostationary state, the solid state hinders this motion and the isomerization rate becomes insignificantly small. Based on a relatively large number of studies, Schmidt et al. suggested that photodimerization may only occur when the two double bonds are parallel to each other and with a center-to-center distance shorter than ca. 4.1 Å. The centrosymmetric arrangement of the α-form leads to centrosymmetric cyclobutanes, the mirror-symmetric arrangement of the β-form leads to mirror symmetric dimers, and the relatively long displacement between double bonds present in the γ-form prevents a reaction from happening. Since the work of Schmidt, numerous examples of topochemical control have been documented (Scheme 2); noteworthy among them are the reactions of several butadienes (**1**) [25,26], coumarins (**2**) [27], and benzylidene-cyclopentanones (**3**) [28].

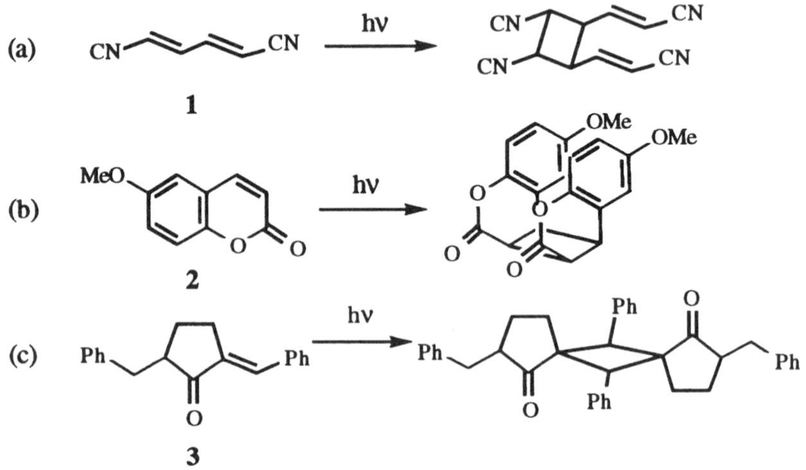

SCHEME 2

2. Nontopochemical Reactions

Schmidt's generalizations are based on a set of crystallographic and chemical data that helps predict or rationalize whether or not a topochemical photodimerization may occur in a given crystal. However, structural information from x-ray diffraction represents a time and space average of a macroscopic sample. An x-ray structure may not capture certain dynamic aspects, chemical impurities, or defect sites that are always present in real crystalline samples. These and other factors affecting the structure of molecules in crystals may result in chemical reactivity not anticipated by topochemical expectations obtained from the average crystal structure. Not surprisingly, several exceptions to Schmidt's topochemical rules have been found (Scheme 3) [2]. Some olefins with packing structures that should lead to cyclobutanes are known to be stable (**4**) [29] reactions that are predicted not to occur based on the distance or orientation of the two double bonds sometimes observed (**5**) [30], and selectivities that are different from those predicted by the packing arrangement can be obtained (**6**) [31,32]. Photochemical reactions have been observed with structural requirements just beyond those delineated by Schmidt's limited data set, suggesting that motions of a certain amplitude may occur in some crystals within the relatively short lifetime of the excited state [33]. Unexpected stability has been correlated with packing or molecular structures that prevent the two alkene centers from approaching one another close enough to

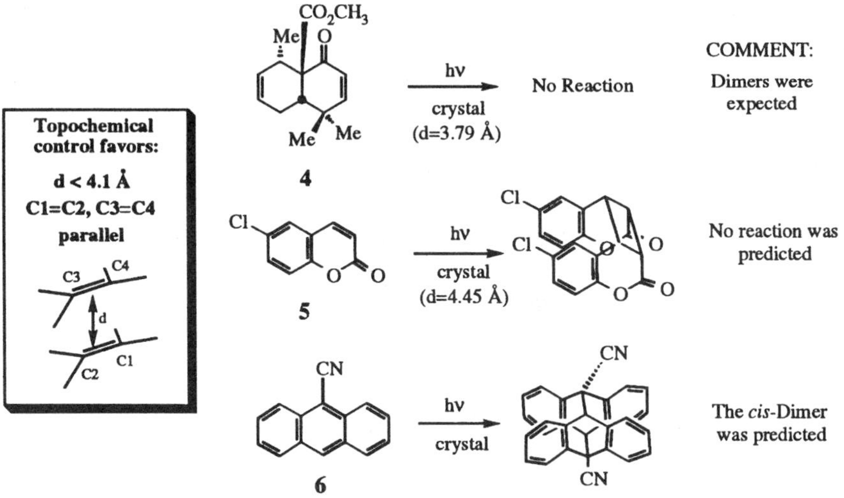

SCHEME 3

form the product. For instance, steric interactions between the substituents in the hexahydronapthalene derivative **4** prevent the cyclobutane ring from forming [29]. Reactivities that are different from those predicted by the packing structure have been explained in terms of defect sites where the distance and orientation requirements for the reaction may be locally fulfilled. Nontopochemical reactions that occur at defect sites show an induction period after which the apparent reaction rate is increased. The unexpected dimerization of several 9-substituted anthracenes such as **6**, which topochemical considerations predict should give the syn-isomer but gives the anti-isomer instead, is one of the most thoroughly documented examples [31,32,34–36]. Differences in geometry between ground and excited states and the role of lattice relaxation have also been cited as a source of unexpected products [33].

3. Unimolecular Rearrangements

Despite recent efforts directed toward the study of two-component crystalline complexes with well-defined stoichiometries [37–41], most bimolecular reactions in crystals studied to date (including those shown in Schemes 1–3) occur between identical molecules. The restrictions on the number of components allowed in crystalline phases and the precise distances and orientations required for bimolecular reactions are among the main limitations imposed by crystals to organic chemical reactivity. However, unimolecular reactions do not depend on the nature and location of their closest neighbors, and photochemical rearrangements in crystals may be investigated with less concern for packing arrangements [4,6,42]. The effect of crystal lattices on unimolecular reactions is profound; bond-breaking and bond-making reactions that cause the structure of the reactant to change toward that of product can be limited by strong repulsive nonbonded interactions which determine whether or not the reaction is energetically feasible.

The "tailormade" space occupied by the reactant in a close packed crystal lattice is determined by its own size and shape. Its boundaries are defined by the van der Waals radii of the closest neighbors which, as proposed by Cohen in 1975, make up a "reaction cavity" (Scheme 4) [43]. In rigid crystals the reaction cavity

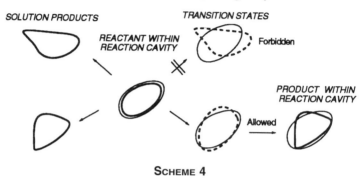

SCHEME 4

is expected to resist deformation, and only those reactions that occur with the least geometrical change are likely to proceed from reactant to transition state to product. This is indicated in Scheme 4; two reaction pathways that give products with very different shapes may be seen in solution, with only one reaction path allowed in the solid state. These intuitive and general guidelines re-state the topochemical postulate with no requirement for defined packing arrangements and apply equally well to unimolecular or bimolecular reactions.

Expectations based on the reaction cavity concept may be tested computationally by comparing the volume of the reactant and the product, and by comparing the structural overlap between reactants and products [9,44–49]. The reaction cavity concept has been confirmed with examples of dimerizations, intramolecular hydrogen abstractions, fragmentation reactions, reactions of biradicals and radical pairs, carbenes and several electrocyclic rearrangements. Figure 1 illustrates this concept with a reaction recently studied in our laboratories which involves the solid state decarboxylation of *cis*-cyclohexanone **7** to *cis*-cyclopentane *cis*-**8** rather than *trans*-cyclopentane *trans*-**8** (see Section IV). Changes in shape and volume accompanying the solid state reaction are illustrated with the van der Waals surfaces of the reactant of the reactant and the two alternative photoproducts.

B. Structure-Reactivity Correlations in Crystals

Reactions that occur in the crystalline solid state offer an opportunity to document the structure of the reactant and its environment to a level of detail that is not

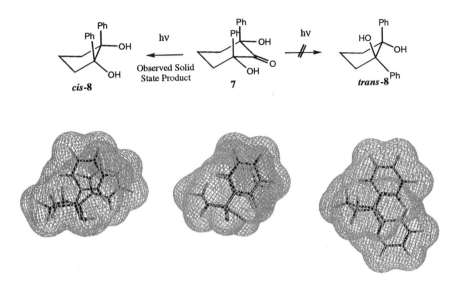

FIGURE 1 Representations of the van der Waals surfaces of a reactive cyclohexanone and its two possible photoproducts illustrate the excellent geometric congruence between the crystalline reactant and the observed photoproduct.

possible in any other reaction medium. Most studies described in the literature are based on product analyses. The selectivity of a reaction in the solid state is compared to that observed in solution and the role of the crystal lattice on selectivity is evaluated. The basic questions that one may answer experimentally with this approach are (1) whether or not the reaction occurs in the solid state, (2) whether or not the reaction is selective and specific, and (3) whether or not the products are topochemically controlled. Unfortunately, *this approach is very ineffective because it yields only binary results from a very rich information source!*

For instance, out of many studies carried out with crystalline olefins that possess slightly different interbond distances and orientations one would only answer *Yes* or *No* to questions of reactivity and selectivity. An oversimplified and pictorial representation of this analysis is illustrated in Fig. 2 with a reaction coordinate for the excited state dimerization of a hypothetical set of crystalline alkenes. The reaction coordinate in this case would be given primarily by the displacement between the two π-systems which, for simplicity, may be assumed to have the required parallel arrangement. In the figure, different crystals can be though of as positioning the prospective reactants at different distances. Each crystal represents a point along the reaction coordinate [50]. Some molecules may

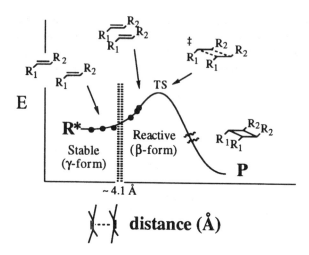

FIGURE 2 The reaction potential shown could be studied in detail with a set of olefin crystals exhibiting a range of molecular separations. In practice, however, product studies usually only reveal whether or not a certain crystal is reactive. This is indicated in the figure by the line at 4.1 Å separating the "stable" and "reactive" regions of the plot.

be categorized as belonging to the γ-form and some may be categorized as belonging to the β-form depending on the translation between molecules in neighboring stacks. Different molecules will require different displacements to reach the transition state and will have different effective activation barriers. However, although different reaction rates and different quantum yields are to be expected from each crystal, the only information that is obtained from product analysis is just *Yes* or *No* with respect to the reaction and only a qualitative boundary that separates examples of the photoreactive and photostable type can be proposed.

Unfortunately, only a few studies have addressed the effect of the crystal lattice on quantum yields of reaction or on absolute rate constants of elementary reaction steps [19,51,52]. These measurements are experimentally very challenging because they are extremely sensitive to minor impurities and rheological factors affecting optical properties. Furthermore, accumulation of the product causes perturbations that alter the local structure, along with absorption coefficients and photochemical rates (see below). Since the mechanistic potential of solid-state reactivity is limited to chemical reactions that occur in rigid, well-defined crystalline environments, chemical reactions must be documented with care and under experimental conditions that guarantee the structure of the original crystal lattice [53–55]. Strategies that may help satisfy this requirement include studies to very low conversion values and studies where the effect of product accumulation is documented. Despite the difficulties and challenges involved in absolute rate measurements in crystals, it seems clear that the rich information obtainable from those measurements, when combined with x-ray diffraction analysis, is valuable and worth pursuing.

II. SOLID-TO-SOLID REACTIONS

An ideal solid-state photochemical reaction would offer high selectivity and specificity with good quantum and chemical yields. However, while high selectivity is frequently achieved in solid state reactions, there may be limitations on the quantum efficiencies and the chemical yields that may be obtained. One of the main obstacles to efficient solid-to-solid reactions comes from changes in reactivity that often occur as the product accumulates. Changes in the composition of a crystalline solid frequently create defect sites which may cause phase [56] changes if the perturbation is high. Phase changes are expected to affect the selectivity and efficiency of further reaction. In the sections that follow, we will analyze the effect of product accumulation in crystals of a reactant, the role of solid solutions or mixed crystalline systems, and the general types of solid state reactions that one may expect based on the types of phases that are involved as the reaction proceeds. Then we will analyze selected literature examples that illustrate these issues.

A. Product Accumulation

1. A Nonlinear Problem

The consequences of product accumulation on the structural and chemical properties of a reacting crystal have been examined by several authors. These can be analyzed from a perspective that focuses on molecular changes or from a crystal phase point of view. A molecular analysis focuses on local structural properties that affect the reaction mechanism [55,57,58]. A solid-phase analysis focuses on how a crystalline ensemble reacts to the changes in composition that occur as the reaction proceeds [59–61]. However, as suggested in Scheme 5, a separation

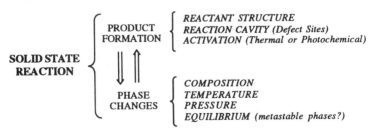

SCHEME 5

of molecular and phase change mechanisms is somewhat artificial since accumulation of the product affects the reactant phase while changes in the medium may have an effect on product formation. A close look reveals a very complex, nonlinear problem.

Most solid state work published in recent years has dealt primarily with a molecular analysis of product formation that seems to arise from the intuitive appeal of the topochemical postulate. Problems associated with phase changes can sometimes be neglected if reactions are carried out to sufficiently low conversion values. However, since preferential reactions at defect sites may be a problem, the involvement of nontopochemical reactions at defect sites should be experimentally documented and avoided. Changes in reaction rates and product selectivity have also been associated with internal stress [54], with sample melting, or with surface effects [62]. In contrast, the mechanisms and consequences of phase transformation have been studied much less. Phase changes depend on the properties of the ensemble and, as suggested in Scheme 5, they are affected by composition, temperature, pressure and whether or not equilibrium is achieved throughout the reaction.

2. Mixed Crystals

As pointed out by Schmidt in the context of his study of cinnamic acid dimerizations, the very early stages of any solid-state reaction which is not initiated at a defect site can be described by a dilute mixed crystal of the product in the crystal phase of the reactant (Scheme 6) [63]. Mixed crystals or solid solutions are

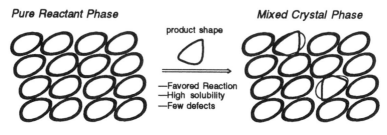

SCHEME 6

structurally regular solid state phases with a crystal structure characteristic of one of the two components. Mixed crystals can be prepared by co-crystallization of the components from solution or from the melt. They require the components to be chemically compatible and structurally similar. Although it has been generally assumed that the solid solute should be randomly distributed in the lattice of the crystalline "solvent," it has been shown that molecular recognition may facilitate deposition on certain crystal faces such that segregation may occur over certain crystal sectors [64,65]. Mixed crystals do not have precise stoichiometric compositions like inclusion complexes or clathrates, but they frequently have limited co-solubilities with compositions beyond which they may not exist.

The formation of mixed crystals by means of a solid-state reaction occurs under kinetic control and may be subject to different constraints than the formation of mixed crystals grown closer to equilibrium conditions from the melt or from solvents. However, whether or not mixed crystallization is possible, and to what extent, is still determined by the structural similarity of the reactant and the product. Kitaigorodskii suggested guidelines based on a volume analysis which indicate that chemically compatible compounds which have optimized overlapping volumes larger than ca. 75% are likely to form solid solutions [66].

It is interesting to note that structural factors facilitating solid state reaction (such as complementarity between the product and the reaction cavity) may also facilitate the mixed crystallization of the product in the lattice of the reactant. We recently postulated that solid state reactions having a reactant ideally predisposed to form a given product will be favored from both molecular and phase transformation points of view [61]. *Products that cause small perturbation to the crystal lattice of the reactant should be tolerated and allow for high selectivities and high conversion values.* Conversely, solid-state reactions which proceed reluctantly may do so because of poor structural similarity between the reactant and the product. Limited solid-state solubility may cause severe crystal lattice perturbations leading to large numbers of defect sites. Such reactions are likely to proceed to low conversion limits or to lead to rapid loss of selectivity.

3. Phase Transformations

Following the formation of a dilute, reactant-like mixed crystal phase, the progress of a solid-state reaction occurring throughout the bulk of a crystal should depend on (1) the solubility of the product in the crystal phase of the reactant, (2) their phase separation mechanisms, and (3) the influence of intermediate and final solid phases on the reactivity of the starting material. While the number of possible scenarios may be large, one may distinguish three general phase transformation pathways that depend on the overall changes that occur as the composition of the crystal changes from reactant to product. The participation of more than one mechanism and more than one intermediate phase may be possible but Fig. 3 aims to give a simplified starting point for a more detailed analysis of solid-state reactions.

As a first approximation it may be expected that solid-to-solid reactions should be guaranteed if they are carried out below the liquid regions of the phase diagram of the reactant and product mixtures. For instance, binary systems with limited solid state solubility may possess phase diagrams such as the one represented in Scheme 7a. The composition axis in this case may be identified with reaction progress from 0 to 100% conversion. As suggested by the dotted line, solid-to-solid reactions may be expected only if the reaction is carried out below the eutectic point of the two-component system. Scheme 7b represents an ideal scenario where the reactant and product may share the same crystal phase through continuous solid solutions. Unfortunately, phase diagrams are generally not

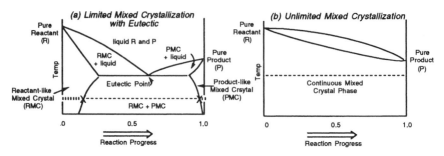

SCHEME 7

known in advance and reactions may be inadvertently carried out in liquid phases. Furthermore, the number of stable and metastable phases increases with the number of components (products) in the mixture so that reactions with low selectivity may show complex phase diagrams and be more likely to proceed through liquid phases. An analysis such as that presented here assumes an infinitely slow reaction which allows equilibration of the reactant and product at a given conversion. This is never the case in practice. In practical terms, it is valuable to explore solid-state reactions at various temperatures and to investigate the phase diagram of the reactant and product(s) by following calorimetric changes as a function of reaction progress. Examination of as-reacted crystals allows for consideration of kinetic effects on phase stabilities.

B. Homogeneous and Heterogeneous Reactions

The analysis which we have presented above presumes that initial product formation occurs randomly throughout the bulk of the crystal, but this need not always

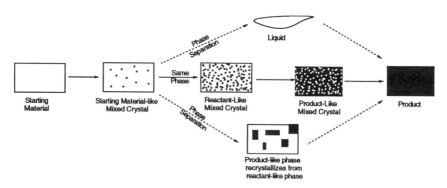

FIGURE 3   Accumulation of product can occur in a continuous mixed crystal, or by phase separation to a liquid (top) or a bi-phasic (bottom) solid.

be the case. Wegner has introduced a useful distinction between *heterogeneous* and *homogeneous* solid-state reactions [67]. Heterogeneous reactions show preferential product formation at distinct nucleation sites defined by defects or a surface. The resulting product distribution leads most often to the separation of a new phase when the product locally reaches the limit of its solid-state solubility, as shown in Scheme 8.

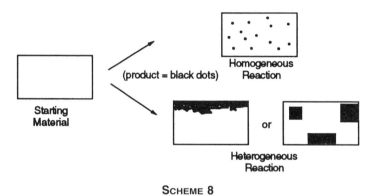

SCHEME 8

Homogeneous reactions are those in which product is distributed randomly in the crystal throughout the reaction. These proceed by way of a solid solution of product in reactant to transform the crystal without ever undergoing phase separation. The strict requirement of random product distribution implies that no *photo*-reaction can be truly homogeneous since absorption at the surface will always be greater than that in the interior. Here we use "homogeneous" to mean by way of a continuous solid solution, emphasizing the lack of phase transformation rather than the requirement of entirely random spatial distribution of product. Homogeneous reactions, then, uniquely preserve the mixed-crystalline nature of the reacting crystal throughout the transformation.

The wavelength chosen for irradiation has been shown to be extremely important for maintaining the homogeneity of a reaction [68,69]. Although proper selection of excitation wavelength should be of obvious concern, it has been given surprisingly little attention. Most solid-state photochemical reactions reported in the literature describe the use of white light spanning the entire UV and visible regions. Excitation with wavelengths that are strongly absorbed by the reactant (high molar extinction coefficient) result in high conversion near surface sites. Locally high concentrations of product favor phase separation mechanisms. To maintain a homogeneous reaction it is desirable that reaction occur in the bulk and not at the surface. It is also preferable that the product not absorb the incident light in order to avoid filter effects which can shield the inside of a crystal and prevent high conversion. This issue will be discussed further below. For both of these reasons, it is always desirable to irradiate crystalline samples at wavelengths where the molar extinction coefficient of the reactant is as small as possible and that of the product is zero. Frequently this occurs at the end of the UV-Vis spectrum of the reactant (Scheme 9) [70]. Several examples will be given below

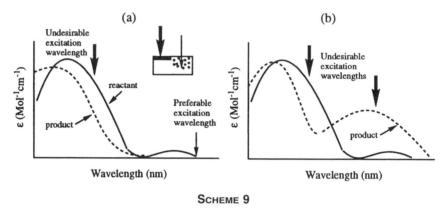

SCHEME 9

where reactions have been observed to be heterogeneous under broadband irradiation but homogeneous when the photolysis conditions were properly adjusted.

Homogeneous reactions maintain the structure of the reacting crystal, and so occur in a single crystal-to-single crystal fashion with phase diagrams similar to that depicted in Scheme 7b. Such reactions can be followed by x-ray crystallography and the structure of the as-formed product can be determined in situ. Reconstructive heterogeneous reactions, which involve recrystallization of a distinct product phase, can also lead to single crystal-to-single crystal behavior. In the sections which follow we mention methods which can be used in the study of solid state reactions, particularly those which can give information about changes in the structure of the reacting crystal, before going on to discuss literature examples of single crystal-to-single crystal transformations as well as reactions which occur with phase separation.

C. Methods of Analysis

1. X-Ray Crystallography

X-ray crystallography is, of course, unparalleled in its ability to yield structural information about a crystal. This technique is only useful for investigating reactions to the extent that long-range order is preserved, however. For ideal single crystal-to-single crystal reactions a high degree of order is maintained, allowing determination of the structure of the as-formed product. This is valuable since in many cases this structure is different from that of the recrystallized product. Crystallography can also be used to detect the formation of new phases, and to determine their crystallographic orientation relative to the reactant phase. This procedure allows *topotactic* reactions, discussed below, to be characterized. In cases where the single crystal nature of the reactant is not preserved, x-ray powder diffraction patterns can often be used to follow changes in unit cell dimensions.

Solid-state reaction in all cases implies the introduction of disorder into a crystal. Even for a perfectly homogeneous reaction the initial product sites are randomly distributed. Thus it is frequently not possible to refine atomic positions in the structures of crystalline reaction intermediates. In rare cases where atomic refinement of intermediates can be done it gives extremely precise information about how a solid-state reaction occurs. Examples of such cases will be discussed below [70,71].

One potential pitfall in using x-ray crystallography is that under x-ray irradiation the crystal may react further, or react too quickly and thus crack. Several examples of solid-state reactions have been discovered in this way, when it was found that an expected structure gave a poor refinement, leading to the realization that the crystal had been transformed on the diffractometer [60,72].

2. Raman and FT-IR Spectroscopies

Prasad, Swiatkiewicz, and Eisenhardt wrote an excellent review of work in the late 1970s using laser Raman spectroscopy as a tool for investigating solid state reactions [73]. Raman spectroscopy is a powerful method for studying solid state reactivity because the wide spectral range which is accessible allows the spectroscopist to simultaneously gather information about the structure of reacting molecules and the bulk properties of the crystal environment surrounding them. We have already mentioned the conceptual distinction between very local changes in the structure of a reacting molecule and greater, more diffuse structural or phase changes that occur in a crystal as a reaction progresses. The utility of Raman spectroscopy stems from the fact that vibrations which arise from intramolecular modes of an isolated molecule fall in a region of the spectrum (200 to 3500 cm^{-1}) distinct from the area where delocalized vibrations involving the bulk of the crystal occur (10 to 200 cm^{-1}). Prasad et al. point out that it is possible, using this technique, to distinguish "physical" changes like phase transformations from "chemical" changes (reactions) that change molecular composition. This method has been used to distinguish homogeneous solid state reactions from heterogeneous ones since it can detect the phase changes which occur in the latter [73]. Closely related to Raman spectroscopy, but usually more readily available, infrared spectroscopy is also a powerful yet simple technique that gives valuable fingerprint and chemical information about solid phases. Chemical reactions can be documented by following intensity changes in bands corresponding to reactants or products. Recent examples of the use of FT-IR spectroscopy for detailed mechanistic work in single-crystalline specimens come from the work of Hollingsworth and McBride in the decarboxylation of diacyl peroxides [55]. Since different polymorphs of a given compound give rise to different spectra, different crystal phases and their transitions can be documented by FT-IR. These technique can be conveniently employed with very small amounts of sample dispersed in KBr pellets or Nujol mulls. We have recently explored the use of FT-IR detection in dilute KBr matrices to determine the relative quantum yields of polycrystalline samples exposed to similar UV irradiation doses [19].

3. Solid State NMR

The potential of solid state NMR is organic solid-state chemistry has been widely recognized, and this technique is likely to increase in importance. Information available from solid state NMR includes aspects of molecular structure, the identification of different solid phases of a given compound (polymorphs), and information on molecular motion. Solid state NMR has also been used to follow the progress of several solid-state reactions [17,19,74,75]. High-resolution spectra of dilute spins such as ^{13}C and ^{15}N can be achieved with cross polarization and rapid (ca. 5 KHz) magic angle spinning (CP-MAS) techniques [76]. Complications may sometimes arise but high-resolution solid-state NMR can be used much in the same way that it is used in solution [77]. Although additional information regarding molecular motion may be obtained from relaxation measurements such as T_1, T_{1p}, T_{CP} [78], the most detailed dynamics information is generally available from analysis of the broad spectra of static samples with line shape analysis and 2D techniques [79,80]. Such experiments can give information about motions that span time scales from 10^{12} to 10^{-1} s^{-1}. Additional examples of the use of NMR in solid-state reactions include the use of imaging techniques to monitor the interface of heterogeneous solid-gas reactions [81].

4. Tunneling Electron Microscopy

In the mid-1970s Jones and Thomas applied electron microscopy to the study of organic solid-state reactions, and some of their work is covered in their 1979 review [82]. Although very useful for the study of inorganic materials, transmission electron microscopy (TEM) has not been widely acknowledged as a useful tool for the investigation of organic crystals. This is largely because of the tendency for high-energy electron beams to destroy samples before adequate information can be gathered. Jones and Thomas have found, however, that electron beam damage can be controlled by stabilizing the substrates with aromatic or chlorine groups, and by optimizing the experimental conditions through variation of the temperature and beam voltage [82].

The advantage of TEM is that it allows a simultaneous measurement of the physical appearance of a very small region of sample and of the electron diffraction pattern of that spot. Although the diffraction pattern cannot usually be resolved to give molecular structures, it can yield information about unit cell parameters and packing arrangements. This is valuable for investigating defect sites in small single crystals, for monitoring phase transformations, and for tracking changes that occur in a crystal as a reaction takes place. For example, the technique has been useful for determining the orientation of linear and planar faults in *p*-terphenyl [83], measuring phase changes in pyrene [84] and observing the presence and orientation of anthraquinone product needles on the surface of an anthracene crystal irradiated in the presence of water vapor [82].

5. Calorimetry

Differential Scanning Calorimetry (DSC) is a sensitive way of detecting phase transformations of a bulk material [85,86]. Monitoring the thermal behavior of a crystal or a powder as a function of its conversion to product can give important information. This technique can verify whether a reaction occurs in a purely solid phase or whether there may be liquid phases involved at a given temperature. Melting point depression can be monitored as product appears, and the characteristic melting of a new phase can be detected if one is formed. DSC can reveal whether or not a eutectic transition attributable to a mixture of phases is present. We have also used DSC in our lab to monitor the thermal stability of reactive crystals.

6. Holography

Köhler, Novak, and Enkelmann have shown that holography can be used as a tool to quantify the rates of solid-state reactions [87]. In some crystalline photoreactions it is possible to irradiate selectively using an optical interference pattern and so to cause conversion of only portions of the reactant crystal. If the product formed has an index of refraction different from that of the reactant then the resulting pattern can create a holographic grating, and characteristics of the reacting crystal can be determined by monitoring the diffraction intensity of the grating. At low conversion, and for systems which show a large change in the index of refraction, the diffraction efficiency is proportional to the square of the extent of the reaction. If this efficiency can be measured as a function of time, then rates and consequently activation energies, can be measured.

Holography is most useful for single crystal-to-single crystal reactions, since phase separation can lead to degradation of the sample. Because of this, the greatest successes have been realized for topotactic, homogeneous reactions, which can often be favored by performing experiments using wavelengths where the reactant is only weakly absorbent, as discussed above.

III. SINGLE CRYSTAL-TO-SINGLE CRYSTAL REACTIONS

Most remarkable among solid state reactions are those which transform a single crystal of starting material into a single crystal of product, maintaining crystallinity throughout the reaction. Such reactions impose severe demands on the structural similarity between the reactant and product and are consequently quite rare. In the cases where they do occur they can be studied by x-ray crystallography which allows a detailed analysis of structural changes. Single crystal-to-crystal reactions are usually, but not always, homogeneous.

Another criterion frequently applied to solid-state reactions yielding crystalline products is that of whether the reaction is *topotactic*. If a reaction occurs so as to retain one or more of the crystallographic axes of the reactant, a definite orientation of the product lattice is obtained relative to that of the reactant and the reaction can be called topotactic. The precise definitions of topaxy found in the solid-state literature vary somewhat in their particulars and the more casual usage of the term varies even more widely. Most authors agree, however, that "product molecules assume preferred orientations relative to crystallographic directions of the reactant structure" [8]. Wegner emphasizes a requirement that the reaction occur throughout the bulk of the reactant crystal, and also points out that topotaxy does not guarantee a topo*chemical* reaction [67].

Several reviews of early work on topotactic polymerizations and isomerizations are available, and the reader is referred to the summaries of Morawetz [88] and Gougoutas [8] for a more complete account. The earliest study of a topotactic reaction appears to have been the observation, in 1932, of the polymerization of trioxane to poly-oxy-methylene [89]. Similar polymerizations of tetraoxane [90] and of trithiane [91] have also been reported to show retention of crystallographic axes from the monomer lattice. Other examples are discussed below. The topotacticity of a reaction can be determined solely by x-ray crystallographic analysis at the reactant and product endpoints. Thus a simple classification of a reaction as topotactic tells very little about how the structure of the crystal lattice changed in the course of reaction.

In this section we look briefly at some well-studied examples of single crystal-to-single crystal reactions. We include reactions which are only homogeneous and single-crystalline under certain conditions. Many of the observations and discoveries made studying cinnamic acid dimerizations and distyrylpyrazine and diacetylene polymerizations offer tremendous promise for carrying out homogeneous crystalline reactions in the future. Progress in understanding homogenous solid-state reactions has also encouraged the study of such systems for applications in specially engineered materials and in data storage devices. A few such applications will be mentioned here.

A. Polymerizations

1. General Background

One class of homogeneous solid-state reactions consists of polymerizations. The realization that monomers can be polymerized in a crystal to give rigidly ordered polymers of high molecular weight with minimal branching has attracted the attention of both polymer chemists and x-ray crystallographers. Baughman has studied factors which are important to the synthesis of large, defect-free polymer crystals [92]. Single-phase polymerization is highly desirable for such purposes since phase separations can lead to strain which in turn can fragment the crystal. (Although defect-free single crystals can in theory result from phase-separation mechanisms for some specific cases.) Baughman outlines a number of criteria which are necessary to maintain a single-phase solid solution throughout a polymerization reaction. These are:

1. *Least atomic motion.* Single-phase polymerizations will in general be more favorable for topochemical reactions than for nontopochemical reactions. That is, reactions which require only small degrees of atomic translation or reorientation are more likely to proceed without phase separation.

2. *Phase stability.* The free energy of a reacting crystal is decreased by the formation of polymer bonds, but may be raised if the polymer is formed in a strained, nonoptimal conformation. Baughman describes an intramolecular free energy (between atoms of the forming polymer) which decreases steadily throughout reaction as bonds are formed, and an intermolecular free energy (between atoms in different chains) which can increase or decrease. If the total free energy, which is the sum of these two terms, can decrease continuously throughout reaction, then phase separation need not occur. If, however, there is an increase in the total free energy that would accompany reaction in a single phase, the product phase will crystallize out.

3. *Reaction uniqueness.* The structure of the monomer lattice must uniquely determine which monomer molecules will react, and in what fashion. Further, symmetry elements of the monomer lattice must be preserved for molecular subunits of the polymer. If this is not the case, local disorder can result. This criterion severely limits the number of monomer crystal structures which can be expected to polymerize via single-phase mechanisms. External forces of some kind which reduce the symmetry of the monomer lattice can be used to help meet the uniqueness criterion.

2. Distyrylpyrazine

One of the best-studied solid-state reactions is the photopolymerization of distyrylpyrazine (**9**) and related compounds to give crystalline polymers containing cyclobutane rings (Scheme 10). This reaction is reminiscent of Schmidt's early work on cinnamic acids, although the presence of two double bonds per monomer can lead to oligomeric or polymeric rather than solely dimeric products. The four-center reaction of **9**, and other related polymerizations, have been reviewed in detail by Hasegawa, who has played a central role in the study of these systems

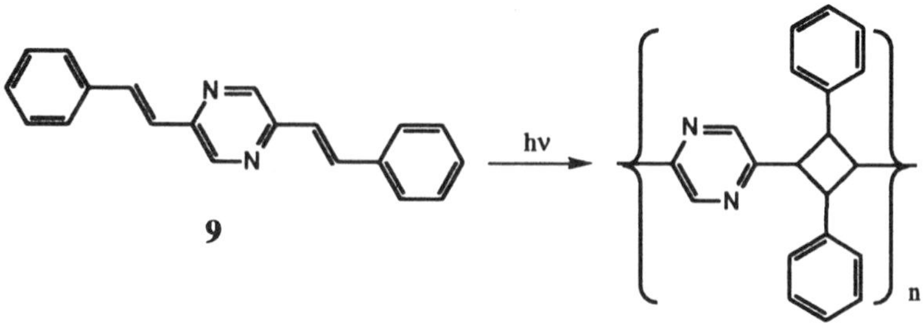

SCHEME 10

[7,93]. The reaction proceeds via a 2 + 2 cycloaddition of two double bonds separated by roughly 3.91–3.97 Å, and in many cases shows retention of the space group of the monomer crystal. Interest in the high structural similarity observed between monomer and polymer crystals prompted extensive study of this system over the last 25 years or so by Hasegawa, Nakanishi, and Wegner. Although this reaction as originally observed occurred heterogeneously, later results have indicated that it can be homogeneous under the appropriate conditions.

Distyrylpyrazine itself can be crystallized as plates or as needles, but only the plates undergo photopolymerization. The monomer packs in the space group P_{bca} with a reactive C–C distance of 3.939 Å [94]. Irradiation at room temperature (well below the melting point of 236°C) with $\lambda > 400$ nm gives rise to oligomers while broadband irradiation containing shorter wavelengths yields polymer. Initially formed oligomer (with $\lambda > 400$ nm) can be converted to polymer by further irradiation below 400 nm [95]. Powder x-ray diffraction studies in 1969 showed that the crystalline monomer reflections convert smoothly into those of the polymer [96]. Later, x-ray examination of single crystals, which were irradiated to generate fragmented polymer crystallites, showed that polymer reflections ("diffraction streaks") grow in superimposed on reflections of the monomer [97]. This revealed that the **a** and **b** axes of the monomer are parallel to those of the polymer and thus that the reaction is topotactic. The space group P_{bca} was maintained throughout the transformation while the **a** axis contracted by 11%, **b** elongated by 13%, and **c** contracted only 2%. It was observed that the center of gravity of the monomer barely moves during reaction. Nakanishi et al. termed the reaction a "lattice duplicating" polymerization due to (1) the control the monomer lattice exerts upon the formation of the polymer and (2) the structural similarity between the arrangement of molecules in the monomer and polymer crystals [97].

Different aromatic residues have been substituted for the pyrazine and phenyl groups in **9** in order to undertake a systematic investigation of this type of compound. Nakanishi et al. have examined the solid-state photochemistry of 1,4-bis[β-pyridyl-(2)-vinyl]benzene (**10**), *p*-phenylenediacrylic acid dimethyl ester (**11**), diethyl ester (**12**) and diphenyl ester (**13**), and *p*-phenylenedi[α-cyanoacrylic

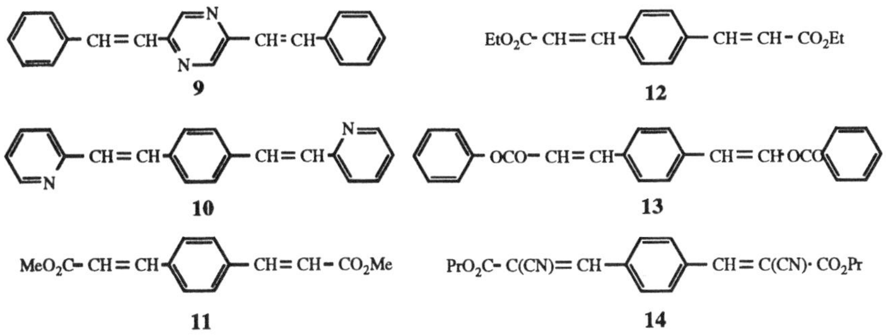

C_6H_5—CH=CH—(pyrazine)—CH=CH—C_6H_5 **9**

EtO_2C-CH=CH—⟨ ⟩—CH=CH-CO_2Et **12**

(pyridine)—CH=CH—⟨ ⟩—CH=CH—(pyridine) **10**

C_6H_5—OCO—CH=CH—⟨ ⟩—CH=CH OCO—C_6H_5 **13**

MeO_2C-CH=CH—⟨ ⟩—CH=CH-CO_2Me **11**

PrO_2C-C(CN)=CH—⟨ ⟩—CH=C(CN)·CO_2Pr **14**

SCHEME 11

acid] di-*n*-propyl ester (**14**) [98]. The corresponding formulas are shown in Scheme 11. X-ray structures were determined after varying amounts of light exposure and the crystals were observed to become opaque over time, although they all retained their shape at low conversion (except for **14**, which cracked). Polymer products, as in the case of **9**, were not formed as single crystals, but rather as aggregates of crystallites aligned along the direction of the polymer chain.

These six compounds (including **9**) were classified by Nakanishi et al. according to three subdivisions [98]: (I) The crystal system, space group, and orientation of all three axes are preserved from monomer to partial polymerization; (II) the crystal system, space group, and direction of a unique axis are preserved; (III) only the crystal system and the space group are preserved. These classes can be related to morphological changes observed in representative examples. Class I reactions have been observed to occur with the formation of large cracks along the polymerization axis, while classes II and III often lead to finer cracks accompanied by fibrillization [98]. According to this scheme, **9** and **10** fall into class I; **12**, **13** and **14** fall into class II; and **11** falls into class III. Thus all of the reactions considered maintain a structural relationship between the monomer and polymer lattices and can be classified as topotactic. One measure of the complementarity of the monomer and polymer lattices is the change in the periodicity of the monomers along the direction of chain growth relative to this periodicity along the polymer chain. For the class I compounds this change is small: 1.8% for **9**, 3.0% for **10**. It is larger, 8.8, 9.3, 10.3, and 11.1%, respectively, for **11**, **13**, **12** and **14**. Nakanishi et al. have hypothesized that *when too much motion is required the reactant lattice will deteriorate*, as was observed for **14**. Clearly, however, even in such cases structural information can be transmitted to the product from the reactant lattice.

Transmission of directional properties from the monomer lattice reveals that many photopolymerizations of the four-center type are topotactic, but this information is insufficient to determine whether the reactions occur via homogeneous or heterogeneous phase-separation mechanisms. Powder x-ray studies done in

1969 revealed a diffraction pattern due to the product phase at conversions as low as 30%, and thus it was concluded that the reaction was heterogeneous, proceeding inward from the surface of the crystal [96]. Later Wegner found that fragmentation of monomer crystals can be prevented and the reaction carried out in a homogeneous fashion under suitable photolysis conditions; Braun and Wegner reported that irradiation with $\lambda > 450$ nm leads to production of oligomeric product as a single crystal [69]. Wavelengths below 450 nm lead to crystal fragmentation because light is absorbed selectively near the surface, generating a nonuniform product distribution. This causes local stress at the interface between regions high in monomer content and those high in polymer, and this stress leads to phase separation and crystal fragmentation. Preservation of the single-crystal structure by using long wavelength irradiation has allowed better x-ray studies to be carried out on partially reacted crystals. From such studies Braun and Wegner saw no evidence for the simultaneous presence of oligomer and monomer phases which would indicate phase separation [69]. The space group of the monomer, P_{bca}, was retained in the crystal of the oligomer and the lattice dimensions of the monomer converted smoothly into those of the oligomer.

The role of defects in the reaction of single crystals of **9** was further probed using dark-field imaging with TEM and etch-pit studies. Etch-pit studies can be used to locate defects since monomer molecules near these sites can be preferentially dissolved away from the rest of a crystal. In investigations of crystals of **9** using these techniques, Braun and Wegner found that although irradiation does lead to defects, at sufficiently long wavelengths the oligomerization reaction does not occur preferentially at these sites, but rather takes place throughout the bulk of the crystal [69].

The polymerization of the **9** oligomer to give the final polymer product shows a behavior different from the monomer oligomerization and has been determined to be heterogeneous. X-ray observation of a crystal of **9** that had been exposed to 365 nm light showed a pattern consistent with a superposition of the oligomeric phase and a distinct phase of the polymer [69]. This implies that the polymer and oligomer undergo phase separation, presumably because polymerization begins at a nucleus and is followed by preferential reactivity at this site. Prasad et al. have used Raman spectroscopy to confirm the heterogeneous nature of the photopolymerization of **9** [73]. They observed changes in the Raman spectrum of **9** between 860 and 1635 cm^{-1} which could be assigned to the disappearance of the double bonds in the monomer and the growth of the cyclobutane ring in the polymer. The lower-energy part of the Raman spectrum showed a superposition of the spectrum of the polymer and the monomer, with the polymer bands growing in as distinct peaks as the reaction progressed, as typical for the appearance of a new phase. They concluded that the polymerization is heterogeneous, and that the polymer, once formed locally in sufficient concentration, phase-separates from the monomer. Thus the polymerization of **9** forms an interesting system in which a two-step reaction begins in a homogeneous manner and then becomes heterogeneous in a later step. The physical characteristics of the crystal can be controlled by wavelength in at least the first step.

A more quantitative Raman study of **9** was undertaken by Eckhardt and Peachey [99]. The homogeneous single crystal-to-single crystal nature of the oligomerization, when carried out at appropriate wavelength, allows Raman modes to be identified and followed throughout the reaction. The smooth changes observed for several strong modes confirm the homogeneous nature of the photo-reaction. Most modes showed only small changes in frequency below 60–70% conversion, at which point larger shifts of 8 to 30 cm^{-1} were observed. The conversion at which the largest frequency shifts were observed corresponded well to the point at which crystal cracking was frequently seen, and was inter-preted as marking the point where an abrupt change occurs in the crystal from the potential of the reactant to the potential of the product. By fitting a lattice potential to the experimental frequency data, specific lattice modes could be correlated with particular molecular motions. The modes which show the greatest changes for the oligomerization of **9** are those involving librations of the molecule about its long axis. This motion brings the reactive double bonds into better alignment for cyclization and also becomes restricted as 2 + 2 dimerization rigidifies the monomers.

Hasegawa et al. have taken advantage of the high order attending photo-polymerization of **9** to produce a novel crystal consisting of an ordered arrange-ment of two polymers [100]. Compound **9** and ethyl 4-[2-(2-pyrazinyl)-ethenyl]-cinnamate (**15**, Scheme 12) form co-crystals from benzene with a composition

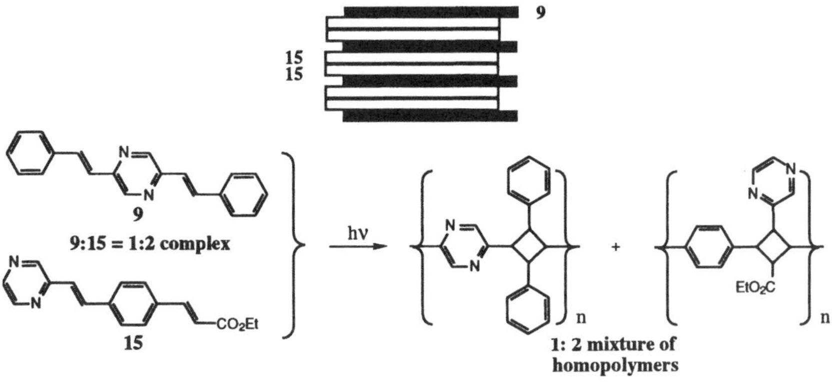

SCHEME 12

ratio of 1:2. In these crystals the monomers stack in alternating planes, with the C−C distance between reactive double bonds much closer for same monomer pairs than for mixed monomer pairs. Irradiation of the co-crystals as a powder gives crystalline polymer with complete conversion of monomer into dimer or higher order oligomers. The authors hypothesize that they have generated an ordered layered polymer formed of alternating sheets of single stranded poly-distyrylpyrazine or poly-ethyl 4-[2-(2-pyrazinyl)-ethenyl]cinnamate, as shown in Scheme 12.

3. Diacetylene Polymerizations

Diacetylenes of the type shown in Scheme 13 can react under thermal or photo-chemical conditions to give crystalline polymers [80]. The change from colorless monomer to brightly colored polymer crystals of the same shape can occur

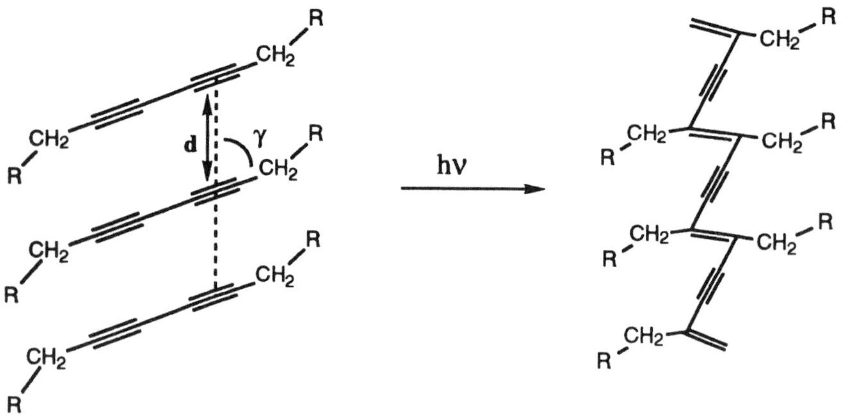

SCHEME 13

rapidly, over a period of less than an hour, probably by a radical mechanism [101]. Reaction is most effective in systems where the monomers are connected via a hydrogen bonded network [102], but other polar R groups can also give high reactivity [103].

Baughman has quantified several necessary conditions for diacetylene poly-merization [92]. The first is a set of geometrical criteria defined as terms of two parameters, the monomer spacing, d, and inclination, γ, along the polymerization axis (see Scheme 13) [92]. Monomers must be close enough together to react and must be able to give product via a least motion path that changes the molecular spacing minimally. Baughman has also pointed out that the possible crystal symmetries from which a single-phase (homogeneous) diacetylene polymeriza-tion can occur are limited. In particular, only a few site symmetries are allowable for the monomer molecules. These restrictions arise from the requirement of reaction uniqueness, which prevents disorder at the molecular level.

Numerous diacetylene polymerizations have been characterized. To select a few that have been well studied, 2,4-hexadiyne-1,6-diol(bis-(p-toluene sulfo-nate)) (**16**), bis-(phenylurethane) (**17**) [101], and 3,5-octadiyne-1,8-diol (**18**) [104]

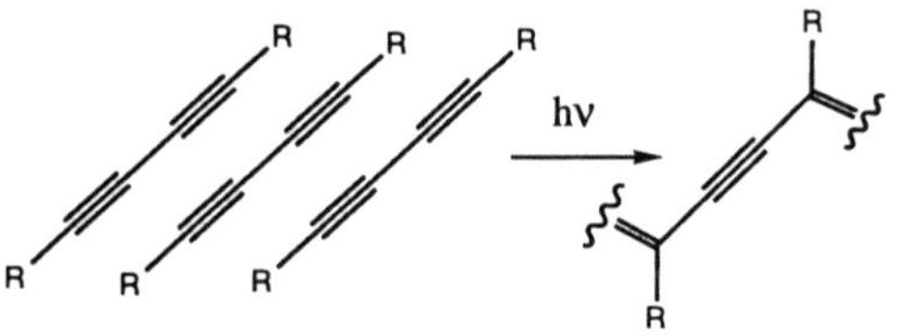

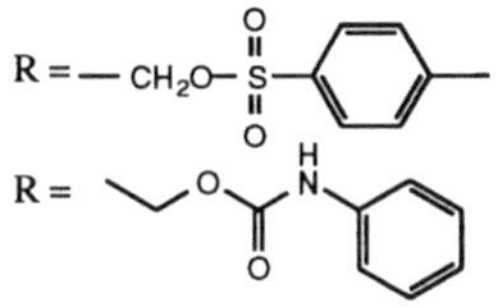

SCHEME 14

(Scheme 14) polymerize to give crystal structures isomorphous with the monomer. Furthermore, **16** has been reported to polymerize to quantitative conversion entirely in a single phase [101,102]. Prasad and Swiatkiewicz have looked at changes in the Raman phonon spectrum upon thermal polymerization of **16** [73] and observed that, according to this criterion, the reaction proceeds in a single phase at least until 30% conversion. Above 30% conversion they observed more complex behavior which they were unable to interpret. This reaction is especially well suited for Raman spectroscopy because the absorption of the acetylene polymer provides a condition for resonance enhancement, giving much greater sensitivity.

Baughman has studied the polymerization of 2,4-hexadiynediol (**19**) by γ-rays with x-ray crystallography [105]. He measured the change of unit cell dimensions as a function of conversion and found smooth changes in lattice parameter **a**, **b**, and **c** of up to 0.3 Å, for crystalline conversions up to 50%. All reflections could be assigned by assuming a single monoclinic phase, and the reaction was concluded to be homogeneous. Phase separation was observed, however, when samples at intermediate conversion were thermally annealed for several days. In this case, amorphous scattering attributable to a separate product phase appeared superimposed on the monomer lattice diffraction peaks. The as-formed crystals show a decrease in the unit cell volume, which then expands upon annealing.

Additional insight into how polymerizations occur can come from conversion vs. time curves. For photochemical polymerizations of **16** these show rapid growth of product upon exposure to light which levels off as the reaction decelerates at higher conversion, possibly because of filtering effects which prevent light from reaching unreacted regions of the crystal [103]. Thermal polymerization shows a considerable lag time prior to rapid polymerization, and proceeds to higher conversion, forming polymer products with very high viscosities. Measurements of the viscosity of the product as a function of time extrapolate to a finite value at the time origin, which has led Wegner to hypothesize the importance of oligomeric nuclei in initiating polymerization [103]. This is not inconsistent with a homogeneous mechanism since oligomeric product accumulation at early times may be distributed as shown in Scheme 15.

The single-crystal nature of the polymerization of **16** has been utilized by

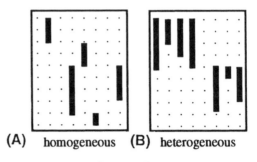

(A) homogeneous   (B) heterogeneous

SCHEME 15

Richter et al. to produce a high-resolution holographic diffraction grating [106]. By illuminating a single crystal of **16** with an interference pattern generated by two 257-nm beams, a series of lines can be written on the substrate. Up to 1000 lines per millimeter can be recorded and instantly read back using a HeNe laser with a wavelength of 633 nm, where the monomer is unreactive and the polymer only weakly absorptive. The advantage of the diacetylene crystal for this purpose is the high resolution (at least 0.6 μm) that can be achieved.

4. Polymerization of NiBr$_2$[P(CH$_2$CH$_2$CN)$_3$]$_2$

Although it represents a thermal reaction of an inorganic complex, we include the polymerization of NiBr$_2$[P(CH$_2$CH$_2$CN)$_3$]$_2$ in our discussion of photochemical organic solid-state reactions because it provides a novel example of a heterogeneous single crystal-to-single crystal reaction [107]. Four notable features characterize this reaction. The polymerization gives rise to only one product—it does so stereospecifically, it retains much of the lattice structure of the monomer, and it demonstrates reaction through a "front" which moves across the single crystal reactant, as shown in Scheme 16. The polymer chains are formed specifically within two distinct sets of Ni sites, with no cross-linking of polymer observed between growing chains. Reaction in this manner gives rise to the same phase as is obtained for polymer grown from monomer in solution.

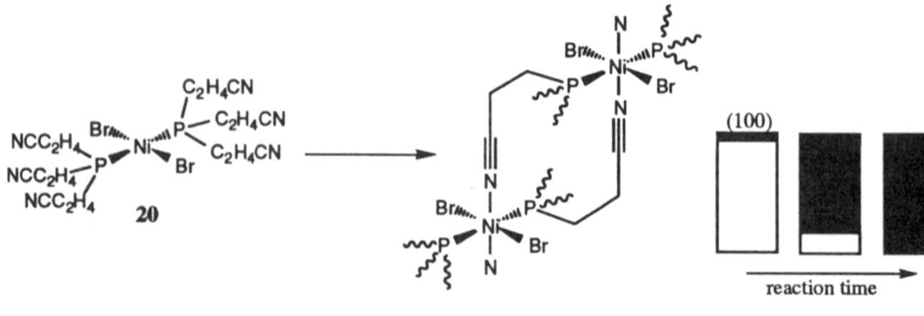

SCHEME 16

Over six months, red single crystals of monomer develop a blue reaction front which traverses the crystal, beginning from a (100) face, as shown on the right in Scheme 16. X-ray analysis at intermediate points has shown the presence of both monomer and polymer crystal lattices, and revealed a topotactic reaction where the **a** and **b** axes of the two lattices are aligned. The **c** axis was found to be rotated by 7.5° relative to its original position in the monomer crystal, leading to a change in space group from P_{bca} to $B2_1/c$. Only one new orientation of the **c** axis was observed; the axis along which the (100) face crystallization originally occurs determines the orientation of the entire single crystal of product.

B. Dimerizations

1. Cinnamic Acids

G. M. J. Schmidt, with his ground-breaking work on the solid-state reactions of cinnamic acids, was one of the first to look at intermediate stages of a reacting organic solid (Scheme 17) [63,108,109]. He did this, in the early 1960s, with the

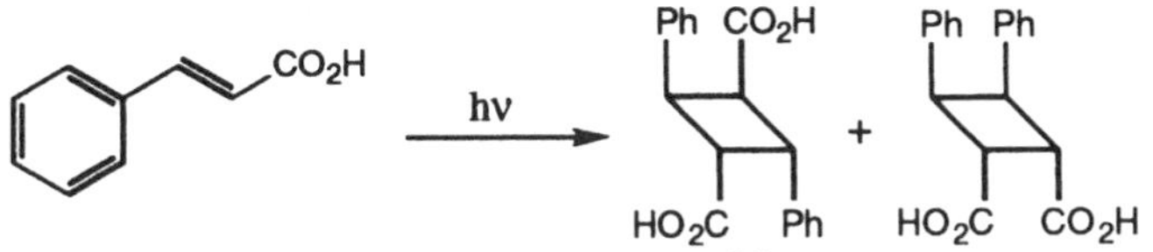

| **21** | **22** | **23** |
| cinnamic acid | α-truxillic acid | β-truxinic acid |

SCHEME 17

use of powder x-ray crystallography. In experiments investigating cinnamic acids Schmidt and co-workers recorded x-ray spectra periodically during the photolysis of powdered samples, which enabled them to determine the time at which a product phase distinct from that of the reactant appeared. They found a variety of behaviors, examples of which are given in Table 1.

Schmidt et al. found that a distinct product phase appeared at less than 10% conversion in some cases (indicating heterogeneous reaction), while in others none was evident at greater than 70% conversion [63]. Certain derivatives, β-truxinic acids in particular, were formed in a modification different from that into which they could be recrystallized. Schmidt interpreted these results in terms of a phase separation of the forming product which occurs when it reaches a limiting solubility in the lattice of the reactant. He further pointed out that recrystallization may be necessary to improve yields in the dimerization of β-cinnamic acids in certain instances where monomer molecules may be stranded in nonreactive sites such as $M_2 .. M .. M_2$ (Section IV.B.5.). 5-Bromo-2-hydroxy-cinnamic acid is an example of a case where there is no evidence for phase separation and reaction is found to be slow and proceed in low yield [63,108].

TABLE 1   Dimerization of Cinnamic Acids as Monitored by Powder X-ray Diffraction*

Compound/conditions (crystal form)	Powder pattern observed	Product
Cinnamic acid (α)	Product pattern after 2 days	α-Truxillic acid
Cinnamic acid (β)	α-Cinnamic acid pattern quickly appears	α-Truxillic and β-truxinic acid
Cinnamic acid (β)	Gradual replacement of reactant with product over 2–11 days	β-Truxinic acid
o-Ethoxycinnamic acid (α)	Changes within 2 days	2,2'-Diethoxy-α-truxillic acid
o-Ethoxycinnamic acid (β) ambient temperature	Diffuse α-monomer after 2 days, no reflections from β-remain	36% 2,2'-diethoxy-α-truxillic acid 32% 2,2'-diethoxy-β-truxinic acid
o-Ethoxycinnamic acid (β) cooled	After 2 days β-monomer after 6 days mainly 2,2'-diethoxy-β-truxinic acid	5% 2,2'-diethoxy-α-truxillic acid 90+% 2,2'-diethoxy-β-truxinic acid

*See structures in Scheme 17.
Source: Reference 109.

The importance of recrystallization to the progress of a reaction was demonstrated again in the solid-state photoreactivity of o-methoxycinnamic acid (24). Cis-o-methoxycinnamic acid (cis-24), when crystallized and irradiated at room temperature, gives rise to the cis and trans acids and to trans/trans dimers (25), as shown in Scheme 18. When the reaction is carried out at low temperature, how-

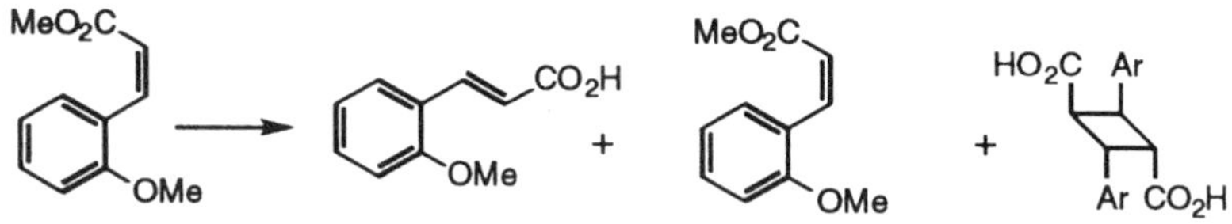

cis-24   trans-24   cis-24   25

SCHEME 18

ever, only trans-24 accumulates. Dimer is only formed upon thermal annealing followed by further irradiation [110]. Presumably this is because the trans isomer is not reactive in the phase of the cis acid in which it is formed, but must recrystallize into a different phase for dimerization to occur. The absence of mixed cis/trans dimers further supports the idea that the reaction does not occur in mixed phases. Similar behavior was seen for β-methyl-trans-cinnamic acid (26) [110]. In this case the trans acid initially partitions between dimerization and isomerization to the cis acid (cis-26). The mixed phase does not give rise to mixed dimers, however, but recrystallization rather "refreshes" the trans phase, allowing further homodimerization to occur.

With co-worker Osaki, Schmidt also observed evidence for a topotactic cinnamide dimerization [111]. While monitoring the x-ray powder diffraction pattern of *trans*-cinnamide after exposure to sunlight, they observed a gradual broadening of the peaks of the reactant lattice with concomitant growth of lines due to a product phase. The new phase was shown to have the same symmetry as the reactant lattice, and a tentative assignment of a reciprocal lattice to the new diffraction lines suggested that the relative positions of monomers in the reactant and the dimer were very similar. Osaki and Schmidt hypothesized that a network of two types of hydrogen bonds between the amides fortified the lattice and prevented the loss of reactant structural information which they had previously observed for the isomerization of 1-bromo-*cis*-cinnamic acid and the dimerization of trans-cinnamic acid. Their observations suggest that this is a heterogeneous topotactic reaction.

Some 30 years after Schmidt's original work, Wegner and co-workers found that cinnamic acid dimerizations can be *homogeneous* under certain reaction conditions. Studying the dimerization of α-*trans*-cinnamic acid (**21**) to truxillic acid (**22**), Enkelmann et al. discovered that irradiation of single crystals at the tail end of their absorption spectrum leads to single-phase reactivity, in up to 100% conversion [70]. Critical to maintaining a single-phase is a uniform distribution of light in the crystal, which is ensured by using a wavelength where the monomer and product absorb very weakly. As revealed by x-ray crystallographic studies at 28, 40, 67, and 100% conversion, a homogeneous crystal was maintained by a compromise of molecular conformations in the monomer and dimer which allowed phenyl rings of both to inhabit the same crystal site. A phenyl ring "tilt" (away from the axis defined by the bond connecting the ring to the reactive double bond, Φ in Scheme 19) and rotation (around this axis, Ω) varied continuously

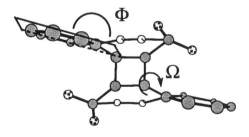

SCHEME 19

during reaction to allow a fit of the product in the reactant lattice (at low conversions) and the reactant in the product lattice (at high conversions). The phenyl tilt varied from 0 to 32.7° and the rotation from 81.5° in the monomer to 100.9° in the

pure dimer crystal. Clearly packing energies must be favorable enough to overcome the conformational energy cost of distorting the molecular geometry.

2. Benzylidenecyclopentanones

The first example of using x-ray diffraction techniques to record the structure of
partially reacted crystals was reported by Nakanishi et al. in 1980. They monitored
the dimerization of 2-benzyl-5-benzylidenecyclopentanone (**28**) and 2-benzyl-
5-*p*-bromobenzylidenecyclopentanone (**29**) [112,113]. Both compounds were
photolyzed to yield single crystals of product which were isostructural with
recrystallized dimer [114]. Compound **29** (Scheme 20), which has a particularly

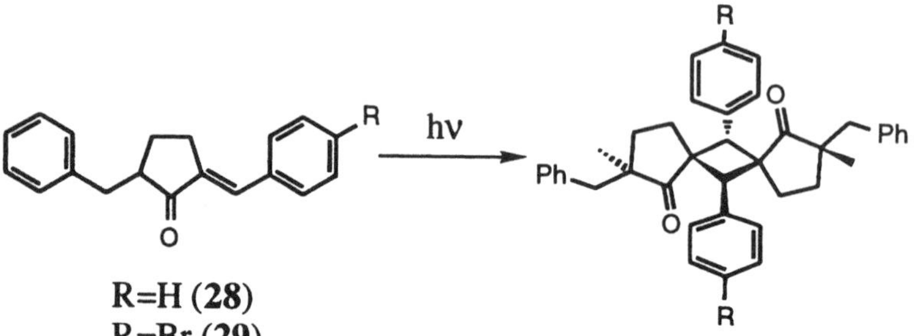

R=H (**28**)
R=Br (**29**)

SCHEME 20

short distance between the reactive double bonds in the monomer crystal, was
very reactive and crystals of this compound showed a tendency to crack. Even
when irradiated extremely slowly (ambient light was sufficient) the structure of
this compound could not be followed to completion in situ.

Complete structures were determined for the monomer and dimer crystals of
both compounds, and unit cell dimensions and intensity data were obtained for
intermediate points in the reaction of **28** by measuring 43 reflections [113,115].
All three crystal axes of monomer and dimer were coincident for **28**, and the space
group P_{bca} was maintained for the photodimerization of both **28** and **29**. Tracing
changes in the unit cell parameters which occur during reaction revealed a smooth,
tough not monotonic, change for both compounds [113]. The magnitude of the
changes observed in the unit cell dimensions is very small, the biggest difference
being about 2% along the **c** axis in **29** (before it cracked). Complete structures
were determined for **28** at 30 and 70% conversion by using phase information
based on atomic positions of the major component (monomer or dimer). Refinement was done using a variable occupancy factor for two of the carbon atoms.
Successful refinement using this method to R values of 0.15 and 0.14 reflects
the extreme structural similarity of the monomer and dimer lattices.

Thermal ellipsoids for the 30 and 70% conversion structures did not show
any consistent alignment, which was taken as an indication that the environment

throughout the crystal was not totally uniform. The authors have speculated that variations in the concentration of dimer throughout the crystal (possibly more at the edges, for example) could give rise to small local variations in molecular orientation; this could account for the averaged ellipsoid shape observed. The overall creation is almost certainly homogeneous since no sign of phase separation was observed, while x-ray structures at intermediate stages could be determined with only moderate broadening of reflection peak widths.

The dimerization of **28** has also been studied by Prasad, who used Raman spectroscopy to monitor both changes in intermolecular vibrations and lattice phonon modes [73]. The Raman spectrum shows the disappearance of alkene stretches at 997, 1180, 1593, and 1625 cm^{-1} as expected, and the appearance of cyclobutane modes at 878, 979, and 1001 cm^{-1}. Phonon modes broadened as the reaction progressed, and bands around 15–40 cm^{-1} showed a shift in frequency. Between about 50 and 66% conversion it was difficult to define distinct bands, but after that point product bands grew in distinctly. This "amalgamation" behavior is good evidence for a homogeneous reaction mechanism.

3. Dimerization of Acridizinium Salts

Wang and Jones have investigated the solid-state reactivity of the series of acridizinium salts shown in Scheme 21 [116]. Many of these undergo single crystal-to-single crystal dimerization when irradiated and can thus be monitored by x-ray

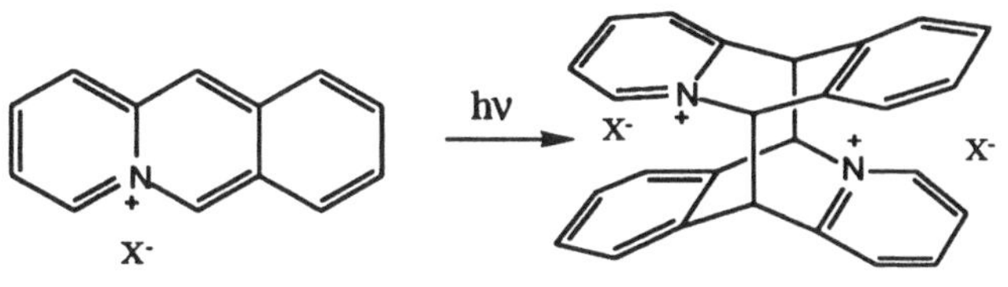

$$X = Br\ (\mathbf{30}),\ I\ (\mathbf{31}),\ ClO,\ BF_4$$

SCHEME 21

crystallography. Wang and Jones have reported the structure of the as-dimerized crystal for the bromide and iodide salts. The bromide salt monomer (**30**) crystallizes into either the $P\bar{1}$ or the $P2_1/a$ space group. Photodimerization in the monoclinic $P2_1/a$ crystal leads to product formation in a $P2_1$ lattice with the same structure that is obtained from recrystallization of the dimer salt. Unit cell parameters at intermediate conversion show a gradual change from the $P2_1/a$ values to those for the $P2_1$ product, although data regarding phase changes was not reported. Irradiation of the triclinic $P\bar{1}$ monomer crystal gave product within the $P\bar{1}$ space group. The iodide salt (**31**) undergoes photodimerization in the $P\bar{1}$ space group with an accompanying increase in unit cell **a** (+0.306 Å) and **b** axes (+0.134 Å) and a decrease in the length of the **c** axis (−0.322 Å).

The dimerization occurs with significant molecular motion, mostly due to one-half of each acridizinium monomer rotating toward its reaction partner by about 19°. There is also a readjustment of the position of the counter ion. The motion of these groups does not occur with perfect regularity as disorder is observed in intermediate structures. Interestingly, reaction of **30** (and possibly that of other salts as well) requires the presence of water of crystallization in order to proceed to complete conversion. The single-crystal nature of the reaction allowed the position of the water to be identified: The oxygen of the water appears to be coordinated to the bromide anion, although why it is essential to reaction has not been discovered. The large motion required, the readjustment of the cation and the intermediate disorder suggest that although it proceeds in a single crystal-to-single crystal manner, this is not likely to be a homogeneous reaction.

4. Styrylpyrilium Triflate Dimerizations

Irradiation of the styrylpyrilium salt (E)-2,6-di-*tert*-butyl-4-[2-(4-methoxyphenyl)-ethenyl]pyrylium-trifluoromethansulfonate (**32**, Scheme 22) at its absorption maximum of $\lambda = 420$ nm results in a heterogeneous reaction in which yellow

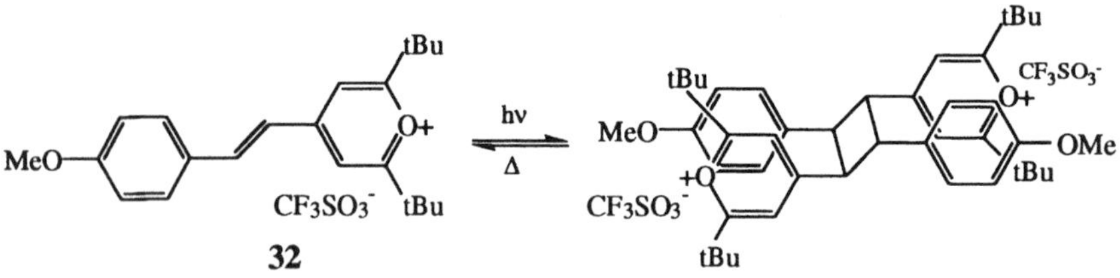

SCHEME 22

microcrystals form on the surface of a red monomer crystal. Novak et al. have shown, however, that irradiation with $\lambda > 570$ nm leads to a single-crystal reaction that is homogeneous within planes of constant irradiation [117]. Thus, this represents another example of a system in which a homogeneous transformation can be maintained if strain due to localized product accumulation can be avoided.

X-ray analysis at intermediate stages of this reaction shows a gradual deformation of the monomer lattice into that of the as-formed dimer. Separate atomic positions could be resolved for carbon atoms directly involved in bond making in product molecules, but the positions of the rest of the atoms in the co-crystal of dimer in monomer showed uniform positions. The aromatic rings were observed to twist through a fairly large angle as the reaction progressed, with their position in intermediate mixed crystals lying between their positions in the monomer and in the as-formed dimer lattice. The as-formed product represents a metastable state, and recrystallization leads to a different modification. Another interesting aspect of the metastable state is that it undergoes thermal reversion to monomer without losing its single crystal nature. This has allowed the system to be used as a holographic grating that could be written photochemically and then erased thermally as a prototype for data storage devices [87].

Köhler, Novak, and Enkelmann have used diffraction by such a holographic grating to study the characteristics of this reaction, to measure its rate, and to estimate an upper limit on the activation energy of both the photodimerization and the thermal reversion to monomer [87]. In their experiment, holographic gratings were written into the crystal by irradiation with an interference pattern created by two 633-nm beams. The same wavelength, with reduced exposure times, was used to read the hologram. Prolonged exposure (about 2000 s) to the reading beam led to photobleaching when the entire crystal was converted to product. After exposure for about 1500 s the crystal loses its homogeneity, as reflected by variations in the intensity of the diffraction pattern, but for irradiation times less than this a single crystal can undergo several write/read/erase cycles without degradation.

At low conversions the diffraction efficiency of the holographic grating can be monitored to determine the rate of photodimerization. Köhler et al. have measured an upper limit on the activation energy for photodimerization of 3.6 kcal/mol, which is very close to a value of about 3.1 kcal/mol determined by Nakanishi et al. for the dimerization of distyrylpyrazine [96]. Rates of disappearance of the diffraction pattern can be used to measure the rate of reversion to starting monomer which occurs upon thermal annealing above 100°C. Köhler et al. have measured the activation energy for this process to be 23.7 kcal/mol.

C. Unimolecular Rearrangements

1. Anthracene Peroxide

X-ray irradiation of single crystals of anthracene peroxide (**33**, Scheme 23) gives mixed crystals of anthraquinone (**34**) and anthrone (**35**), with retention of the shape of the original crystal [118]. This transformation has been monitored by

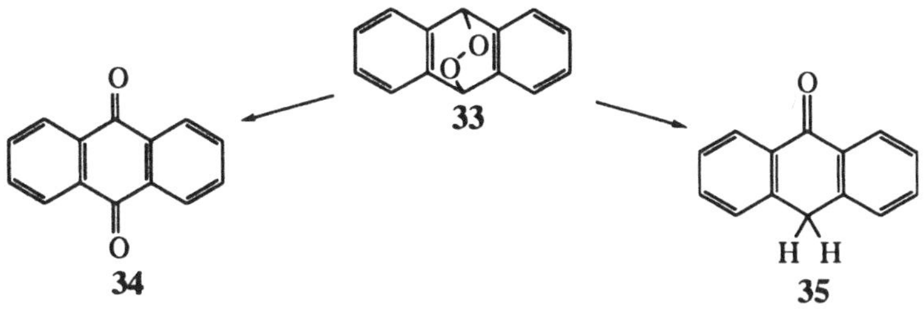

SCHEME 23

x-ray diffraction. While complete structures of intermediate points could not be solved, these studies revealed the creation and destruction of symmetry elements in the crystal. The reaction appears to take place after columns of molecules along the **b** axis of the parent crystal first disorder with respect to each other (while remaining parallel to the **b** axis and retaining the molecular spacing along this axis) and then contract with the intermolecular spacing along the **b** axis changing from 5.8 Å to 4 Å, presumably as reaction occurs. Product molecules with the new spacing recrystallize, leading to sharpened diffraction spots characteristic of ordered mixed crystals. The transformation is single crystal-to-single crystal, and seems to occur throughout the crystal, but does not occur in a single phase. Significant amounts of molecular motion are required to pass from reactant to either product, and while distinct relationships between adjacent molecules in the reactant are not necessarily preserved in the product, the reaction proceeds with high structural integrity and can be monitored throughout by x-ray crystallography. It is, therefore, difficult to classify strictly as homo- or heterogeneous. The preservation of the crystallographic orientation of the **b** axis, however, along which columns of molecules are stacked in both reactant and product, typifies a topotactic reaction.

2. p-Bromophenylacyl Ester of Hirsutic Acid

The *p*-bromophenylacyl ester of hirsutic acid (**36**, Scheme 24) shows a dramatic topotactic rearrangement induced by the x-ray irradiation [72]. The reactant lattice is preserved in the conversion of **36** to a 2:3 ratio of **36** and its keto form, **37**.

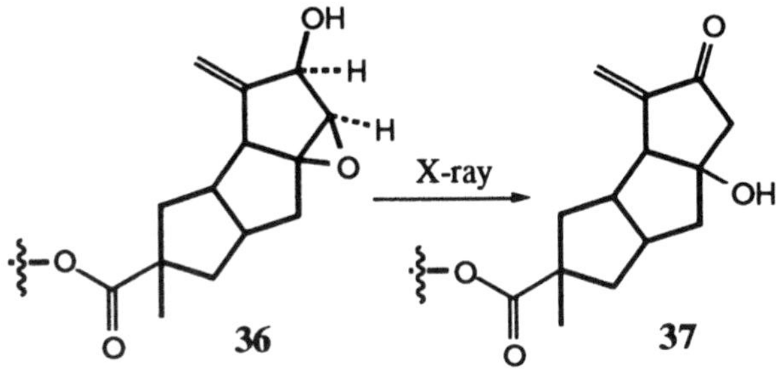

SCHEME 24

Cell parameters change by only about 1% and most atoms remain in their original positions despite migration of 2 hydrogens, the formation of the ketone and the opening of the epoxide. The reaction is driven by the formation of hydrogen bonds between **36** and **37** that cannot occur between molecules of **36** alone. No phase separation of the product was observed, and indeed the unintended reaction was only recognized after it proved impossible to satisfactorily solve the crystal structure assuming it was composed of **36**.

3. Bis-(ortho-iodobenzoyl)-Peroxide Rearrangement

Upon standing at room temperature, heating to 110° or exposure to x-ray radiation, single crystals of bis-(ortho-iodobenzoyl)peroxide (**38**, Scheme 25) give product **39** without change in the morphology of the crystal [119]. Gougoutas has provided a summary of his study of this reaction in a 1971 review article [8]. The product is formed in the space group Cc (from the Pc reactant) which is one of

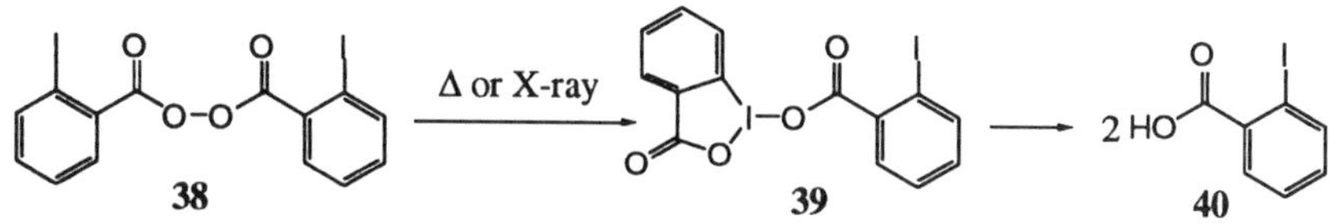

SCHEME 25

two possible polymorphs which can be crystallized from solution. A product phase appears to be formed directly from the reactant without the intervention of intermediate phases. The crystals formed are opaque, and appear to have amorphous regions, although they still diffract. If heated to 120°, a polycrystalline sample results. The as-formed crystal of **39** also reacts to give single crystals of iodobenzoic acid (in its native crystal, $P2_1/c$), with no change in the shape of the crystal. This transformation does not occur in freshly crystallized **39**, however, and it is assumed that defects from imperfections in the as-reacted crystal are necessary to allow access of the water required for hydrolysis in the last step. Both reactions to form **39** and iodobenzoic acid are topotactic with one crystallographic axis preserved along the short-repeat axis.

Interestingly, analysis shows that one-half of the phenyl rings in **38** are required to flip by 180° to give the product structure from the reactant. The ability of the reaction to occur within an intact crystal without fragmentation or loss of crystallographic orientation is remarkable in light of the significant amount of molecular motion required to reach the geometry of the product from that of the reactant. It seems unlikely that this reaction proceeds in a single phase, given the large amount of motion required.

4. Cobaloxime Isomerizations

Several different types of reactions have been observed to occur in single crystals of cobaloxime complexes [120,121]. The axial ligand $C(CH_3)(H)(CN)$ shown in Scheme 26 for chiral cobaloxime **41** racemizes in a single crystal when exposed

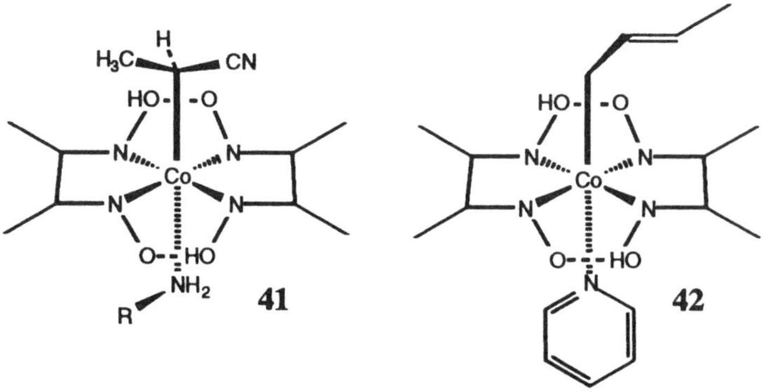

SCHEME 26

to x-ray irradiation. ESR evidence suggests that it is the Co−C bond that is homolytically cleaved [122]. X-ray crystallographic studies at intermediate points indicate that the reaction of **41** occurs with first order kinetics, as determined by monitoring the unit cell dimensions. These change smoothly and continuously, with **a** and **c** decreasing very slightly and **b** increasing (by about 2%) as reaction progresses, preserving the reactant space group (P2$_1$). The electron density changes significantly only at the site of the newly racemic center.

Ohashi and co-workers have carried out a systematic study of derivatives of **41** (Scheme 26) to determine factors which allow racemization of some compounds but not others. For example, they find that for R = (R) or (S) methylbenzylamine or R = pyrrole, racemizations occur with a rate constant of about $1.5–3.0 \times 10^{-6}$ s^{-1}. For other compounds (with R = triphenylphosphine, tributylphosphine, diphenylethylphosphine, and diethylphenylphosphine) no reaction was observed at room temperature. The reactivity (or lack thereof) correlates well with the cavity volume accessible to the reacting group [123]. For compounds that crystallize with two molecules per unit cell the effect of cooperative racemization can be observed and can also be related to cavity size [124–126]. Possibilities available to reactive chiral compounds have been classified by Ohashi as follows [120].

If there is one molecule/unit cell, (I) the molecule may racemize if the cavity size is greater than 11.5 Å, resulting in some disorder in the crystal structure. If there are two molecules/unit cell, (II) one molecule may invert and one may stay the same, (III) both molecules may racemize but equilibrium conditions lead to one molecule reverting and one inverting, (IV) both molecules racemize, but to different extents, or (V) both molecules racemize equally.

Complete inversion of one of the centers, as in II and III, can be driven by creation of a crystallographic inversion symmetry which allows tighter packing of the molecules, reduces the unit cell volume, avoids bad close contacts, and lowers the energy of the crystal [124,125]. Preferential isomerization of one molecule in the unit cell over the other has been correlated with the relative available reaction volumes. In cases where the reaction volume for each of two molecules is large enough for facile racemization it has been speculated that entropic factors may favor a final state which allows disorder in both sites. Computational studies which account for the potential packing energies of various structures have been carried out to study these reactions. The calculations confirm the existence of possible low-energy paths for racemization in some instances where it has been observed [127,128]. These criteria can be used to rationalize changes in the space group and cell dimensions observed for these single crystal-to-single crystal transformations.

Another type of isomerization, that of *cis-trans* isomerization of an axial Co substituent has also been observed to occur in a single crystal-to-single crystal manner in cobaloximes. Compound **42** was monitored by x-ray crystallographic determination of the unit cell dimensions at 9 points over 2 h [121]. Molecular structures at the beginning and end of the photolysis were identical except for disorder that was introduced due to the newly formed cis substituent. Changes in the unit cell dimensions occurred gradually and smoothly and indicated first-order kinetics, at least at early stages of the reaction. The **a** axis expanded by 0.253 Å, while the **b** and **c** axes contracted by 0.016 and 0.385 Å, respectively. These changes could be rationalized based on the orientation of the *cis*-butane substituent along the **a** axis, and the alignment of the *trans* along the **c**. The smooth changes and relatively small motions involved suggest the possibility of a homogeneous reaction.

5. A Metastable Phase in the Reactivity of 1,2-Diphenyldiazopropane

In our laboratories we have shown that the crystalline solid state may exert a very high degree of control on the selectivity of several highly reactive intermediates. The reaction in Scheme 27 illustrates the transformation of a diazo compound

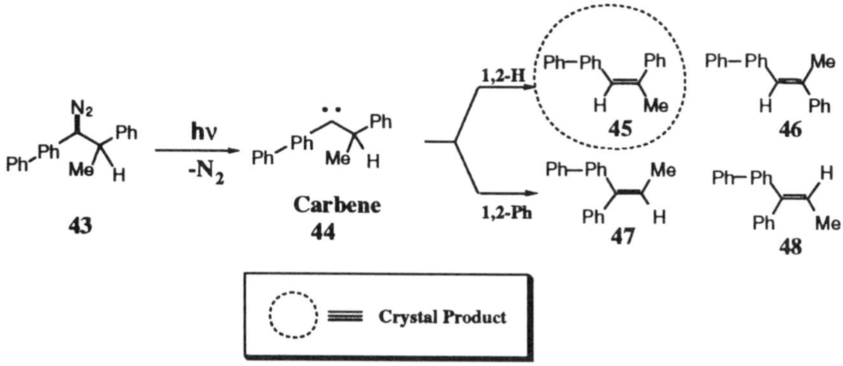

SCHEME 27

into four stereoisomeric olefins arising from 1,2-H shifts and 1,2-Ph migrations by the intermediacy of an arylalkyl carbene with a triplet ground state [129,130]. When the reaction is carried out in solution, all possible products are observed in yields that vary with temperature and solvent polarity. Reactions carried out in the solid state with λ ≥ 450 nm proceed with complete stereoselectively to give the *cis*-stilbene type product, **45**, with conversions up to 95% (Scheme 27). Reactions carried out under conditions where the photoproduct absorbs (e.g., λ ≥ 300 nm) result in lower selectivities that change as a function of irradiation time.

The mechanism for solid-state control in this case relies on the rigidity of the crystal lattice and on a conformation that predisposes the reactant toward the observed product. In the case of carbene 1,2-R shifts, it is known that the migrating group must be aligned with the empty *p*-orbital of the *sp*²-hybridized singlet-state carbene [131,132]. As shown in Scheme 28, the solid state conforma-

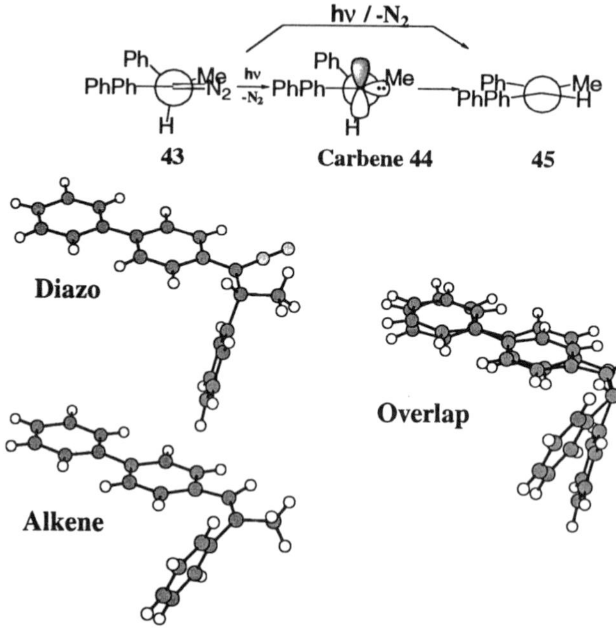

$$hv / -N_2$$

43   Carbene 44   45

Diazo

Overlap

Alkene

SCHEME 28

tion of diazo **43** determined by x-ray diffraction aligns the small hydrogen group for migration, while the bulky phenyl group is minimally displaced when *cis*-alkene **45** forms.

Although the packing structures of the reactant and recrystallized product are quite different, there is excellent overlap between the reactant and product molecular geometries (Scheme 28, bottom). FT-IR and ^{13}C CPMAS NMR analyses of the solid-state reaction give spectra that change continuously with the extent of conversion to give a final appearance very different from those of the pure reactant or product phases [16]. It is readily apparent that the solid phase of the "as-reacted" product is quite different from that obtained after recrystallization. This observation suggests that reaction proceeds in the original phase of the reactant, or that phase separation occurs to give a different polymorph of the photoproduct. A distinction between these two possibilities was accomplished by thermal analysis (DSC). The lack of a eutectic point and a progressive broadening and lowering of the melting endotherm, which was followed by the thermal denitrogenation of the diazo compound, suggests that the two component mixture exists throughout the reaction as a continuous solid solution. Although a rigorous characterization of the final phase and its relation to the phase of the reactant are still to be determined, it is clear that the product does not phase-separate but adopts a metastable phase that seems closely related to that of the reactant. It was shown that nitrogen remains trapped in the crystal lattice but it is not known whether it adopts a preferred crystallographic position or if it remains disordered. The features of this reaction described above suggest that it may indeed be a single crystal-to-single crystal transformation. Arguing against this are the observations that the crystal becomes opaque during irradiation, and that the surface shows evidence of disruption by small eruptions, which give rise to a fine powder at the surface at some sites. It is important to stress that any changes in packing arrangement and molecular structure occurring as product accumulates in this reaction do not have a deleterious effect on the observed selectivity.

IV. RECONSTRUCTIVE SOLID-TO-SOLID REACTIONS

A. Phase Separation Reactions

Single crystal-to-single crystal reactions such as those discussed above are exceedingly rare, as most solid-state reactions disrupt the lattice hosting the reaction to greater extents. Some reactions destroy the crystallinity of the medium to give liquid or amorphous phases and some reactions lead to new crystalline phases by not-well-understood reconstructive mechanisms. Although some reactions occur by phase separation to the stable form of the product, others may lead to metastable phases that are not available by crystallization from solution, sublimation, or the melt. All of the examples analyzed in this section, which illustrate phase-separation mechanisms, lead to severe optical and mechanical degradation of the reacting crystal or to the formation of polycrystalline products.

In principle, the evolution of a reacting system, from a phase transformation point of view, may be followed according to the phase diagram of the reactant and the product(s). A solid-to-solid reaction along the dotted line in the hypothetical phase diagram of Scheme 7 would start with formation of reactantlike mixed crystals (RMC). In the example in Scheme 7, the reaction would continue within the RMC phase until the solubility limit of the product is reached, at about 20% conversion. After this, the RMC no longer tolerates the product and phase separation would occur. Segregation of the reactant and the product would result in recrystallization into the RMC phase and a "productlike" mixed crystal phase (PMC) containing about 10% of starting material. Overall, conversion values between ca. 20 and 90% would involve a weighted mixture of crystals with compositions given by the solubility limits of the two allowed solid phases and, at the end, the reaction would reach completion within the PMC phase. A cartoon representation of these changes is shown in Scheme 29.

The hypothetical phase diagram in Scheme 7 assumes that the reactant and the product are at equilibrium, which is probably not true in most solid-state reactions. Since diffusion, rotation, and conformational motions are highly re-

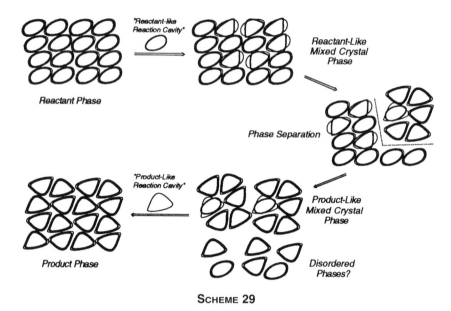

SCHEME 29

stricted in the solid state, it is expected that relaxation rates will be very slow. It is likely that many solid-state reactions will involve metastable phases and the phase separation and recrystallization will be highly dependent on whether the reaction is carried out in large crystalline specimens or in fine powdered samples. Since phase separation and recrystallization require a great deal of molecular motion, solubility limits in a solid-state reaction may exceed those observed by equilibrium co-crystallization of the components. It is likely that efficient solid-to-solid reactions will require very high solubility of the product in the crystal phase of the starting material. Although it is expected that disordered phases should result in loss of reaction control, it is possible that productlike phases may have a similar influence on the reaction as the phase of the reactant.

Although there are very few documented examples of solid-state reactions that proceed with high control and that occur by phase-separation mechanisms, it is likely that this will prove to be the most common type of solid-to-solid reaction. A few examples from the literature are described below.

1. Decarbonylation of cis-2,6-Diphenyl-2,6-dihydroxycyclohexanone

Using an excitation wavelength of $\lambda = 350$ nm Choi et al. recently observed that the solid-state reaction of the cis-ketone **7** may be carried out to completion to give the cis-photoproduct cis-**8** with no loss of stereoselectivity in a remarkably clean solid-to-solid reaction. A summary of the reaction in Scheme 30 gives the melting points of the pure reactant (146°C) and product (104°C) phases and illustrates some mechanistic details.

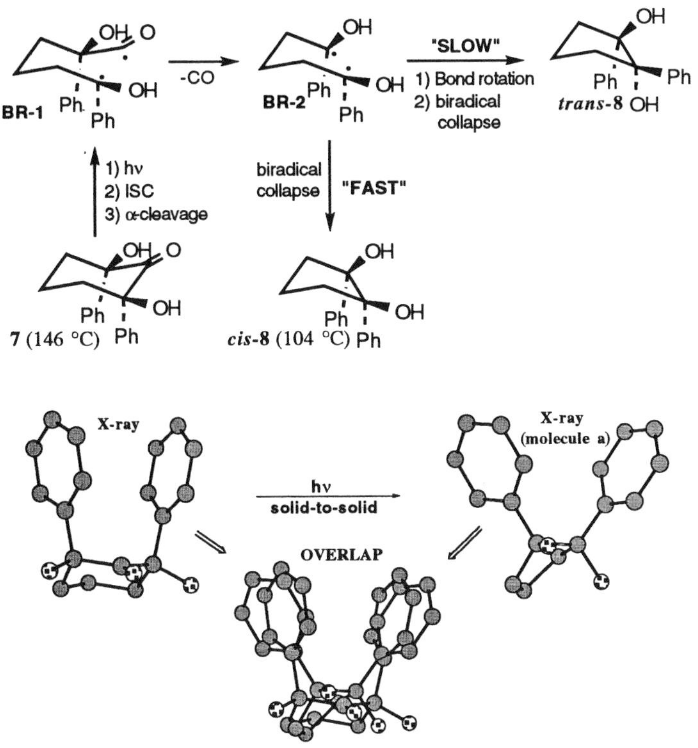

SCHEME 30

As depicted in Scheme 30, the reactant crystallizes with the phenyl groups in axial positions. Equatorial hydroxyl groups are involved in hydrogen bonding with centrosymmetric and translationally related molecules to form an extended dimeric ribbon. The reaction is proposed to proceed from the triplet excited state of the ketone to give acylalkyl biradical BR-1 which decarbonylates to give dialkyl biradical BR-2. In solution, BR-2 adopts extended conformations and undergoes bond rotation so that both *cis* and *trans* isomers are formed in a 1:1 ratio. A study of the reaction in solvents of increasing viscosity and in glassy sucrose matrices revealed that rigidity alone is not enough to achieve the selectivity observed in the crystalline solid state. In the crystal, excellent structural overlap between the reactant and the product is consistent with a topochemical reaction and suggests a good solid-state solubility of each molecule within the crystal lattice of the other. The x-ray structure of compound **7** gives r.m.s. deviations of 0.874 Å and 0.947 Å with each of the two molecules per asymmetric unit present in crystals of *cis*-**8** (see also Fig. 1).

^{13}C CPMAS NMR and FT-IR spectral analysis of the pure reactant, the recrystallized product and several partially irradiated samples demonstrates that the reaction proceeds through crystalline phases. Spectral studies also implicate a

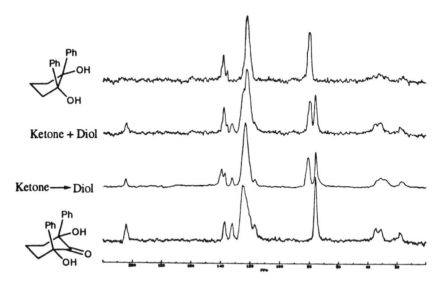

FIGURE 4 Solid-state ^{13}C CPMAS NMR (TOSS) spectra of (from top to bottom) pure photoproduct *cis*-**8**, weighted addition of the spectra of *cis*-**8** and starting material **7**, a partially reacted sample containing **7** and *cis*-**8**, and a sample of pure **7**.

phase-separation mechanism. No phases distinct from those of the pure reactant and pure product were observed, and the spectra of partially photolyzed samples could be efficiently simulated by the weighted sum of reactant and product spectra (Fig. 4). This suggests that the crystal structure of the as-reacted product is indistinguishable from that obtained upon crystallization from solvents, and it clearly demonstrates the involvement of a phase-separation mechanism. Thermal analysis by DSC showed that compounds **7** and *cis*-**8** have sharp endotherms at their respective melting points (146°C and 104°C, respectively) while partially reacted samples reveal a eutectic transition characteristic of a two-phase system from the lowest detectable conversion values.

While product analysis shows reaction control that is characteristic of a highly organized environment, spectral, thermal, and diffraction data reveal two different phases. The x-ray structures of the reactant and the product have two entirely different packing structures, and no other polymorphs were obtained under various crystallization conditions. In addition, single crystals became opaque upon irradiation and x-ray analysis showed the loss of long-range order. The rate of formation of the *cis* isomer, however, remains much higher than the rate of formation of the *trans* isomer at all conversion values, and kinetic analysis gives a remarkably clean (apparent) first-order reaction. We speculate that, while it is clear that the reaction is highly heterogeneous from a macroscopic point of view, it may occur microscopically in a more homogeneous manner. Although the solubility of the product in the reactant phase may be high, because of the intrinsic heterogeneity of photochemical illumination a reaction front may develop due to preferential reaction near the surface. That phase separation occurs within these microscopic mixed fronts with very high conversion values is suggested by sequential dissolution and analysis of crystal layers which clearly show a concentration gradient. As indicated in Scheme 31, most of the sample at any given time

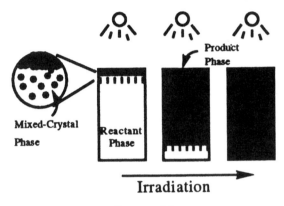

SCHEME 31

remains in the phase of the pure reactant or in the phase of the "as-formed" product so that only these two phases are observable with standard analytical techniques. The reaction, however, may occur primarily in a small volume of mixed crystalline character that exists at the reaction front. Reactions like this may be described as occurring via a phase-separation mechanism that involves steady-state mixed crystalline phases (Scheme 31).

2. Photorearrangement of tert-Butyl Succinimide

The photochemical reaction of crystalline *tert*-butyl succinimide recently studied by Fu et al. constitutes another interesting and well-documented example of an efficient solid-state reaction that proceeds through mixed crystalline phases and which involves a phase-separation mechanism [133].

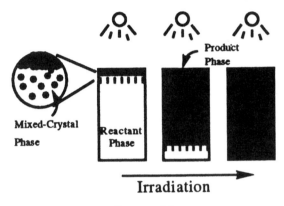

SCHEME 32

N-tert-Butyl succinimide (**49**, Scheme 32) crystallizes in the monoclinic space group P2$_1$/n with two molecules per asymmetric unit. The final product, tetrahydro-1-H-azepine-2,5-dione (**50**), crystallizes in the space group P2$_1$/c with only one molecule per asymmetric unit. Remarkably, the reactant melts at 44°C and the product at 175–176°C. With excitation at 253.7 nm, **49** reacts by intramolecular hydrogen abstraction and cyclization of the corresponding 1,4-biradical to yield an unstable aminoalcohol, which rapidly tautomerizes to the tetrahydro-1-H-azepine-2,5-dione (**50**). While irradiation of single crystals levels off at ca. 60% conversion, finely powdered samples lead to 100% conversion with chemical yields of 80%. Analysis of the reaction by solid-state FT-IR showed no accumulation of the aminoalcohol intermediate. X-ray powder diffraction at low-conversion values revealed the presence of a dilute solid solution of the product in the crystal phase of the reactant, which was followed by the formation of a new crystal phase between 6 and 14% conversion. Although it was suggested that these values may represent the solubility limit of the product in the reactant, it was not conclusively shown that they represent the composition of the crystal. It was also observed that the powder pattern obtained at high conversion is slightly different from that of the recrystallized phase, indicating that the terminal solid solution of the reactant in the crystal phase of the product is very similar, but not identical, to that of the recrystallized product. It was demonstrated that reaction occurs from a single crystal to a polycrystalline phase through two distinguishable solid solutions.

3. Incorporation of Solvents During Reaction

Several solid-state photochemical reactions have been investigated with polycrystalline samples suspended in solvents. Solvents such as water, where the reactant and the product are likely to be insoluble, are usually chosen and a surfactant is added to maintain the suspension. There are at least two apparent advantages to this method. First of all, photochemical equipment commonly used for fluid samples can be readily adopted to solid-state reactions. Secondly, it is expected that all microcrystals in a powdered sample will be homogeneously exposed to the incident light in a well-stirred reactor. Interestingly, while several examples of solid-to-solid reactions in suspended crystals have been documented, there are some cases where the solvent is incorporated into the phase of the final product. In a report by Nakanishi et al. [134] it was shown that *p*-formyl cinnamic acid (**51**, Scheme 33) forms mirror-symmetric dimers. While irradiation of crystals suspended in hexane gave amorphous cyclobutanes in 85% yield, suspension of the crystals in water gave a 100% yield of a crystalline photodimer with one water molecule of crystallization.

In a later example [135], it was shown that photodimerization of 3,4-dichlorocinnamic acid (**52**) proceeds in high yields in various hydrocarbons to give similar mirror-symmetric products with solvent of crystallization. The reaction does not occur in the fluid phase and dimerization of the suspended monomer

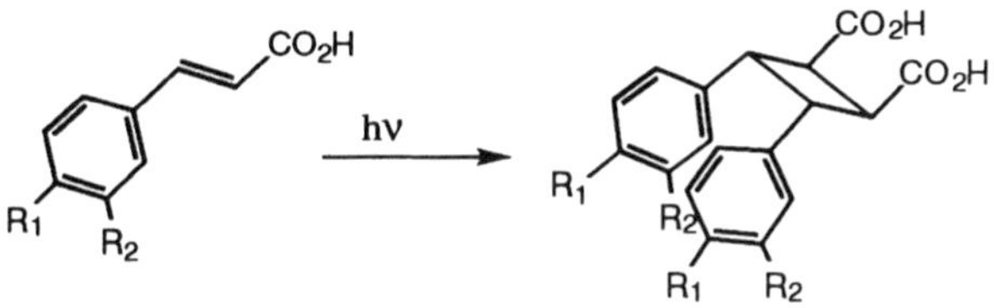

a) $R_1 = CHO$, R_2=H (**51**)
a) $R_1 = R_2$= Cl (**52**)

SCHEME 33

is more efficient than in the dry solid. As in the case of *p*-formyl cinnamic acid (**51**), incorporation of solvent was shown to affect the crystallinity of the dimer. While the dry dimer was amorphous the solvated dimer was crystalline. The solvent incorporation was temperature dependent, and could be prevented at $-10°C$. While the detailed mechanism for these reactions is yet to be elucidated, the critical role of water in maintaining crystallinity suggests the operation of a reconstructive phase transition.

B. Factors That Limit Conversion

1. Excitation Wavelength

There are numerous literature examples of photochemical solid-state reactions that proceed to limited conversion values. This is also a common problem of photochemical reactions in fluid media. Explanations are usually trivial and the problem sometimes, but not always, avoidable. The most common reason for a solid-state photochemical reaction to halt occurs when the product competes for incident light and acts as a filter. This limitation can be avoided in systems where the reactant and the product have nonoverlapping electronic spectra (Scheme 9a), but it is unavoidable when the product absorbs more strongly than the starting material at all wavelengths.

Solid-state reactions that conform to Scheme 9a include those with photo-products that have excitation energies much higher than those of the reactant. Reactions with photoproducts that absorb at longer wavelengths and with high extinction coefficients, as in Scheme 9b, are likely to be inefficient in the solid state. For instance, the photochemical dehydration of hydroxytriaryl methanols such as **53** (Scheme 34) to yield diphenylmethylene-1,4-benzoquinones (**54**), studied by Lewis et al. [136], is a well-documented example of the latter. It was shown that a thermal reaction proceeds to completion through a solid solution in well-demarcated zones of prismatic monoclinic crystals. In contrast, the photo-chemical reaction stops at less than 1% conversion with most of the product located on a thin film near the surface. The efficiency of the solid photochemical reaction in this case is very low because the product absorbs light at the same

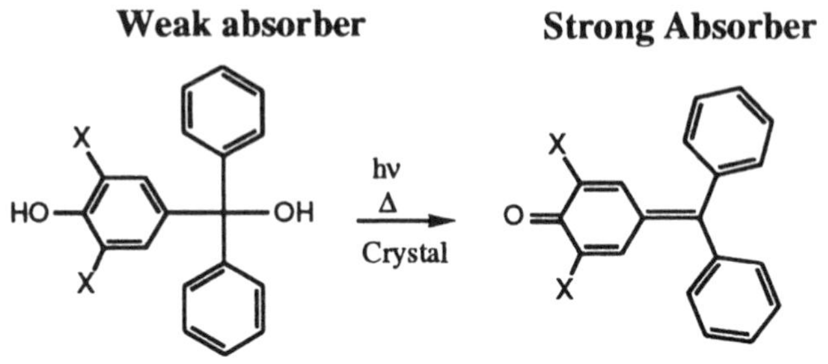

Weak absorber   **Strong Absorber**

SCHEME 34

wavelengths as the starting material, forming an effective filter that prevents illumination inside the crystal. Nonetheless, it was shown in this case that the thermal reaction may be nucleated by photochemical initiation.

Situations where the absorption spectra of reactants and products are not conducive to high yield solid-state reactions may occur in classic photochromic systems. In a recent study on the crystalline photochromism of 1,2-bis(2,4-dimethyl-3-thienyl)perfluorocyclopentane (**55**, Scheme 35) by Irie et al., it was

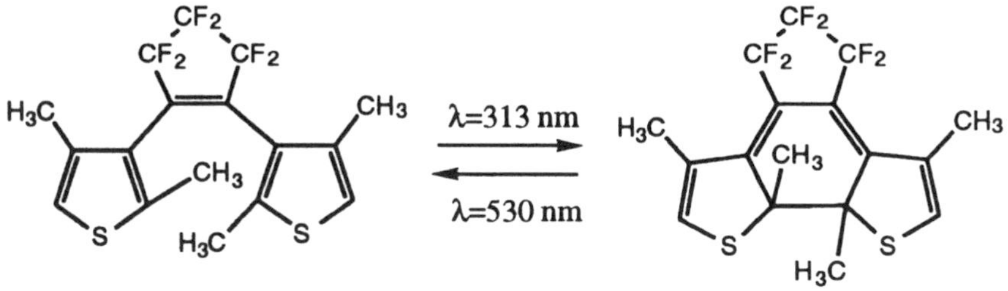

SCHEME 35

elegantly shown that reaction occurs under control of the crystal lattice [137]. The colorless open form of **55** was irradiated at 313 nm to give the closed form **56** which displays an intense red color. Although it was shown that the red form is thermally stable, analysis of irradiated crystals showed limiting conversions of only 7.8%. The low conversion was explained in terms of the overlapping absorption of the two isomers. In fact, the closed form has a larger extinction coefficient at the excitation wavelength of 313 nm and gives rise to a thin filtering layer. In contrast, the reverse reaction, from the closed to the open form, proceeds to completion by excitation at 530 nm.

Another example that stresses the importance of excitation wavelength on a solid-state reaction can be illustrated with recent work from our group. This involves the photochemical denitrogenation of diazo compound **43** to yield stilbene derivatives **45** and **46** [16]. As discussed before, selective excitation of the n,π^* diazo band in the visible ($\lambda > 380$ nm) yields stilbene **45** in almost 100% selectivity and with almost 100% conversion. Shorter wavelength radiation ($\lambda > 350$ nm) is strongly absorbed by the product, leading to photoisomerization of the stilbene, and a photostationary state favoring **46**. Maintaining crystal control over the reaction, therefore, depends critically upon selection of an appropriate wavelength.

2. Optical Degradation

The efficiency of solid-state reactions may also be decreased by degradation of the optical properties of the medium caused by accumulation of the product. This limitation is likely to be of concern in large crystalline samples where light must penetrate through reacted regions. From a preparative point of view this potential problem may be circumvented by using polycrystalline samples.

3. Product Acting as a Trap

Photochemical reactions in the solid state may also go to limited conversion when the product does not compete for incident light but may act as an efficient energy quencher. This is likely to be of importance in triplet state reactions when the product may act as an energy trap. Although this is a factor that may also limit the conversion of photochemical reactions in solution, because of efficient energy delocalization by dipole-dipole and exchange mechanisms it is likely to be more important in the crystalline solid state.

4. Accumulation of Internal Strain

Accumulation of the product may increase the rigidity of the medium, diminish the rate of the reaction, and ultimately limit the chemical yield of a solid-state reaction. In a recent study on the photocyclization of three chloro-substituted α-oxoamides (**57–59**, Scheme 36) by Hashizume et al. it was speculated that limited chemical yields may be attributed to increased internal strain caused by the

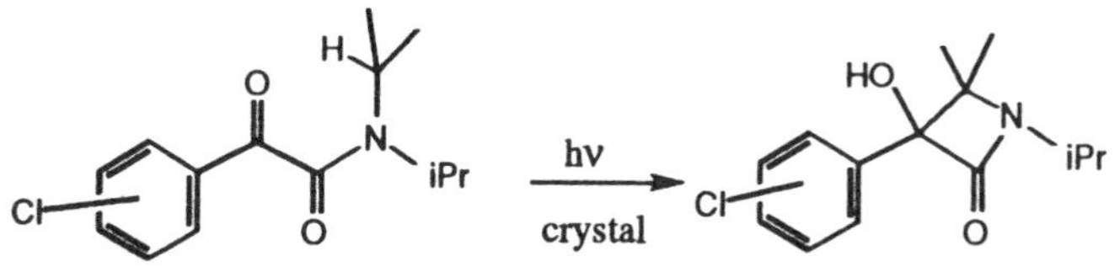

	Chemical Yield	Optical Yield (% ee)
a) o-Cl (**57**)	42	—
b) m-Cl (**58**)	75	100
c) p-Cl (**59**)	50	—

SCHEME 36

torsion of the bulky phenyl group required for a 1,4-biradical intermediate to make the final product [138]. The highest chemical yield was obtained with a *meta*-chloro-substituted oxoamide. This compound crystallizes in the chiral space group $P2_12_12_1$ and reacts 100% enantiospecifically to give optically active products with no loss of enantioselectivity as a function of conversion. Similar observations have been made for other oxoamides which undergo remarkably similar solid-to-solid reactions with good chemical yields and little or no loss of steric control [139].

5. Statistical Isolation

Irradiation 1-aza-diene **60** (Scheme 37) at ambient temperature with a Hanovia medium-pressure mercury lamp led to an interesting solid-state reaction [140]. The products obtained after 9 h of irradiation included a topochemical dimer **61** formed in 65% and a 1:1 mixture of epimeric bicyclo[3.1.0]hexenes **62** and **63**

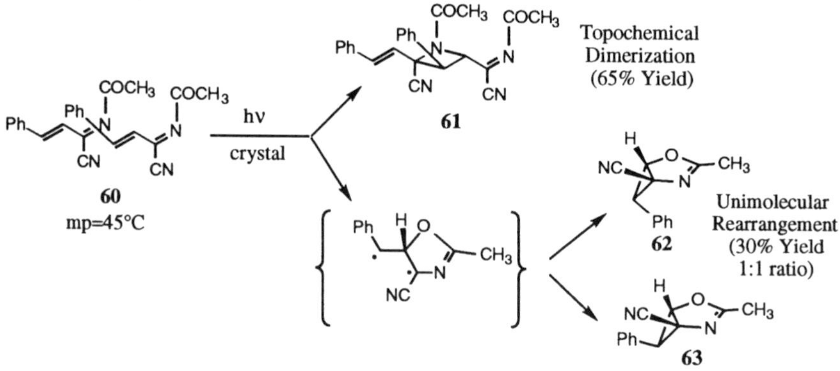

SCHEME 37

formed in 30%. An x-ray analysis of **60** showed that both types of reactions are topochemically allowed. Dimerization occurs between the imine and an alkene double bond of a neighboring aza-diene to yield an azetidine, while the unimolecular reaction is proposed to occur via a resonance stabilized benzyl-α-cyano 1,3-biradical formed by addition of the carbonyl oxygen to one of the alkene carbons. Interestingly, it was shown that the product ratio varies strongly as a function of conversion. While dimerization is favored at low-conversion values, rearrangement becomes increasingly important as the reaction proceeds. To explain the variation in the product ratio, it was suggested that changes in reaction selectivity as a function of conversion may be explained in terms of a statistical model. In this model, the rate of dimerization is highest at low conversions, and the azetidine dimer **61** is the dominant product. However, as the reaction proceeds, the local lattice my be altered so that potential reaction partners may no longer exist or may not have the required orientation. The authors explored various models of lattice disruption. A relatively good agreement with experiment was found by (1) simulating photochemical events with randomly selected molecules, (2) assuming that reaction occurred preferentially along a given direction of the stack, (3) choosing a molecule and, if an appropriate reaction partner was available assuming dimerization and if not, assuming rearrangement. The simulation was continued until all molecules were chosen in this way.

Although this statistical model can qualitatively explain the observed results, it is by no means the only viable explanation. With crystals that melt at 45°C and a reaction carried out at room temperature (20°C), there should be some concern that the sample goes through liquid phases below the eutectic of the four-component system.

V. CONCLUDING REMARKS

A great deal of interesting work has been carried out in the field of solid-state chemistry in the last few years. The potential of this area both for practical applications and for basic scientific purposes is slowly receiving broader recognition. Organic chemistry inside crystals provides promise for materials applications, for carrying out selective synthetic transformations and for use as a tool in studying fundamental aspects of organic reactivity. It is clear, however, that this field is still young, and that much remains to be done. In particular, the study of how microscopic reactivity is related to macroscopic phase changes is important, but has been addressed only sporadically in the past. In this review we have pointed out that there are only a few examples where detailed studies of the effects of conversion on crystal quality have been carried out. The influence of conversion on reaction rates and selectivities has not traditionally been well documented. Phase-separation mechanisms, in particular, have in the past been largely avoided or ignored. Understanding these effects is an important goal if the full potential of solid-state chemistry is to be realized. In closing, here we summarize a few observations based on the examples discussed in this paper which may prove to be generalizable upon further study of a larger data set.

In order to preserve reaction selectivity to high conversions it appears desirable (although not *always* necessary) to avoid phase-separation mechanisms. Phase separation is more likely to occur if a reaction involves considerable molecular motion, and so topochemical least-atomic-motion mechanisms are the most promising candidates for homogeneous transformations. Several experimental variables have also been discussed which can further influence how a solid state reaction proceeds. The irradiation wavelength is crucially important since an inappropriate choice can lead to high local concentrations of product, favoring phase separation. Such accumulation may further give rise to filtering effects which limit conversion. Reactions should in general be carried out slowly to avoid cracking and at temperatures considerably below the melting points of the components involved, in order to avoid liquid phases. Small particles may have a greater chance of receiving uniform irradiation, and where surface effects are not important, crystalline powdered samples may prove preparatively useful. A number of other techniques to favor homogeneous reactions no doubt remain to be discovered. For example, the observation that solvent molecules or counterions can play an important role in the phase behavior of some transformations suggests that such "additives" might be systematically investigated as a means of holding a crystal together during a reaction [141].

In conclusion, we maintain that it is important to the development of better models in this field that experimentalists carefully characterize both the molecular *and* the phase changes that occur during solid-state reactions. Reports on product selectivity alone are useful but incomplete. It is desirable to document solid-state reactivity as a function of conversion, as a function of temperature, and as a function of irradiation wavelength. In our own laboratory, we are systematically approaching problems with these issues in mind using FT-IR, CP-MAS NMR, and DSC. Our results, many of which have been outlined here, are interesting and indicate that there is a great deal that can be learned using such standard techniques. We hope that this chapter, while necessarily incomplete, will motivate some discussion and analysis of these issues in the growing solid-state photochemistry community.

ACKNOWLEDGMENTS

We gratefully acknowledge the National Science Foundation, the Petroleum Research Fund, and the University of California for their generous support of this research. The National Science Foundation is also acknowledged for a graduate research fellowship to A.E.K.

REFERENCES AND NOTES

1. Roth, H. D. *Angew. Chem. Int. Ed. Engl.* 1989, *28*, 1193–1207.
2. Ramamurthy, C.; Vankatesan, K. *Chem. Rev.* 1987, *87*, 433–481.
3. Lamartine, R. *Bull. Soc. Chim. Fr.* 1989, 237–246.
4. Scheffer, J. R.; Garcia-Garibay, M.; Nalamasu, O. *Org. Photochem.* 1989, *8*, 249–347.
5. Venkatesan, K.; Ramamurthy, V. In *Photochemistry in Organized and Constrained Media*; V. Ramamurthy, ed.; VCH: New York, 1991, pp. 133–184.
6. Scheffer, J. R.; Pokkuluri, P. R. In *Photochemistry in Organized and Constrained Media*; V. Ramamurthy, ed.; VCH: New York, 1991, pp. 185–246.
7. Hasegawa, M. *Adv. Phys. Org. Chem.* 1995, *30*, 117–171.
8. Gougoutas, J. Z. *Pure and Appl. Chem.* 1971, *27*, 305–326.
9. Gavezzotti, A.; Simonetta, M. *Chem. Rev.* 1982, *82*, 5220.
10. Desiraju, G. R. *Organic Solid State Chemistry*; Elsevier: Amsterdam, 1987, p. 550.

11. Desiraju, G. R. *Endeavour* 1984, *8*, 201.
12. Bonner, W. C. *Top. Stereochem.* 1988, *18*, 1–98.
13. Scheffer, J. R.; Garcia-Garibay, M. A. In *Photochemistry of Solid Surfaces*; T. Matsuura and M. Anpo, eds., Elsevier: Amsterdam, 1989.
14. Sakamoto, M. *Chem. Eur. J.* 1997, *3*, 684–689.
15. Bloor, D.; Chance, R. R. *Polydiacetylenes*; M. Nijhoff: Dordrecht, 1985; vol. 102.
16. Shin, S.; Cizmeciyan, D.; Keating, A. E.; Khan, S.; Garcia-Garibay, M. A. *J. Am. Chem. Soc.* 1997, *119*, 1859–1868.
17. Shin, H. S.; Keating, A. E.; Garcia-Garibay, M. A. *J. Am. Chem. Soc.* 1996, *118*, 7626–7627.
18. Mahé, L.; Izouoka, A.; Sugawara, T. *J. Am. Chem. Soc.* 1992, *114*, 7904–7906.
19. Choi, T.; Peterfy, K.; Khan, S. I.; Garcia-Garibay, M. *J. Am. Chem. Soc.* 1996, *118*, 12477–12478.
20. Tomioka, H.; Watanabe, T.; Hirai, K.; Furukawa, K.; Takui, T.; Itoh, K. *J. Am. Chem. Soc.* 1995, *117*, 6376–6377.
21. Kohlshutter, H. W. *Anorg. Allg. Chem.* 1918, *105*, 121.
22. Schmidt, G. M. J. *Solid State Photochemistry*; Verlag Chemie: New York, 1976.
23. Liebermann, C.; Ilinski, M. *Ber. Dtsch. Chem. Ges.* 1885, *18*, 3193–3201.
24. Liebermann, C. *Ber. Dtsch. Chem. Ges.* 1877, *10*, 2177–2179.
25. Green, B. S.; Lahav, M.; Schmidt, G. J. M. *J. Chem. Soc. B.* 1971, 1552–1564.
26. Lahav, M.; Schmidt, G. J. M. *J. Chem. Soc. B.* 1967, 239–243.
27. Gnanaguru, K.; Ramasubbu, N.; Venkatesan, K.; Ramamurthy, V. *J. Org. Chem.* 1985, *50*, 2337–2346.
28. Theocharis, C. R.; Jones, W. In *Organic Solid State Chemistry*; G. Desiraju, ed.; Elsevier: Amsterdam, 1987, pp. 47–67.
29. Ariel, S.; Askari, S.; Scheffer, J. R.; Trotter, J.; Walsh, L. *J. Am. Chem. Soc.* 1984, *106*, 5726–5728.
30. Gnanaguru, K.; Ramasubbu, N.; Venkatesan, K.; Ramamurthy, V. *J. Photochem.* 1984, *27*, 355–362.
31. Stevens, B.; Dickinson, T.; Sharpe, R. R. *Nature* 1964, *204*, 876–877.
32. Cohen, M. D.; Ludmer, Z. *J. Chem. Soc., Chem. Commun.* 1969, 1172–1173.
33. Murthy, G. S.; Arjuna, P.; Venkatesan, K.; Ramamurthy, V. *Tetrahedron* 1987, *43*, 1225–1240.
34. Heller, E.; Schmidt, G. M. J. *Isr. J. Chem.* 1971, *9*, 449–462.
35. Ludmer, Z. *Chem. Phys.* 1977, *26*, 113–121.
36. Ebeid, E.-Z., M.; Bridge, J. *J. Chem. Soc., Faraday Trans. I* 1984, *80*, 1131–1138.
37. Suzuki, T.; Fukushima, T.; Yamashita, Y.; Miyashi, T. *J. Am. Chem. Soc.* 1994, *116*, 2793–2803.
38. Koshima, H.; Matsuura, T. *J. Heterocyclic. Chem.* 1994, *31*, 121–142.
39. Koshima, H.; Ding, K.; Chisaka, Y.; Matsuura, T. *J. Am. Chem. Soc.* 1996, *118*, 12059–12065.
40. Koshima, H.; Maeda, A.; Masuuda, N.; Matsuura, T.; Hirotsu, K.; Kada, K.; Mizutani, H.; Ito, Y.; Fu, T. Y.; Scheffer, J. R.; Trotter, J. *Tetrahedron: Asymmetry* 1994, *5*, 1415–1418.
41. Meng, J.; Wang, W.; Wang, H.; Matsuura, T.; Koshima, H.; Sugimoto, I.; Ito, Y. *Photochem. Photobiol.* 1993, *57*, 597–602.
42. Scheffer, J. R. In *Geometric Requirements for Intramolecular Photochemical Hy-*

drogen Atom Abstraction: Studies Based on a Combination of Solid State Chemistry and X-Ray Crystallography; G. R. Desiraju, ed., VCH: Amsterdam, 1987, pp. 1–45.

43. Cohen, M. D. *Angew. Chem. Int. Ed. Engl.* 1975, *14*, 386–393.

44. Ariel, S.; Askari, S.; Scheffer, J. R.; Trotter, J.; Walsh, L. In *Steric Compression Control. A Quantitative Approach to Reaction Selectivity in Solid State Chemistry*; chap. 15 ed.; Fox, M. A., ed., American Chemical Society: Washington, D.C., 1985; Vol. ACS Symp. Ser. 278, pp. 243–256.

45. Ariel, S.; Askari, S. H.; Scheffer, J. R.; Trotter, J.; Wireko, F. *Acta Crystallogr.* 1987, *B43*, 532–537.

46. Scheffer, J. R.; Trotter, J.; Garcia-Garibay, M.; Wireko, F. *Mol. Cryst. Liq. Crys. Inc. Nonlin. Opt.* 1988, *156*, 63.

47. Keating, A. E.; Shin, S. H.; Houk, K. N.; Garcia-Garibay, M. A. *J. Am. Chem. Soc.* 1996, *119*, 1474–1475.

48. Zimmerman, H. E.; Zhu, Z. *J. Am. Chem. Soc.* 1995, *117*, 5245–5262.

49. Zimmerman, H. E.; Zhu, Z. *J. Am. Chem. Soc.* 1994, *116*, 9757–9758.

50. Bürgi, H. B.; Dunitz, J. D. *Acc. Chem. Res.* 1983, *16*, 153–161.

51. Zimmerman, H. E.; Zuraw, M. J. *J. Am. Chem. Soc.* 1989, *111*, 2358–2361.

52. Ito, Y.; Matsuura, T. *J. Photochem. Photobiol.* 1989, *50*, 141–145.

53. Walter, D. W.; McBride, J. M. *J. Am. Chem. Soc.* 1981, *103*, 7074–7084.

54. McBride, J. M. *Acc. Chem. Res.* 1983, *16*, 304–312.

55. Hollingsworth, M. D.; McBride, J. M. *Adv. Photochem.* 1990, *15*, 279–379.

56. As indicated by West in Ref. X, chaps. 9 and 11, a phase can be defined as "a physically distinct and mechanically separable (in principle) portion of a system, each phase being itself homogeneous." It is also pointed out by West that this definition is sometimes difficult to apply in nonstoichiometric solids and mixed crystalline systems.

57. Hollingsworth, M. D.; McBride, J. M. *J. Am. Chem. Soc.* 1985, *107*, 1792–1793.

58. Hollingsworth, M. D.; McBride, J. M. *Chem. Phys. Lett.* 1986, *130*, 259–264.

59. Dunitz, J. *Pure and Appl. Chem.* 1991, *63*, 177–185.

60. Dunitz, J. *Acta Crystallogr.* 1995, *B51*, 619–631.

61. Garcia-Garibay, M. A.; Constable, A. E.; Jernelius, J.; Choi, T.; Cizmeciyan, D.; Shin, S. H. In *Physical Supramolecular Chemistry*; L. Echegoyen and A. Kaifer, eds., Kluwer Academic Publishers: Dordrecht, 1996, pp. 289–312.

62. Pokkuluri, P. R.; Scheffer, J. R.; Trotter, J. *Tetrahedron Lett.* 1989, *30*, 1601–1604.

63. Schmidt, G. M. J. *J. Chem. Soc.* 1964, 2014–2021.

64. McBride, J. M.; Bertman, S. B. *Angew. Chem. Int. Ed. Engl.* 1989, *28*, 330.

65. Vaida, M.; Shimon, L. J. W.; Weisinger-Lewin, Y.; Frolow, F.; Lahav, M.; Leiseirowitz, L.; McMullan, R. K. *Science* 1988, *241*, 1474–1479.

66. Kitaigorodskii, A. I. *Mixed Crystals*, Springer-Verlag: Berlin, 1984.

67. Wegner, G. *Pure Appl. Chem.* 1977, *49*, 443–454.

68. Enkelmann, V.; Wegner, G.; Novak, K.; Wagener, K. B. *J. Am. Chem. Soc.* 1994, *115*, 10390–10391.

69. Braun, H.-G.; Wegner, G. *Makromol. Chem.* 1983, *184*, 1103–1119.

70. Enkelmann, V.; Wegner, G.; Novak, K.; Wagener, K. B. *J. Am. Chem. Soc.* 1993, *115*, 10390–10391.

71. Leibovitch, M.; Olovsson, G.; Scheffer, J. R.; Trotter, J. *J. Am. Chem. Soc.* 1997, *119*, 1462–1463.

72. Comer, F. W.; Trotter, J. *J. Chem. Soc. B.* 1966, 13.
73. Prasad, P. N.; Swiatkiewicz, J.; Eisenhardt, G. *Applied Spectroscopy Reviews* 1982, *18*, 59–103.
74. Scheffer, J. R.; Wong, Y.-F.; Patil, O. A.; Curtin, D. Y.; Paul, I. C. *J. Am. Chem. Soc.* 1985, *107*, 4898–4904.
75. Stitchell, S. G.; Harris, K. D. M.; Aliev, A. E. *Struct. Chem.* 1994, *5*, 327–333.
76. Pines, A.; Gibby, M. G.; Waugh, J. S. *J. Chem. Phys.* 1973, *59*, 569–590.
77. Fyfe, C. A. *Solid State NMR for Chemists*; C.F.C. Press: Guelph, Ontario, 1983.
78. Torchia, D. A.; Szabo, A. *J. Mag. Res.* 1982, *49*, 107–121.
79. Blumich, B.; Spiess, H. W. *Angew Chem. Int. Ed. Engl.* 1988, *27*, 1655–1672.
80. Schmidt, C.; Blumich, B.; Spiess, A. H. *J. Magn. Res.* 1988, *79*, 269–290.
81. Butler, L. G.; Cory, D. G.; Dooley, K. M.; Miller, J. B.; Garroway, A. N. *J. Am. Chem. Soc.* 1992, *114*, 125–135.
82. Jones, W.; Thomas, J. M. *Prog. Solid St. Chem.* 1979, *12*, 101–124.
83. Jones, W.; Thomas, J. M.; Williams, J. O.; Hobbs, L. W. *J. Chem. Soc. Faraday Trans. II* 1975, *71*, 138–145.
84. Jones, W.; Cohen, M. D. *Mol. Cryst. Liq. Cryst.* 1977, *41*, 103.
85. Briehl, H.; Butenuth, J. *Thermochimica Acta* 1990, *167*, 249–292.
86. Cammenga, H. K.; Epple, M. *Angew. Chem. Int. Ed. Engl.* 1995, *34*, 1171–1187.
87. Kohler, W.; Novak, K.; Enkelman, V. *J. Chem. Phys.* 1994, *101*, 10474.
88. Morawetz, H. *Pure & Appl. Chem.* 1966, 12, 201–210.
89. Kohlschütter, H. W.; Sprenger, L. *Z. Phys. Chem.* 1932, *B16*, 284.
90. Hayashi, K.; Nishii, M.; Okamura, S. *J. Polymer Sci.* 1964, *C4*, 839.
91. Lando, J. B.; Stannett, V. *J. Polymer Sci.* 1965, *A3*, 2369.
92. Baughman, R. H. *J. Polymer Sci. Polymer Phys. Ed.* 1974, *12*, 1511–1535.
93. Hasegawa, M. *Chem. Rev.* 1983, *83*, 507–518.
94. Sasada, Y.; Shimanouchi, H.; Nakanishi, H.; Hasegawa, M. *Bull. Chem. Soc. Japan* 1971, *44*, 1262.
95. Tamaki, T.; Suzuki, Y.; Hasegawa, M. *Bull. Chem. Soc. Japan* 1972, *45*, 1988.
96. Nakanishi, H.; Suzuki, Y.; Suzuki, F.; Hasegawa, M. *J. Polym. Sci. A-1* 1969, *7*, 753.
97. Nakanishi, H.; Hasegawa, M. *J. Polym. Sci. Part A-2* 1972, *10*, 1537–1553.
98. Nakanishi, H.; Hasegawa, M.; Sasada, Y. *J. Polymer Sci. Polymer Phys. Ed.* 1977, *15*, 173–191.
99. Peachey, N. M.; Eckhardt, C. J. *J. Phys. Chem.* 1993, *97*, 10849–10856.
100. Hasegawa, M.; Kinbara, K.; Adegawa, Y.; Saigo, K. *J. Am. Chem. Soc.* 1993, *115*, 3820–3821.
101. Wegner, G. *Makromol. Chem.* 1970, *134*, 219–229.
102. Wegner, G. *Z. Naturforsch.* 1969, *24b*, 824–832.
103. Wegner, G. *Makromol. Chem.* 1971, *145*, 85–94.
104. Melveger, A. J.; Baughman, R. H. *J. Polymer Sci. Polymer Phys. Ed.* 1973, *11*, 603–619.
105. Baughman, R. H. *J. Appl. Phys.* 1972, *43*, 4362–4370.
106. Richter, K. H.; Güttler, W.; Schwoerer, M. *Chem. Phys. Lett.* 1982, *92*, 4–6.
107. Cheng, K.; Foxman, B. M. *J. Am. Chem. Soc.* 1977, *99*, 8102–8103.
108. Cohen, M. D.; Schmidt, G. M. J. *J. Chem. Soc.* 1964, 1996–2000.
109. Cohen, M. D.; Schmidt, G. M. J.; Sonntag, F. I. *J. Chem. Soc.* 1964, 2000–2013.
110. Schmidt, G. J. M. *Pure Appl. Chem.* 1971, *27*, 647.

111. Osaki, K.; Schmidt, G. M. J. *Isr. J. Chem.* 1972, *10*, 189–193.
112. Jones, W.; Nakanishi, H.; Theocharis, C. R.; Thomas, J. M. *J. C. S. Chem. Comm.* 1980, 610–611.
113. Nakanishi, H.; Jones, W.; Thomas, J. M.; Hursthouse, M. B.; Motevalli, M. *J. Phys. Chem.* 1981, *85*, 3636–3642.
114. Nakanishi, H.; Jones, W.; Thomas, J. M. *Chem. Phys. Lett.* 1980, *71*, 44–48.
115. Nakanishi, H.; Jones, W.; Thomas, J. M. *J. C. S. Chem. Comm.* 1980, 611–612.
116. Wang, W.-N.; Jones, W. *Tetrahedron* 1987, *43*, 1273–1279.
117. Novak, K.; Enkelmann, V.; Wegner, G.; Wagener, K. B. *Angew. Chem. Int. Ed. Engl.* 1993, *32*, 1614–1616.
118. Lonsdale, K.; Nave, E.; Stephens, J. F. *Phil. Trans. Roy. Soc. London* 1966, *A261*, 1.
119. Leffler, J. E.; Faulkner, R. D.; Petropoulos, C. C. *J. Am. Chem. Soc.* 1958, *80*, 5435.
120. Ohashi, Y. *Acc. Chem. Res.* 1988, *21*, 268–274.
121. Yamada, T.; Uekusa, H.; Ohashi, Y. *Chem. Lett.* 1995, 187–188.
122. Ohashi, Y.; Yanagi, K.; Kurihara, T.; Sasada, Y.; Ohgo, Y. *J. Am. Chem. Soc.* 1981, *103*, 5805–5812.
123. Ohashi, Y.; Uchida, A.; Sasada, Y.; Ohgo, Y. *Acta Crystallogr.* 1983, *B39*, 54–61.
124. Ohashi, Y.; Yanagi, K.; Kurihara, T.; Sasada, Y.; Ohgo, Y. *J. Am. Chem. Soc.* 1982, *104*, 6353–6359.
125. Kurihara, T.; Ohashi, Y.; Sasada, Y. *Acta Crystallogr.* 1983, *B39*, 243–250.
126. Uchida, A.; Ohashi, Y.; Sasada, Y.; Ohgo, Y.; Baba, S. *Acta Crystallogr.* 1984, *B40*, 473–478.
127. Uchida, A.; Dunitz, J. D. *Acta Crystallogr.* 1990, *B46*, 45–54.
128. Gavezzotti, A. *J. Am. Chem. Soc.* 1983, *105*, 5220–5225.
129. Tomioka, H.; Hayashi, N.; Izawa, Y.; Senthilnathan, V. P.; Platz, M. S. *J. Am. Chem. Soc.* 1983, *105*, 5053–5057.
130. Pomerantz, M.; Witherup, T. H. *J. Am. Chem. Soc.* 1973, *95*, 5977–5988.
131. Evanseck, J. D.; Houk, K. N. *J. Phys. Chem.* 1990, *94*, 5518–5523.
132. Schaefer III, H. F. *Acc. Chem. Res.* 1979, *12*, 288–296.
133. Fu, T. Y.; Scheffer, J. R.; Trotter, J. *Can. J. Chem.* 1994, *72*, 1952–1960.
134. Nakanishi, F.; Nakanishi, H.; Tasai, T.; Suzuki, Y.; Hasegawa, M. *Chem. Lett.* 1974, 525–528.
135. Nakanishi, F.; Hirakawa, M.; Nakanishi, H. *Isr. J. Chem.* 1979, *18*, 295–297.
136. Lewis, T. W.; Curtin, D. Y.; Paul, I.C. *J. Am. Chem. Soc.* 1979, *191*, 5715–5725.
137. Irie, M.; Uchida, K.; Eriguchi, T.; Tzuzuki, H. *Chem. Lett.* 1995, 899–900.
138. Hashizume, D.; Kogo, H.; Sekine, A.; Ohashi, Y.; Miyamoto, H.; Toda, F. *J. Chem. Soc. Perkin 2* 1995, 61–66.
139. Sekine, A.; Hori, K.; Ohashi, Y.; Yagi, M.; Toda, F. *J. Am. Chem. Soc.* 1989, *111*, 697–699.
140. Teng, M.; Lauher, J. W.; Fowler, F. W. *J. Org. Chem.* 1991, *56*, 6840–6845.
141. Gamlin, J. N.; Jones, R.; Leibovitch, M.; Patrick, B.; Scheffer, J. R.; Trotter, J. *Acc. Chem. Res.* 1996, *29*, 203–209.

6

Chemical and Photophysical Processes of Transients Derived from Multiphoton Excitation: Upper Excited States and Excited Radicals

W. Grant McGimpsey
Worcester Polytechnic Institute, Worcester, Massachusetts

To review all of the previously reported upper excited state photophysics and photochemistry as well as processes reported for excited reaction intermediates would require substantially more than a single chapter in this volume. For this reason, I have limited the scope of the chapter to include only the chemistry and photophysics of upper excited states and the chemistry and photophysics of excited radicals of organic systems.

While information is given on the reactions of both upper excited singlet and triplet states, emphasis is placed on upper triplet states. This reflects the greater number of recent literature reports involving two-color studies, a method that lends itself more conveniently to transients such as triplet states that have relatively long lifetimes.

Information is also given for excited neutral radicals. Several recent reviews have described the reactions of excited radicals, radical ions, biradicals, carbenes, ylides, and other intermediates. Neutral radicals derived from organic systems represent the most widely studied group of transients. Here, the attempt has been made to draw together the results of earlier low- and room-temperature spectroscopic studies of radicals with the more recent results obtained by laser flash photolysis and laser jet techniques.

In compiling the information in this chapter, I have relied heavily on several very comprehensive reviews that have appeared over the past few years [1–7]. In particular, the 1978 review by Turro et al. [1] is extremely thorough in describing the intra- and intermolecular photophysics and chemistry of upper singlet and triplet states. In fact, rather than reproduce the same details here, I direct the reader to this review for a summary of upper state behavior reported prior to 1978. (A description of azulene and thione anomalous fluorescence is included since these systems are the best-known systems that display upper state behavior.) I also direct readers to the reviews by Johnston and Scaiano [2] and Wilson and Schnapp [3] which focus on the chemistry of both upper triplet states and excited reaction intermediates as studied by laser flash photolysis (one- and two-color methods) and laser jet techniques. Also, Johnston's thorough treatment of excited radicals and biradicals [4] and the review of thioketone photophysics and chemistry by Maciejewski and Steer [5] are excellent sources of detailed information.

Finally, I have attempted to be thorough in describing the decay processes of upper states and excited radicals, but I make no claim that this chapter constitutes a fully comprehensive review of each chemical system. Rather, I have selected examples that are representative of behavior instead of cataloging the results for every system.

I. INTRODUCTION

Kasha's rule states that in condensed phases, fluorescence and phosphorescence emissions are observed only from the vibrationally relaxed lowest electronically excited singlet and triplet states, respectively, irrespective of the state that was originally excited [1,8]. With a few notable exceptions such as the anomalous $S_2 \rightarrow S_0$ fluorescence emission of azulene and some thioketones, by and large this rule has been shown to have general application. There is nothing magic about the rule; it simply implies that the rate constants for internal conversion from upper excited states to the lowest excited state, and subsequent vibrational relaxation are generally much greater than the rate constants for the emissive processes from these upper states.

The lack of many exceptions to Kasha's rule subsequently led to the erroneous inference that no other upper excited state process, be it photophysical or photochemical, intra- or intermolecular, could efficiently compete with internal conversion. However, with the increasingly common use of pulsed lasers in photochemical studies over the past 20 years, it has been possible to show that the upper excited states of many systems have a rich and varied chemistry. Up to this time studies were usually carried out with low-intensity lamps or broadband flashlamps that failed to produce significant populations of upper excited states, with the result that upper state chemical or photophysical processes were below the threshold of detection. The high light intensity provided by pulsed lasers can result in sequential, resonant multiphoton excitation, thereby preferentially producing larger concentrations of upper states, and in the process amplifying the photophysical and chemical effects to a point where they can be detected.

Having recognized that upper states can potentially contribute in a non-negligible way to the photochemical behavior of a molecule, efforts have been made over the past 15 years to probe this behavior using a number of methods; in particular, two-color (two-laser) flash photolysis and the laser jet technique [9].

The former approach uses two lasers with different wavelengths, firing sequentially, to produce the upper state. The first laser produces the lowest excited state of the molecule and the second laser selectively excites the excited state into an upper level. The two-color approach offers a distinct advantage over one-laser multiphoton excitation since both spectral and temporal tunability of the two photons is possible. To successfully use only one laser for both excitation steps it is necessary that the lowest excited state be produced in significant concentrations *during* the laser pulse. Otherwise, there will be no excited state present to absorb the second photon. Obviously, this is not a problem with the lowest singlet state, but if the pulse is very short, large triplet concentrations may not be achieved during the pulse. It is also necessary that the excited state have a considerable extinction coefficient at the laser wavelength so that it can successfully compete with the ground-state molecules for the laser photons. On the other hand, with the two-color approach a variable delay can be introduced between the two photons to allow for the buildup of the excited state concentration or, in the event that two transients with different lifetimes are produced by the first photon, the second photon can be delayed to allow time for one of the transients to decay. In addition, if the laser providing the second photon is tunable (e.g., flashlamp-pumped dye or Nd/YAG-pumped OPO) the wavelength of the second photon can be adjusted to correspond to the absorption maximum of the excited state. It should be noted that the two-color approach is most applicable to a study of upper triplet state behavior since singlet state lifetimes are typically too short compared with the laser pulse duration to allow for efficient upper singlet production.

The laser jet approach typically involves focusing a CW argon ion laser at a high velocity jet of a solution containing the compound of interest. A light amplication effect is produced by this technique which can result in very high concentrations of multiphoton transients.

It should be noted that these two techniques compliment each other. The two-color approach provides spectroscopic and kinetic information about the system but end-product analysis is difficult because of the small amounts of products generated by this method. On the other hand, while the laser jet technique does not provide direct spectroscopic or kinetic information, products can be produced in significantly larger concentrations. It should also be pointed out, however, that while a few recent published studies have combined these two techniques, this powerful combination has yet to be exploited fully.

The development of the two-color and laser jet approaches has also allowed the study of the photochemical behavior of excited states of reaction intermediates, i.e., transient species that are chemically distinct from the original ground or excited state, such as neutral and ion radicals, biradicals, carbenes, and ylides. In fact, the study of excited reaction intermediates has been more comprehensive than the study of upper states. Originally, the short-lived nature of the ground-state transient itself led to the incorrect assumption that the excited transient would be too short-lived to participate in any chemical or photophysical processes other than deactivation to the ground state. However, this is now known not to be the case and some surprising differences between the ground- and excited-state behavior of reaction intermediates have been observed.

II. UPPER EXCITED STATES: PHOTOPHYSICS

A. $S_2 \rightarrow S_0$ Emission

The best-known exception to Kasha's rule is the anomalous fluorescence displayed by azulene and its derivatives (nonalternant hydrocarbons) and some aliphatic and aromatic thioketones.

Beer and Longuet-Higgins first reported the anomalous $S_2 \rightarrow S_0$ fluorescence spectrum of azulene (1) in 1955 (λ_{max} 374 nm) [10]. The large $S_2 - S_1$ energy gap and a poor Franck Condon factor for $S_2 \rightarrow S_1$ relaxation combine to allow $S_2 \rightarrow S_0$ emission to compete with other non-radiative processes [1]. The effect of energy gap on the fluorescence quantum yield for 1 and its derivatives (2–18) has been extensively studied and the data are reproduced in Table 1 [11,12]. It is clear that substitution has the effect of decreasing the S_2 energy but does not have a corresponding effect on the S_1 energy. Thus, substitution provides an energy gap gradient which can be correlated with the fluorescence yield. In addition to simple substitution, benzannulation and phenyl substitution of azulene yields derivatives that also exhibit anomalous fluorescence (19–26), although little photophysical data is available for these compounds. (See Chart 1 [13,14].)

Like azulenes, thioketones are known to display anomalous $S_2 \rightarrow S_0$ fluorescence. The $S_2 \rightarrow S_0$ transition is electric dipole allowed and this, combined with large S_2–S_1 energy gaps in many thioketones, results in relatively large fluorescence quantum yields and nanosecond to picosecond lifetimes [5]. Table 2 gives photophysical information for the S_2 state of a variety of thioketones [15–24]. Most of this data was acquired in perfluoroalkane solvents. Maciejewski and Steer note that the S_2 relaxation dynamics of these compounds are solvent dependent

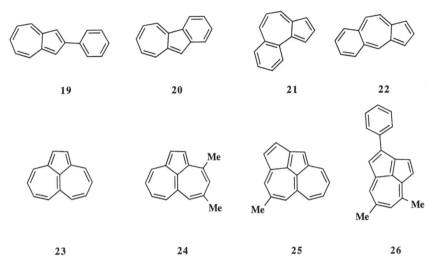

CHART 1 Phenyl Substituted and Benzannulated Azulenes

and therefore the intramolecular relaxation dynamics can be masked by strongly interacting solvents. The use of weakly interacting perfluoroalkanes reveals more intramolecular information [5]. This is demonstrated by the order of magnitude reduction in the lifetime of xanthione (33) in hexane as compared to the lifetime in a perfluoroalkane solvent [5].

When data are obtained in perfluoroalkanes it becomes clear that $S_2 \to S_1$ internal conversion is the dominant S_2 nonradiative deactivation pathway for aromatic thiones and several enethiones while $S_2 \to S_0$ fluorescence is a minor pathway. Deuterium substitution, especially β− to the thioketone, has a pronounced effect on the relaxation and has led to the conclusion that high-frequency C−H vibrations are important accepting modes in this process.

In contrast to the aromatic thiones, alicyclic thiones such as 27–29 exhibit much shorter S_2 lifetimes. Reportedly, this effect is caused by a reversible photochemical mechanism [1,5]. Photochemical decay is also an important S_2 process in 42–46 [1].

B. Triplet–Triplet Fluorescence

Triplet–triplet ($T_2 \to T_1$) fluorescence has been reported for anthracene and a variety of substituted anthracenes in solution at room temperature [25,26]. This work came about as a result of the initial observation of a long wavelength emission band superimposed on the tail of the $S_1 \to S_0$ emission of 9-bromoanthracene (54) and 9,10-dibromoanthracene (55). Table 3 gives the wavelength of the (0,0) band for the emission as well as the measured quantum yields (see p. 259).

TABLE 1 S_2–S_2 Energy Gap and S_2–S_0 Fluorescence Quantum Yields for Azulene and Substituted Azulenes

	Azulene	E(S$_1$) kcal/mol	E(S$_2$) kcal/mol	ΔE [E(S$_2$) - E(S$_1$)]	Φ$_{fl}$	Ref.
1	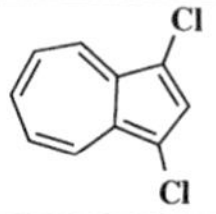	36.3	77.2	40.9	0.37	11,12
2	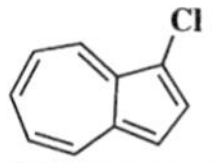	38.3	79.2	40.9	0.23	11,12
3	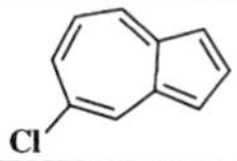	39.2	79.8	40.6	0.20	11,12
4	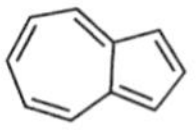	40.9	80.9	40.0	0.20	11,12
5	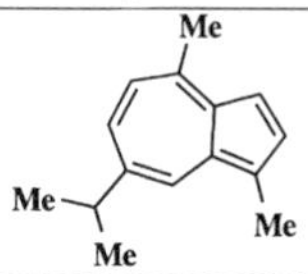	39.5	77.8	38.3	0.092	11,12
6	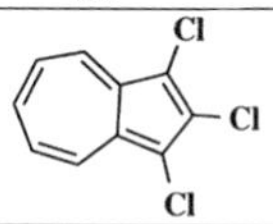	39.2	76.4	37.2	0.076	11,12
7	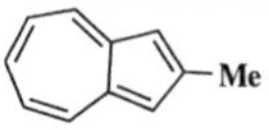	42.6	79.5	36.9	0.052	11,12
8	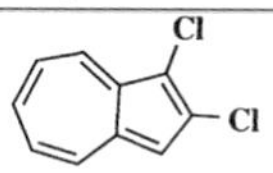	41.5	78.1	36.6	0.040	11,12
9	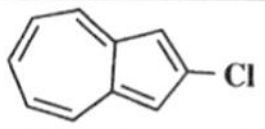	44.0	79.8	35.8	0.043	11,12
10	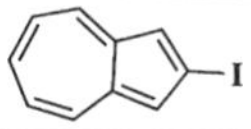	42.9	77.5	34.6	0.024	11,12
11	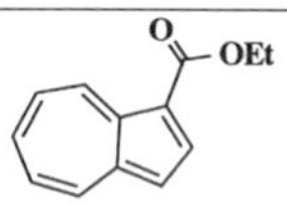	44.3	77.2	32.9	~0.001	11,12
12	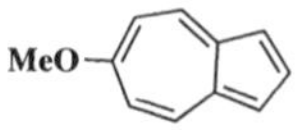	45.2	77.8	32.6	~0.001	11,12
13	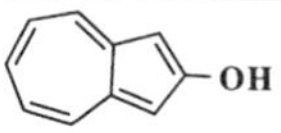	46.6	77.8	31.2	~0.0005	11,12

TABLE 1 Continued

	Azulene	$E(S_1)$ kcal/mol	$E(S_2)$ kcal/mol	ΔE $[E(S_2) - E(S_1)]$	Φ_{fl}	Ref.
14	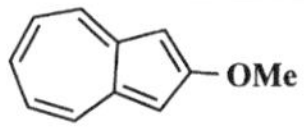 OMe	46.9	76.9	30.0	~0.0005	11,12
15	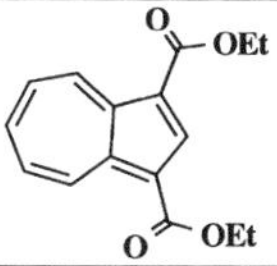	47.8	75.8	28.0	~1 x 10⁻⁵	11,12
16	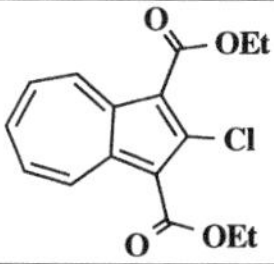	49.5	76.9	27.4	~1 x 10⁻⁵	11,12
17	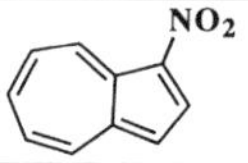	45.8	70.9	25.1		11,12
18	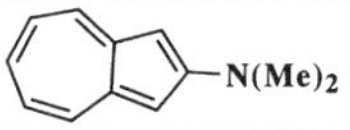 N(Me)₂	47.8	68.4	20.6	~1 x 10⁻⁵	11,12

Prior to this study, photochemical sensitization studies [27–29] and energy gap correlations [30] yielded T_2 lifetimes for meso-substituted (9,- and 9,10-) anthracenes in the 200-ps range. However, the yield values shown in Table 3 suggested considerably shorter lifetimes. More recent energy transfer studies agree with the yield data, giving lifetimes of < 20 ps (vide infra) [31].

The effect of substituents on fluorescence yield is varied and does not lend itself to ready explanation. Generally, deuterium and alkyl substitution have little effect while the presence of halogens increases the yield. However, 9,10-dichloroanthracene does not exhibit any observable emission. While this would ordinarily be attributed to the existence of other efficient decay pathways, the T_2 lifetime of 20 ps [31] is considerably longer than for other derivatives that *do* exhibit fluorescence. No explanation for these contradictory observations has been given.

C. T_n–S_1 Reverse Intersystem Crossing (RISC)

The two-color technique has been used to obtain direct confirmation of $T_n \rightarrow S_1$ reverse intersystem crossing (RISC) for a variety of systems including **54** and **55** [32–37], a variety of cyanine dyes (**56–58**) [38], substituted isoalloxazines (**59–62**) [39], and the phototautomers of several systems produced by photoinduced proton transfer (**63–67**) [40–46]. In these systems, irradiation into the T–T

TABLE 2 S_2-s_0 Fluorescence Emission Parameters and Nonradiative Decay Rate Constants for Thioketones

	Thioketone	$\lambda_{0,0}$	τ_{fl} ps	Φ_{fl}	k_{rad} s^{-1}	k_{nonrad} s^{-1}	Solvent	Ref.
27	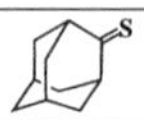		0.3	1.7×10^{-4}	6×10^8	3×10^{12}	perfluoroalkane	15,16
28			0.05	3.5×10^{-5}	7×10^8	2×10^{13}	perfluoroalkane	16
29	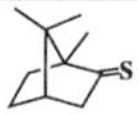		0.1	6.5×10^{-5}	6×10^8	1×10^{13}	perfluoroalkane	16
30	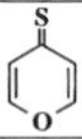	364	<20	1×10^{-4}	$>5 \times 10^6$	$>5 \times 10^{10}$	perfluoroalkane	17,18
31	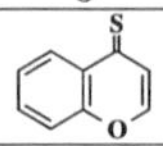	392	210	2.3×10^{-2}	1.1×10^8	4.8×10^9	perfluoroalkane	18,19
32	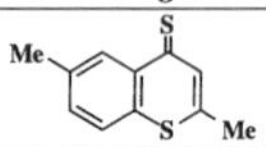	430	101	3.8×10^{-3}	3.8×10^8	9.9×10^9	perfluoroalkane	17,18
33	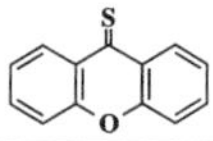	419 / 25	175	1.4×10^{-2}	8×10^7	4.8×10^9	perfluoroalkane / hexane	18-20 / 20
34	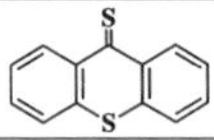	460	64	2.3×10^{-3}	3.6×10^7	1.6×10^{10}	perfluoroalkane	18,19
35	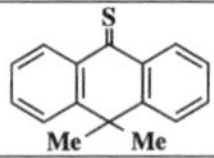	420	410	1.7×10^{-2}	4.1×10^7	2.4×10^9	perfluoroalkane	18,19
36	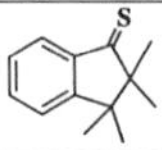	344	880	0.14	1.6×10^8	9.7×10^8	perfluoroalkane	17-19, 21,22
37	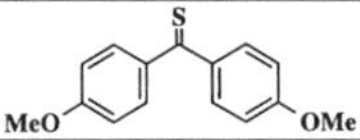	~359	35	5.6×10^{-4}	1.6×10^8	2.9×10^{10}	perfluoroalkane	18
38	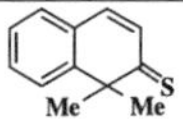		153	2.6×10^{-2}	1.7×10^8	6.4×10^9	perfluoroalkane	23
39	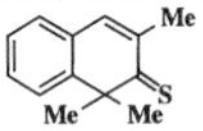		19	3.3×10^{-3}	1.7×10^8	5.9×10^{10}	perfluoroalkane	23
40	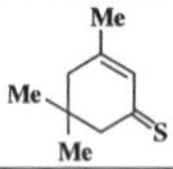		11	1.8×10^{-3}	1.6×10^8	9.1×10^{10}	perfluoroalkane	23
41	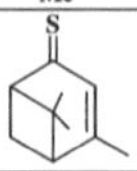		250	4.1×10^{-2}	1.6×10^8	3.9×10^9	perfluoroalkane	23
42	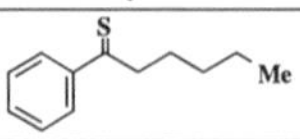		418	0.44			hexane	24

TABLE 2 Continued

	Thioketone	$\lambda_{0,0}$	τ_{fl} ps	Φ_{fl}	k_{rad} s^{-1}	k_{nonrad} s^{-1}	Solvent	Ref.
43	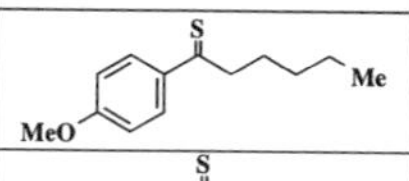		~90	0.42			hexane	24
44	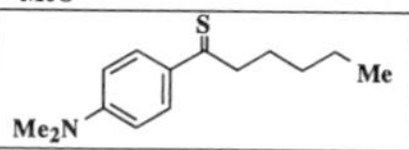			0.47			hexane	24
45	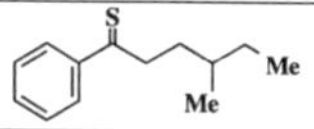		370	0.45			hexane	24
46	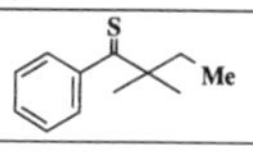		301	0.37			hexane	24

absorption band by a second laser pulse results in $S_1 \to S_0$ fluorescence emission taking place concurrently with the pulse.

A quantitative study of the RISC quantum yield was carried out with **55** [37]. This study involved UV irradiation of **55** in benzene or cyclohexane solution to produce the T_1 state and subsequent photolysis by a second laser tuned to the $T-T$ absorption band. The second pulse was accompanied by depletion (bleaching) of the $T-T$ absorption and $S_1 \to S_0$ fluorescence [the fluorescence was detected and quantified by an optical multichannel analyzer (OMA)]. The quantum yield of RISC, Φ_{RISC}, was calculated using Aberchrome 540, a reversible fulgide, as a two-laser actinometer. The values for Φ_{RISC} obtained by this method were 0.17 and 0.09 for cyclohexane and benzene solvents, respectively. This compares with 0.19 found for **55** in ethanol solvent [36].

Much larger Φ_{RISC} values were found for the cyanine dyes **56–58** [38]. A similar two-color technique was used to measure these yields with the exception that *meso*-tetraphenylporphyrin was used in place of Aberchrome 540 as the two-laser actinometer. As Table 4 shows, there is an inverse relationship between the triplet depletion (bleaching) quantum yield, Φ_{B1}, and the $S \to T$ ISC yield, Φ_{ISC}. Thus a large value for Φ_{B1} was accompanied by a small Φ_{ISC}. This reflects the "cyclic" flow of energy following excitation of T_1 to T_n. Once RISC has occurred, decay of the S_1 state partitions between ISC and fluorescence. If Φ_{ISC} is small, relatively few S_1 states will be cycled back to T_1 and Φ_{B1} will be large. This effect masks the actual efficiency of RISC. Thus the ratio $\Phi_{B1}/(1 - \Phi_{ISC})$ is given as an indicator of RISC efficiency.

RISC has also been observed by a two-color method for the isoalloxazine structures **59–62** shown in Chart 2 [39]. The discovery of RISC in these compounds came about as a result of time-resolved resonance Raman measurements on their triplet states. S_1-S_0 fluorescence generated by the Raman probe pulse

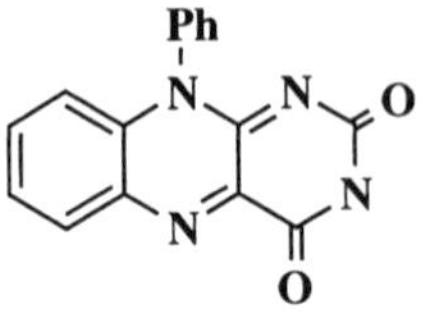

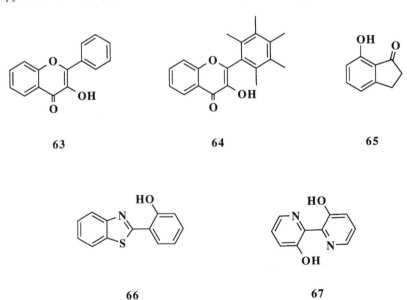

59 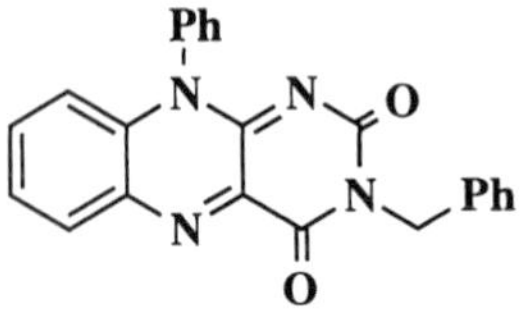   **60** 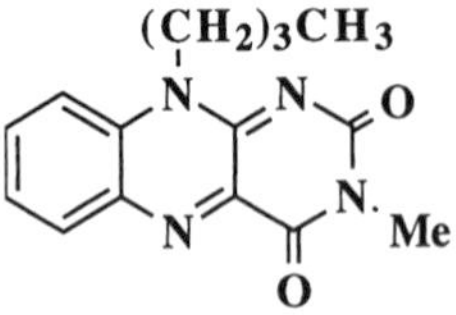   **61** 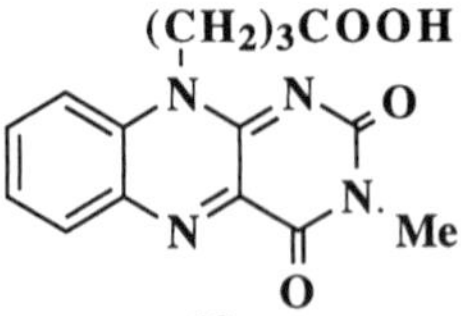   **62**

CHART 2

interfered with the expected Raman scattering. The RISC yield was estimated to be $\Phi_{RISC} \approx 8 \times 10^{-3}$.

RISC has been observed by two-color irradiation for a series of photo-tautomers generated by photoinduced proton transfer [40–46]. [The parent molecules, 3-hydroxyflavone (**63**), 2′,3′,4′,5′,6′-pentamethylphenyl-3-hydroxyflavone (**64**), 7-hydroxyindanone (**65**), 2-(2′-hydroxyphenyl)benzothiazole (**66**), and 2,2′-bipyridine-3,3′-diol (**67**), are shown in Chart 3.] Following proton transfer in the

63   **64**   **65**

66   **67**

CHART 3

TABLE 3 T_2-T_1 Fluorescenec (0, 0) Band Position and Quantum Yield for Anthracene and Substituted Anthracenes

	Anthracene	λ_{0-0} (nm)	$\Phi(T_2 \to T_1)$	Ref.
47	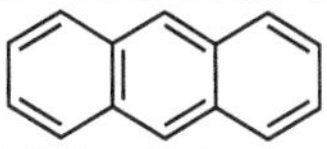	891	5×10^{-8}	26
48	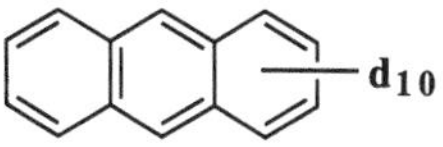 d_{10}	888	5×10^{-8}	26
49	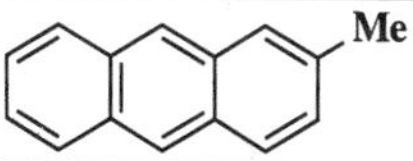 Me	896	4×10^{-8}	26
50	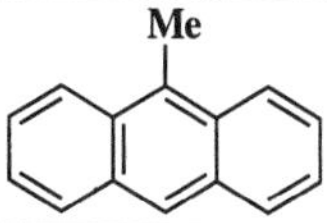 Me	917	$< 2 \times 10^{-8}$	26
51	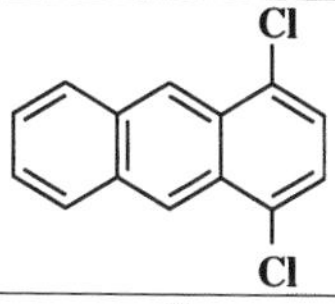 Cl Cl	959	70×10^{-8}	26
52	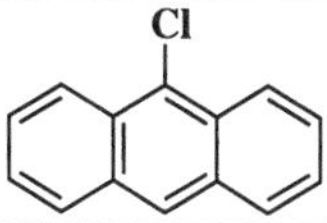 Cl	860	20×10^{-8}	26
53	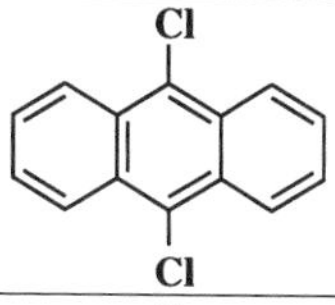 Cl Cl	none observed	none observed	26
54	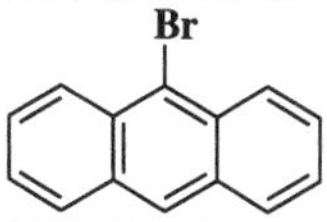 Br	867	30×10^{-8}	25,26
55	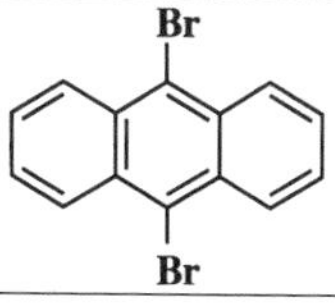 Br Br	841	100×10^{-8}	25,26

TABLE 4 RISC Parameters for Cyanine Dyes

Cyanine		Φ_{ISC}	Φ_{BI}	$\Phi_{BI}/(1-\Phi_{ISC})$	Ref.
56	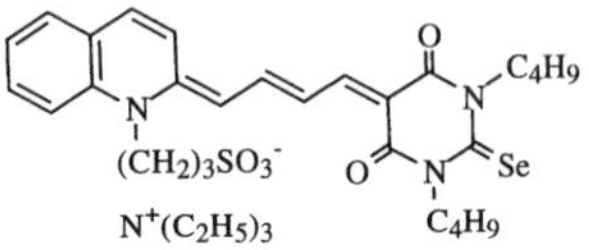	0.13	0.58	0.68	38
57	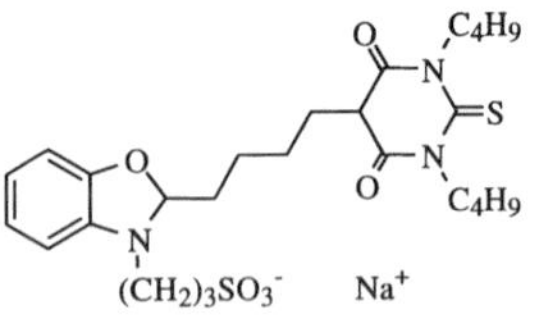	0.04	0.74	0.77	38
58	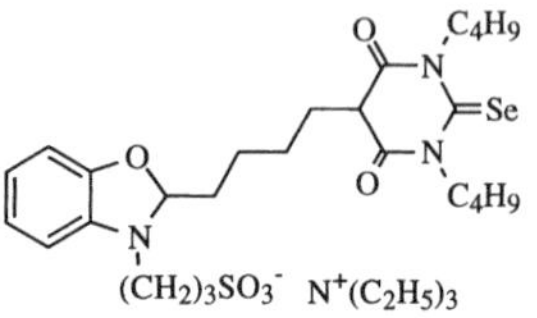	0.80	0.15	0.75	38

singlet manifold, ISC takes place to yield the triplet phototautomer. Excitation of this species causes RISC to the singlet phototautomer and fluorescence.

The reasons why some systems exhibit efficient RISC are not entirely clear, although there appears to be a correlation between efficiency and a large $T_2 - T_1$ energy gap (reducing the rate constant for internal conversion), heavy atom substitution and, possibly, a small $T_2 - S_1$ energy gap. A large $T_2 - T_1$ energy gap is almost certainly partially responsible for the large RISC yield in **55**, and heavy atom substitution may play a role in **55–58**.

D. Triplet–Triplet Energy Transfer

The T_2 lifetimes for anthracenes are relatively long lived because the rate constant for $T_2 \rightarrow T_1$ internal conversion is small relative to other molecules that have smaller $T_2 - T_1$ energy gaps. This allows processes such as $T - T$ fluorescence and RISC to be detected. It also increases the likelihood of bimolecular reactions between T_2 and added quencher molecules. This was first illustrated by anthracene sensitization of the rearrangement of rigid dienes such as **68** and **69** to the di-π-rearranged products [27,47–49]. The participation of anthracene T_2 was inferred from the knowledge that the rearrangement is a triplet process requiring > 68 kcal/mol (anthracene T_2 possesses ~ 74 kcal/mol, whereas the energy of T_1 is 42 kcal/mol).

68   69

Anthracene and substituted anthracenes are also known to sensitize the dimerization of 1,3-cyclohexadiene, another triplet state process [E(T_1) for 1,3-cyclohexadiene = 53 kcal/mol] [28]. In similar experiments, the T_2 state of naphthalene was also shown to undergo energy transfer to quenchers [29].

The first direct time-resolved evidence for energy transfer from an upper excited triplet state in solution at room temperature was published in 1987 [50]. This study made use of the two-color technique to photoexcite the T_1 state of benzophenone, **70**, in benzene solvent. As the extensive (almost quantitative) triplet depletion was not accompanied by any product formation, it was concluded that the excitation energy was transferred to the triplet manifold of the benzene solvent. The energetics of this donor-acceptor system are certainly conductive to this process. The benzophenone T_1 and T_n energies (69 kcal/mol and ca. 120 kcal/mol, respectively—the second photon in the two-color excitation provides roughly 50 additional kcal/mol to the T_1 state) bracket the benzene T_1 energy (85 kcal/mol) and therefore benzene acts in the same way toward benzophenone as 1,3-cyclohexadiene acts toward anthracene, i.e., as an exclusive upper triplet energy accepter.

Since the benzene T–T absorption could not be detected spectroscopically, there was no further transient evidence for this process. However, direct confirmation of upper triplet state energy transfer was obtained using 2-acetylphenanthrene, **71**, as an upper triplet state energy donor and biphenyl as the energy acceptor [51]. The T_1 energy of **71** lies below that of biphenyl by 5 kcal/mol, making energy transfer from T_1 energetically unfavorable. Excitation of the T_1 state of **71** to an upper triplet level in a two-color experiment not only led to depletion of the T_1 state, but also to production of the biphenyl T–T absorption. The transient behavior in this system was different from that of **70** in that the depletion caused by the second laser pulse was temporary rather than permanent. Thus, depletion and recovery in **71** are caused by energy transfer from the upper triplet of **71** to biphenyl, followed by back energy transfer from biphenyl T_1 to the T_1 state of **71**. Apparently, the decay of triplet benzene is too rapid to allow the same back energy transfer to occur in **70**. Figure 1 shows the energetics of this system.

Stern–Volmer analysis (measuring the magnitude of the biphenyl T–T absorption produced by the second pulse as a function of the concentration of added biphenyl) yielded an estimate of ~600 ps for the lifetime of the upper triplet state of **71**.

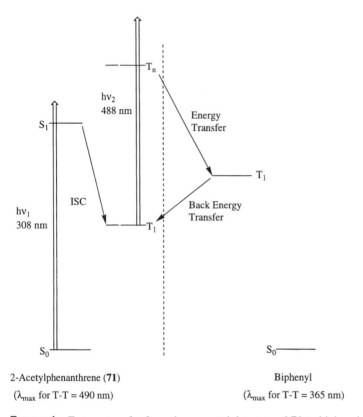

2-Acetylphenanthrene (**71**)   Biphenyl
(λ_{max} for T-T = 490 nm)   (λ_{max} for T-T = 365 nm)

FIGURE 1   Energy transfer from the upper triplet state of **71** to biphenyl.

Since these first time-resolved studies, similar quenching methods (i.e., choosing an exclusive upper state quencher) have been used in conjunction with the two-color approach to demonstrate energy transfer from anthracene T_2 to acrylonitrile, benzonitrile, dimethoxybenzene [52] and from the upper triplets of **70, 71** and p-terphenyl to a variety of quenchers including carbon tetrachloride [53].

Another recent study makes use of the participation of the T_2 state in the $S \rightarrow T$ ISC process in anthracene [31]. 1,3-Octadiene was used to intercept some of the T_2 states before they relaxed to T_1 and the decrease in T_1 yield was used to estimate the T_2 lifetime. Further, this study compensated for the effects of static and time-dependent quenching that comes into play at the relatively large quencher concentrations that are required when quenching sub-nanosecond-lifetime transients. The lifetimes obtained (given in Table 5) were significantly less than previously estimated from other quenching studies and are in line with the lifetimes implied from the T–T fluorescence quantum yields discussed above.

TABLE 5 T_2 Lifetimes for Anthracenes and Substituted Anthracenes

	Anthracene	$\tau\,(T_2)$ (ps)	Ref.
52	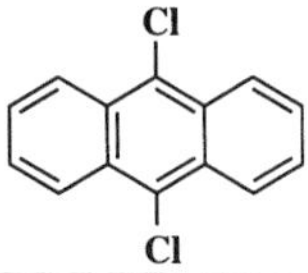	20.8	31
47	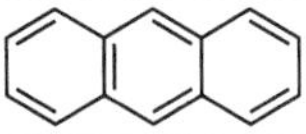	8.8	31
48	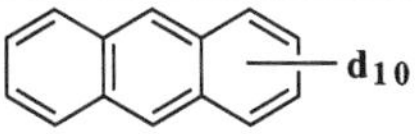	10.4	31
72	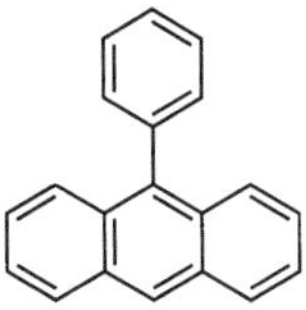	6.2	31
50	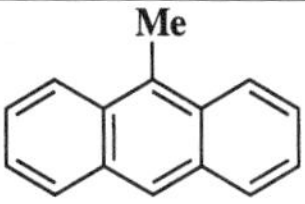	8.8	31
52	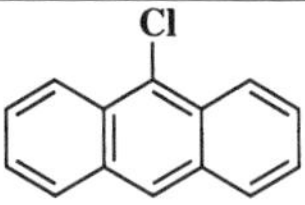	9.1	31
73	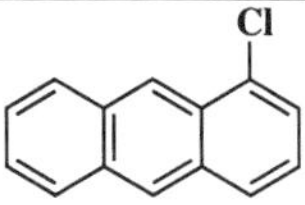	3.5	31
74	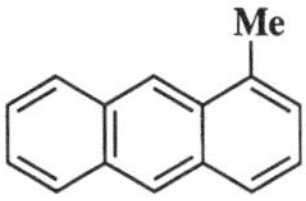	1.6	31
54	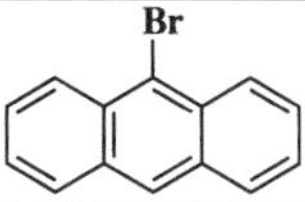	5.5	31
55	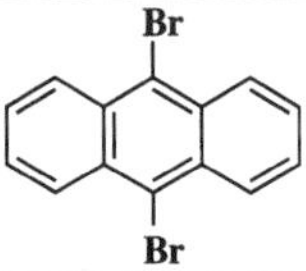	7.0	31

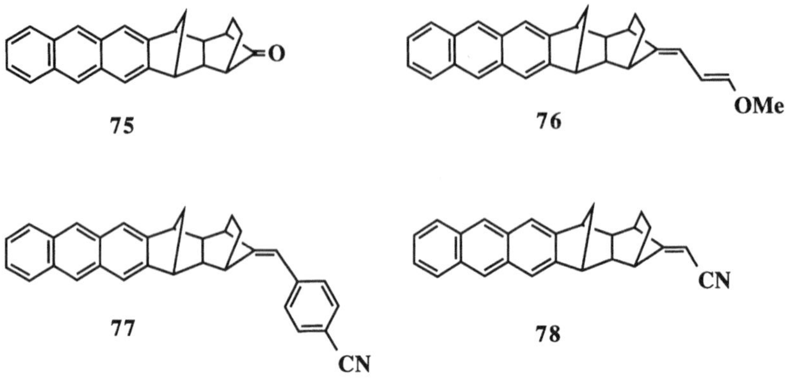

75   76   OMe

77   78

CN

CHART 4

The T_2 state of anthracene has been shown by the two-color technique to participate in intramolecular energy transfer as well [54]. Compounds **75–78** (Chart 4) all contain the anthracene chromophore joined to acceptor chromophores by way of the rigid norbornylogous bridges used extensively by Paddon-Row, Verhoeven, and co-workers [55]. Each of the acceptor chromophores is capable of acting exclusively as an anthracene T_2 quencher. In **75**, energy transfer results in production of the ketone triplet state which decays by a combination of back energy transfer into the anthracene T_1 state and ring opening to form products. In **76–78**, initially, the spectroscopic (planar) triplet state of the acceptor is formed which subsequently relaxes to the perpendicular triplet. From there, decay involves either formation of the perpendicular ground state or back energy transfer to anthracene T_1 coupled with bond rotation back to a planar ground state of the acceptor. Since rotation coupled with back energy transfer involves overcoming an energy barrier, the rate of back energy transfer is decreased and perpendicular ground state formation is favored. This energy transfer behavior is summarized in Fig. 2.

E. Electron Transfer

Given that triplet–triplet energy transfer proceeds via a Dexter (electron exchange) mechanism, it is not surprising that electron transfer can also occur via upper triplet states. Two-color experiments with anthracene in acetonitrile in the presence of ethylbromoacetate, a dissociative electron acceptor, showed that excitation to an upper triplet state led to depletion of the T–T absorption and concurrent production of the anthracene cation radical as a result of electron transfer (Scheme 1) [52].

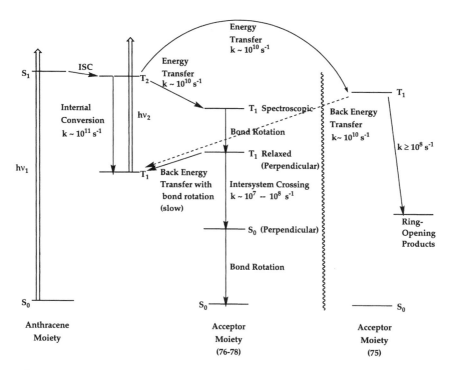

SCHEME 1

III. UPPER EXCITED STATES: PHOTOCHEMISTRY

In their 1978 review [1], Turro et al. thoroughly documented the upper state chemistry reported, up to that time, that was significantly different in mechanism or yield from that observed for the lowest excited states. This earlier work often involved a comparison of products formed under low-intensity CW irradiation in

FIGURE 2   Upper triplet state intramolecular energy transfer processes in bichromophores 75–78.

low-temperature matrices and room-temperature liquid solution. In many cases products that were found in low concentration or were absent in the room-temperature studies were readily observed in the low-temperature experiments. Such behavior was usually a good indication of upper triplet state chemical reaction. Triplet lifetimes at low temperature are sufficiently long to allow high steady-state concentrations to be produced under low-intensity irradiation. Absorption of a second photon by the triplet states can then potentially lead to relatively efficient two-photon product formation (usually triplet production and reexcitation were achieved by one CW source but in some cases two sources were used). At room temperature, under CW irradiation, triplet steady state concentrations are much lower owing to their shorter lifetimes and reexcitation is usually negligible. A few examples of the earlier work are given here, but the reader is directed to Turro's review for an in-depth discussion of these studies. Thus, upper triplet and singlet states were found to be involved in a variety of reactions including: formation of molecules in the pleiadene family, **79** and **80** [56–58] among others, in 77 K matrices (Scheme 2) (the upper triplet formation of **80** has also

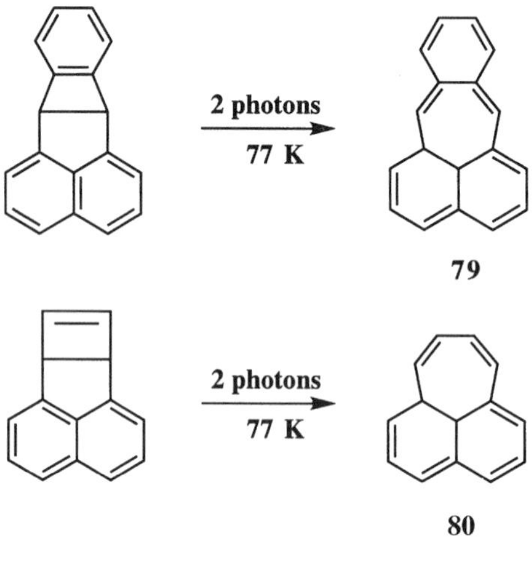

SCHEME 2

been investigated more recently with excimer laser excitation [3]); α-cleavage in ketones such as benzaldehyde, acetophenone, and benzophenone in low-temperature matrices following irradiation with two CW light sources (upper triplet) [59]; α-cleavage of acetophenone in the vapor phase (upper singlet) [60]; H-atom abstraction by some α,β-unsaturated ketones (upper singlet) [61]; cyclo-addition of thioketones with olefins (upper singlet) [62]; intermolecular and intramolecular H-atom abstraction by thioketones (upper singlet) [63,64]. A recent, detailed review and discussion of the mechanisms for these upper singlet thioketone reactions are given by Maciejewski and Steer [5].

The more recent two-color and laser jet studies of upper state chemistry involve bond cleavage and photoionization, H-atom abstraction, cycloadditions, sigmatropic shifts, and "tandem" excited state reactions in which the energy of an upper state promotes a sequence of reactions rather than just one. The recent review by Wilson and Schnapp [3] summarizes many of these reactions. The following sections of this review describe recent reports of carbon–halogen, carbon–carbon, and nitrogen–hydrogen bond cleavage as well as examples of cycloaddition reactions occurring from upper states. Discussion of photoionization is included with the description of N–H bond cleavage in amines. Photoionization by direct excitation into upper singlet levels and following one-source multiphoton excitation is a fairly common process and is not discussed here. Instead, only ionization following two-color excitation in the triplet manifold will be described.

A. Carbon–Halogen Bond Cleavage

Carbon–halogen bond cleavage in aryl halides is believed to involve initial excitation into an upper singlet state that is stable with respect to cleavage, followed by intersystem crossing to an upper triplet state that is either dissociative itself or can cross to an upper dissociative (σ, σ^*) triplet [65–67]. The two-color approach has been used to demonstrate directly that excitation to an upper triplet state results in carbon–halogen cleavage. For example, in 2-bromonaphthalene (**81**) and 9-bromophenanthrene (**82**) [68], promotion of T_1 to an upper triplet by dye laser excitation

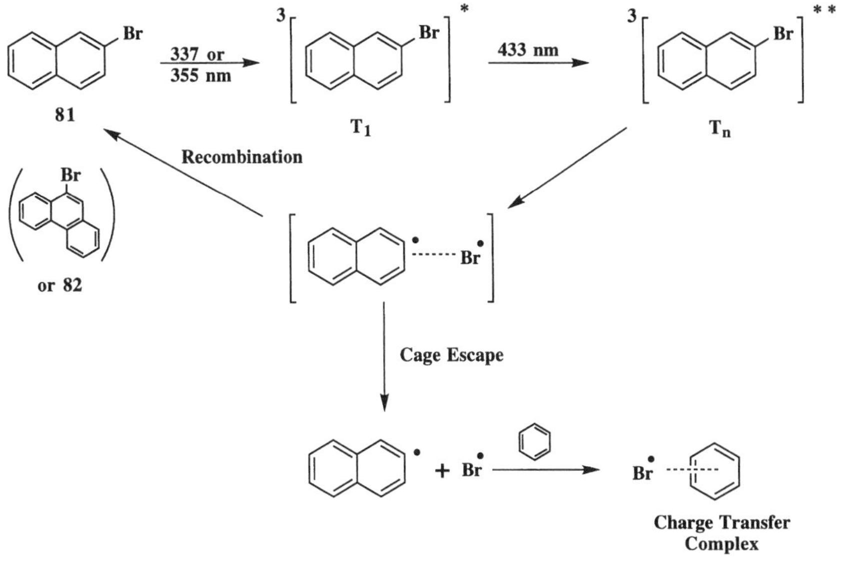

SCHEME 3

in benzene solution leads to cleavage as evidenced by production of the bromine atom–benzene charge transfer complex, itself a transient with $\lambda_{max} \approx 550$ nm and $\tau \approx 1$ μs (Scheme 3). Thus the T–T absorption is depleted and the 550-nm absorption is enhanced concurrent with the dye laser pulse. Quantum yields for debromination were measured using the charge transfer complex and Aberchrome 540 as actinometers and were found to be 0.04 and 0.05 for **81** and **82**, respectively. It was noted that the debromination yields were consistently smaller than the triplet depletion yields (bleaching yields), indicating that not all cleavage events led to complex formation. Since there was no product formation other than that expected from cleavage, it was concluded that a considerable fraction of aryl radical–bromine atom pairs undergo geminate recombination to give ground state starting material.

Carbon–bromine cleavage from upper triplet states has also been observed by two-color photolysis of **55** and for the ketones **83–85** [69]. In the latter cases cleavage is followed by a neophyl-type rearrangement and decarboxylation yielding the arylmethyl radical (Scheme 4).

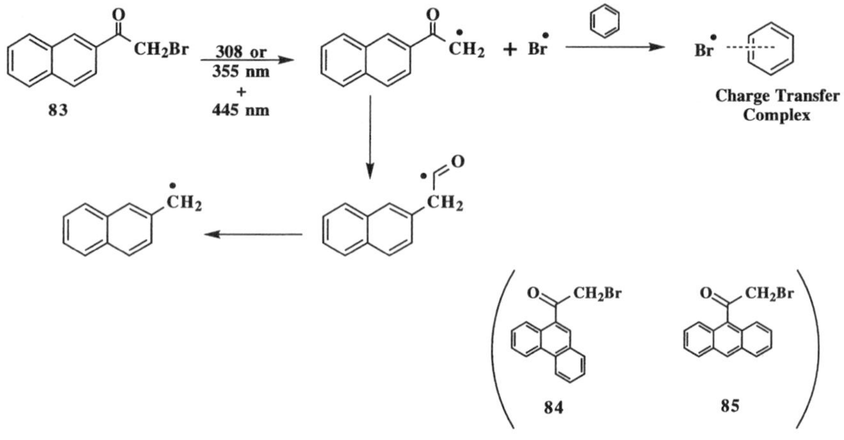

SCHEME 4

Laser flash photolysis and phosphorescence measurements, as well as measurements of the effect of temperature on the quantum yields of triplet sensitized dehalogenation in thiophene derivatives **86–88** (Chart 5) also indicated the participation of an upper triplet state (σ, σ^*) in the carbon–halogen cleavage reaction [70]. Significant enhancement in dehalogenation was observed when using low-energy triplet sensitizers with energies several kcal/mol greater than that of the unreactive T_1 state of the thiophenes (e.g., $E_T = 57$ kcal/mol for chrysene; $E_T = 53$ kcal/mol for **86**). The explanation given for this effect was that the sensitizer gains

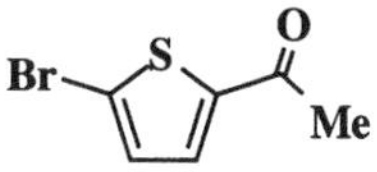

86 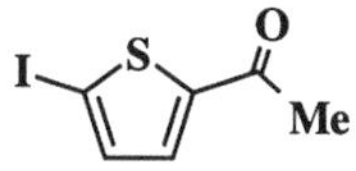 **87** 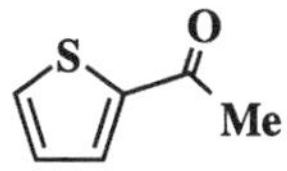 **88**

CHART 5

sufficient activation at higher temperatures to undergo energy transfer to a higher, dissociative triplet. From the thermally induced enhancement of dehalogenation, estimates of the upper triplet state energies were made. Thus, $E(T_n) = 60-70$ kcal/mol for **86** and **87** and ~60 kcal/mol for **88**.

B. Reluctant Norrish Type I Bond Cleavage

The efficiency of Norrish type I cleavage from lowest triplet states is effected by both the electronic configuration of the triplet as well as the amount of energy stored in the triplet. Molecules that possess π, π^* lowest triplet states are generally unreactive toward cleavage as are those that have low triplet energies relative to the energy required for cleavage. It has now been shown in numerous examples that this "reluctance" to undergo cleavage can be overcome by promoting the lowest triplet to a higher energy upper triplet level. In fact, the observation of cleavage from upper triplets represents some of the best illustrations of the application of the two-color approach.

The first time-resolved demonstration of reluctant type I cleavage was reported for benzil (**89**) [71] and di-1-naphthyl-2-propanone (**90**) [72]. **89** possesses a T_1 state with n, π^* character but its energy (54 kcal/mol) is insufficient to promote cleavage. However, dye laser excitation of the intense 480 nm $T-T$ absorption band produces an upper triplet with more than enough energy to undergo fragmentation. Evidence for cleavage was provided by the observation of dye laser-induced depletion of the $T-T$ absorption combined with two-color product studies carried out on solutions containing a mixture of **89** and 4,4'-dimethylbenzil. The major species found in these product studies was the radical cross-coupling product, 4-methylbenzil. Since only a trace of this material was found in one-laser experiments, it is a good indicator of cleavage.

In initial two-color experiments performed with **90**, and other compounds such as benzophenone (**70**), in benzene solvent, extensive $T-T$ depletion was observed. This led to the assumption that upper state chemistry such as type I cleavage was efficient. However, as was pointed out above, benzene and other aromatic solvents can quench upper triplet states by energy transfer and thereby cause depletion. A reexamination of several reluctant type I systems using other solvents (particularly acetonitrile) revealed that energy transfer quenching by aromatic solvent is usually a more important upper triplet decay pathway than cleavage.

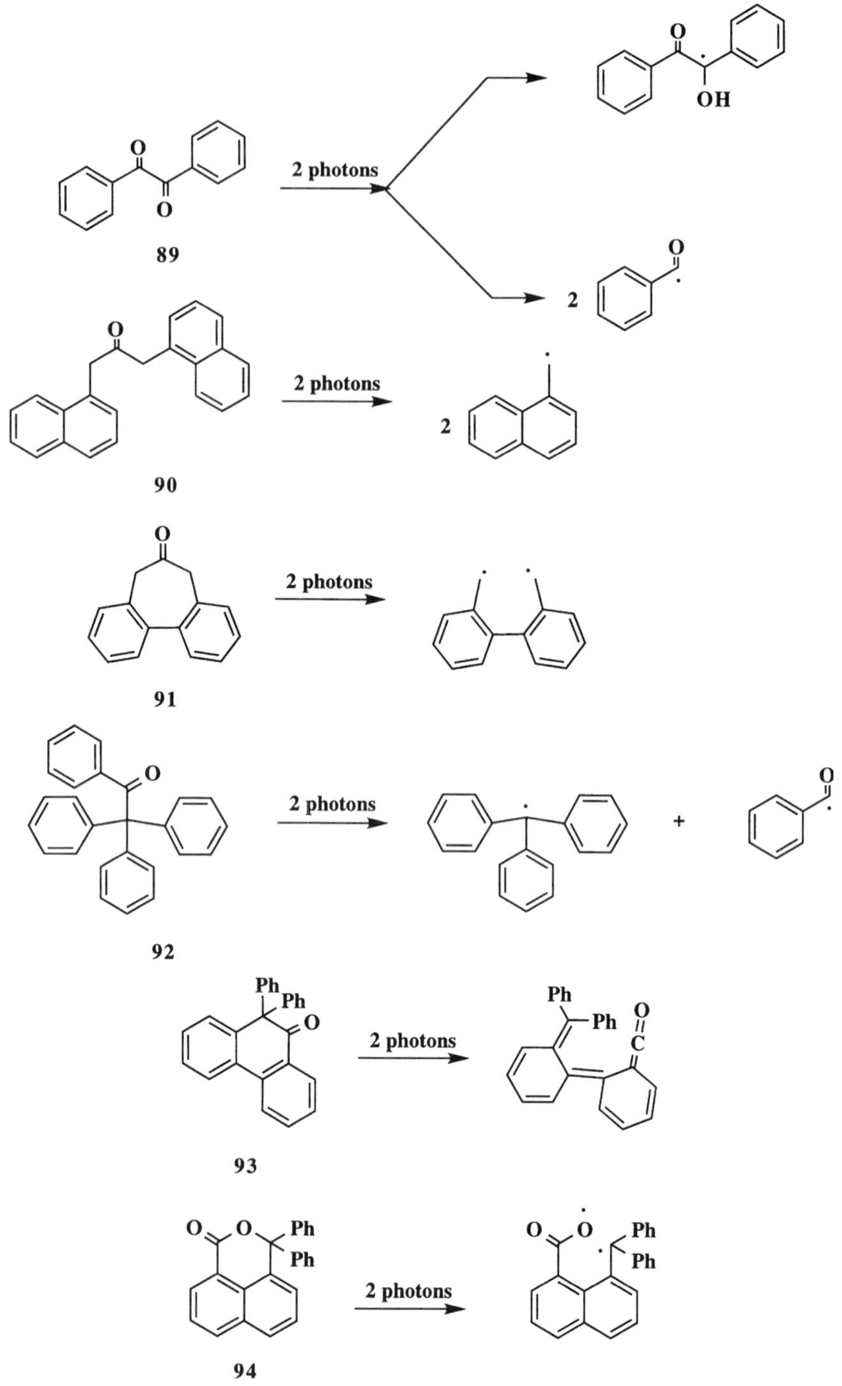

SCHEME 5

Scheme 5 shows several examples of reluctant type I cleavage. In **90**, the lowest triplet state is π, π^* in character but again the upper triplet state undergoes cleavage. In this case, in addition to triplet depletion, product studies and time-resolved observation of the naphthylmethyl radical were indicators of cleavage. Compounds **91** [73] and **93** [3] also have unreactive π, π^* lowest triplets and again undergo cleavage from upper triplet states. β-cleavage via an upper triplet state was observed for **94** [74]. The resulting biradical subsequently undergoes decarboxylation followed by a series of rearrangements.

C. Nitrogen–Hydrogen Bond Cleavage and Photoionization

In general, aromatic amines undergo photoionization in polar solvents yielding a cation radical. In secondary amines, the cation radical can subsequently deprotonate to give a neutral aminyl radical. Photoinduced homolytic N–H bond cleavage can occur in polar and nonpolar media. The participation of upper excited states, particularly triplets, in these processes has been the subject of several recent two-color studies. To date, the upper triplet states of iminodibenzyl [75], carbazole [76], and phenothiazine [77] (**95–97**, respectively) have been shown to undergo cleavage in polar and nonpolar solvents. An upper singlet state has also been implicated in cleavage of **96** [76]. Compounds **95, 97** [75,77], diphenylamine (**98**) [78], promazine (**99**), and chlorpromazine (**100**) [79] all undergo photoionization in polar solvents. (See Chart 6.)

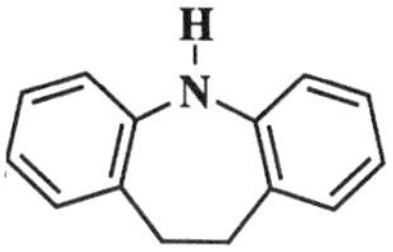

95 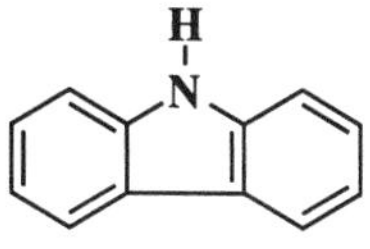   96 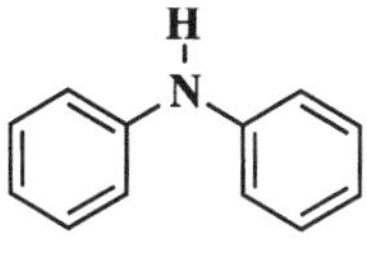   98

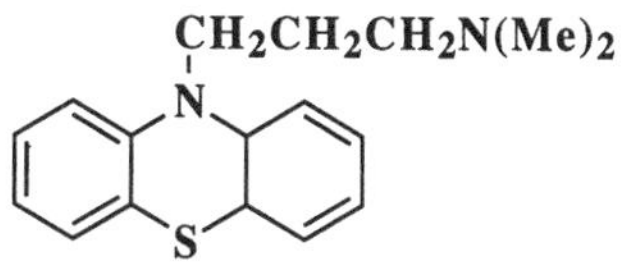

99 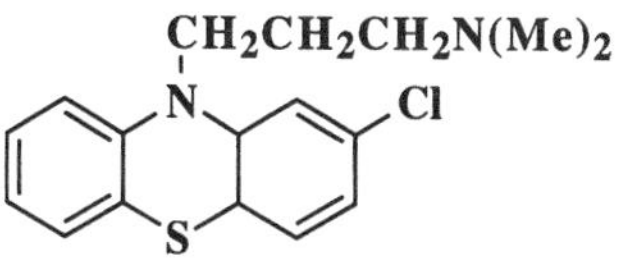   100

CHART 6

Thus, in the case of **97** [77], the upper triplet state undergoes N–H cleavage in nonpolar solvents following 355 + 445 nm two-color irradiation, while in acetonitrile, cleavage and ionization both occur (Scheme 6). These processes were indicated by the concurrent depletion of the T–T absorption and growth of absorp-

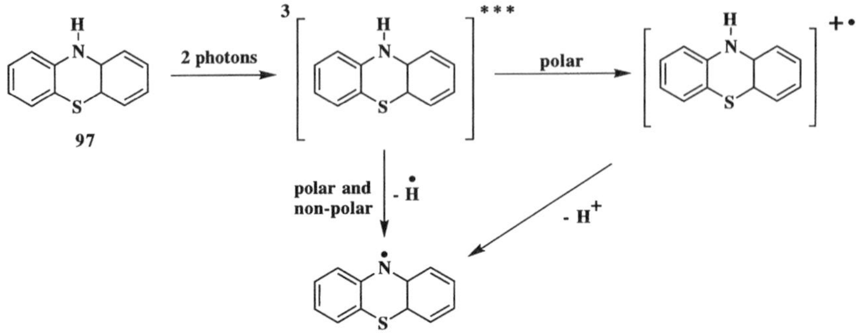

SCHEME 6

tions due to the neutral aminyl and cation radicals. Quantum yields for depletion of the T_1 state were highly solvent dependent: 0.12 and 0.004 in cyclohexane and acetonitrile, respectively. Indeed, the inefficiency of depletion is even more pronounced in alcohols where little or no depletion was observed. This was a general observation for most of the amines studied and is possibly due to specific solute–solvent interactions. It should be noted, however, that hydrogen bonding can be ruled out as a source of this effect since triplet depletion for **99** was observed in aqueous solution. A possible explanation lies in how well the electron is solvated following photoionization. In acetonitrile, the electron undergoes an irreversible reduction of the solvent while in alcohols and water, a physical rather than chemical interaction occurs. However, it has been suggested that the energetics of solvation are different for water and alcohols and that the aqueous electron can be thought of as more stable than an electron solvated by methanol (deeper potential well for the aqueous electron). Thus in alcohols, the efficiency of geminate electron–cation radical recombination may be higher than in water. (Recombination to form the upper or lower triplet states would not lead to triplet depletion.)

The extinction coefficients for the $T–T$ neutral aminyl, and cation radical absorptions of **97** were used to calculate the quantum efficiencies for N–H cleavage and photoionization. The results indicate that in cyclohexane, the efficiency of cleavage is ca. 90%. Thus, roughly 90% of those upper triplet states that do not relax to T_1 undergo cleavage. In acetonitrile on the other hand, the efficiencies for neutral and cation radical production are 0.53 and 0.45, respectively. In other words, of the upper triplets that do not regenerate T_1, half decay to neutral radical and the other half to cation radical. It should be noted that the actual proportion of direct cleavage events may be smaller than indicated from the efficiencies because one of the cation radical decay routes is deprotonation to form the neutral radical.

Similar photoionization results were observed with **99** and **100** although picosecond (355 nm) and two-color measurements on **100** indicated that photoionization is a two-photon process occurring from both an upper singlet and an upper triplet state.

Two-color photoionization through the triplet manifold has also been reported for biphenyl in aqueous micellar solution [80].

D. Cycloaddition Reactions

Although bond cleavage and photoionization are common processes occurring from upper excited states, the energy stored in these states can also facilitate the formation of new bonds. For example, multiphoton excitation of perylene and pyrene in cyclohexane leads to hydrogen abstraction from the solvent and subsequent radical coupling products [81]. Two-color irradiation of anthracene (**47**) in the presence of acrylonitrile leads to a four-fold increase in the amount of cyclo-adduct **101** as compared to that obtained under one-laser irradiation [52] (Scheme 7). That this is an upper triplet process was indicated by the triplet depletion

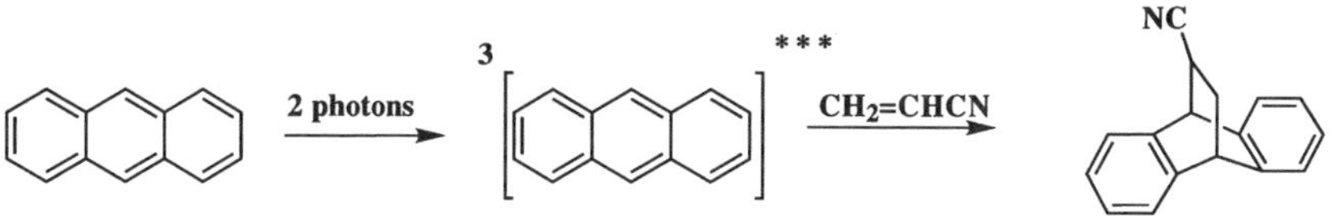

SCHEME 7

observed concurrent with the second pulse. However, it is unclear whether the initial step in the cycloaddition mechanism is due to energy transfer from the upper triplet of anthracene to acrylonitrile or electron transfer (both of which have been shown to occur with anthracene) or some other route.

Intramolecular cycloaddition for 5-phenyltricyclo[5.2.1.0^{2,6}]deca-4,8-dien-3-one, **102**, producing the cage product, **103** has been reported to proceed through an upper singlet state [82] (Scheme 8). The participation of an upper singlet was determined by a comparison of yields obtained using different excitation wave-

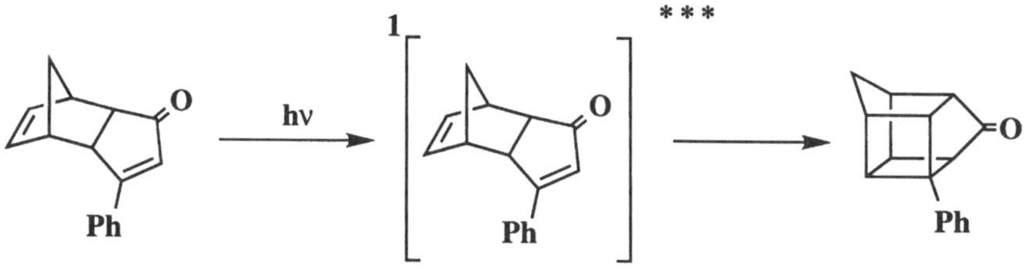

SCHEME 8

lengths and by benzophenone triplet sensitization. Irradiation into the higher energy π, π^* band yielded primarily **103**, while excitation of the lower energy n, π^* band leads to triplet formation and dimerization.

IV. EXCITED RADICALS: FLUORESCENCE

A considerable amount of data exists on the luminescence properties of excited benzyl [6,83–90,94–97], arylmethyl [91–93,100–108], diphenylketyl [7,109–116], and heteroradicals [117–123] in low-temperature matrices and in room temperature solution. Tables 6 through 11 tabulate both spectral and lifetime information. Where data is available, these tables also include information on the absorption spectra of the excited radicals. Due to the fact that numerous radicals listed in the tables are partially or totally deuterated, structures indicate the number of hydrogen atoms and/or deuterium atoms present.

In general, for benzyl and other arylmethyl radicals there is a large spectral red shift between the observed absorption maxima of the ground-state radicals and the emission maxima for the excited states. The origin of this shift lies in the low oscillator strength for the $D_0 \rightarrow D_1$ transition which renders it nearly unobservable for most of the radicals studied. Thus, the observed absorption is due to the $D_0 \rightarrow D_n$ transition. Excitation into this band partially leads to relaxation to the D_1 state followed by emission. The effect of substituents on the emission has been studied extensively and has shown that substitution usually leads to a spectral red-shift.

Room-temperature solution lifetimes for these excited radicals vary substantially from < 1 ns for benzyl radical itself to nearly 0.5 μs for diphenylmethyl radical [84,95]. The efficient deactivation of excited benzyl radical has been suggested to occur by thermal activation of the emissive lowest excited state to a close-lying upper state that decays by a nonradiative route. Thus, in low-temperature glasses, the excited benzyl radical lifetime is over 1 μs, but at higher temperatures the upper state can be accessed thermally. The differences in lifetimes that are observed upon substitution of the benzyl radical can all be rationalized on the basis of subtle changes in the energies of the lowest excited state and this close-lying upper state. The long room-temperature lifetime of excited diphenylmethyl radical and naphthylmethyl radical can be explained by a much larger energy gap between D_1 and D_2 which renders thermal activation inefficient [100,101,106]. Excited triphenylmethyl radical has a lifetime that is similar to that of benzyl. However, in this case deactivation is by a chemical route (ring closure—vide infra) [105].

Unlike the arylmethyl radicals, diphenyl ketyl radicals have substantial extinction coefficients for the $D_0 \rightarrow D_1$ absorption and emission that shows the expected mirror image relationship to the absorption band. Ketyl radical lifetimes are generally in the 10-ns range and show only minor temperature dependence. Deuterium substitution of the hydroxy group has a pronounced effect on lifetime,

TABLE 6 Spectral and Lifetime Data for Fluorescence from Excited Benzyl and Substituted Benzyl Radicals in Low-Temperature (77 K) Matrices

	Benzyl Radical	λ_{max} (emission) (nm)	τ (μs)	Solvent	Ref.
104	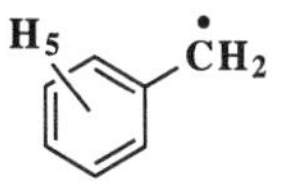 H_5—$\dot{C}H_2$ (benzyl)	461.5	----	cyclohexane	83
105	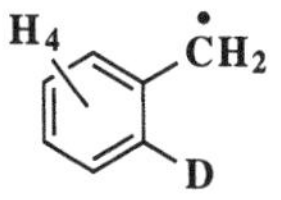 H_4—$\dot{C}H_2$, ortho-D	broad	1.8	3-methylpentane	84
106	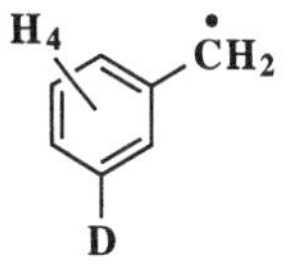 H_4—$\dot{C}H_2$, meta-D	broad	1.52	3-methylpentane	84
107	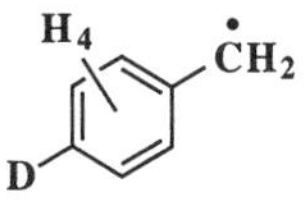 H_4—$\dot{C}H_2$, para-D	broad	1.54	3-methylpentane	84
108	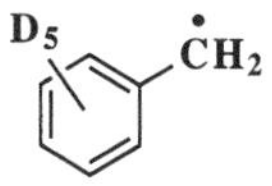 D_5—$\dot{C}H_2$	broad	3.26	3-methylpentane	84,85
109	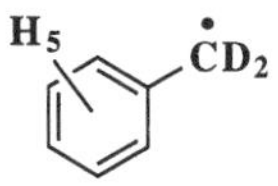 H_5—$\dot{C}D_2$	broad	1.73 1.26	3-methylpentane	84,85
110	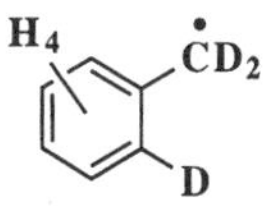 H_4—$\dot{C}D_2$, ortho-D	broad	3.21	3-methylpentane	85
111	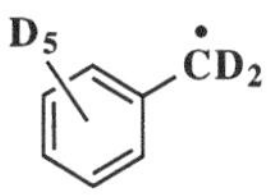 D_5—$\dot{C}D_2$	broad	3.53	3-methylpentane	83-85

TABLE 6 Continued

	Benzyl Radical	λ_{max} (emission) (nm)	τ (μs)	Solvent	Ref.
112	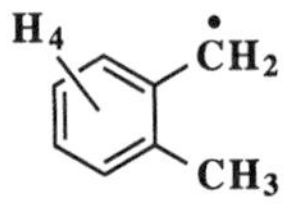 H_4 — $\dot{C}H_2$ / CH_3 (ortho-methyl)	477.5	0.87	ethanol	86,87
113	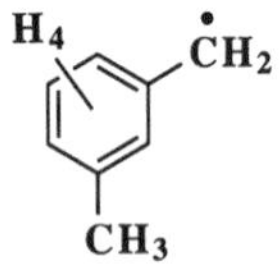 H_4 — $\dot{C}H_2$ / CH_3 (meta-methyl)	476.0	0.80	ethanol	86,87
114	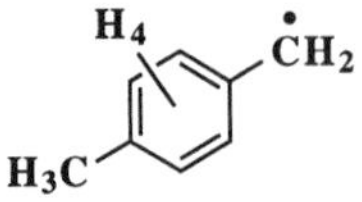 H_4 — $\dot{C}H_2$ / H_3C (para-methyl)	469.0	1.33	3-methylpentane	88,89
115	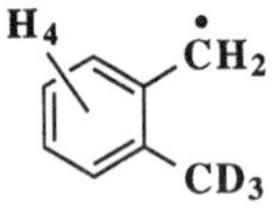 H_4 — $\dot{C}H_2$ / CD_3	broad	1.02	3-methylpentane	85
116	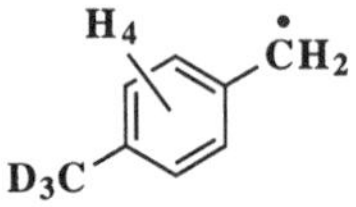 H_4 — $\dot{C}H_2$ / D_3C	broad	2.85	3-methylpentane	85
117	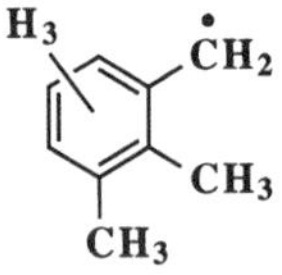 H_3 — $\dot{C}H_2$ / CH_3 / CH_3	477 (hexane)	0.62 (3-methylpentane)		85,90
118	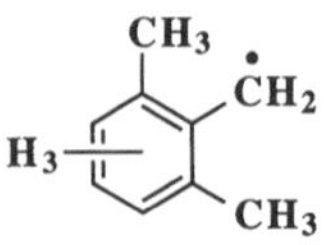 CH_3 / $\dot{C}H_2$ / H_3— / CH_3	494.5	----	hexane	90

	Benzyl Radical	λ_{max} (emission) (nm)	τ (μs)	Solvent	Ref.
119	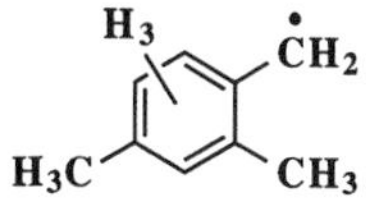 H_3 — ring — $\overset{\bullet}{C}H_2$; H_3C, CH_3	471	----	hexane	90
120	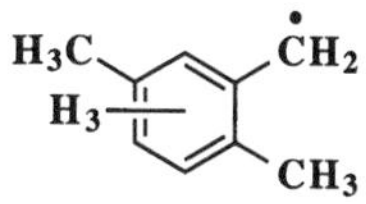 H_3C — ring — $\overset{\bullet}{C}H_2$; H_3, CH_3	496.5	----	hexane	90
121	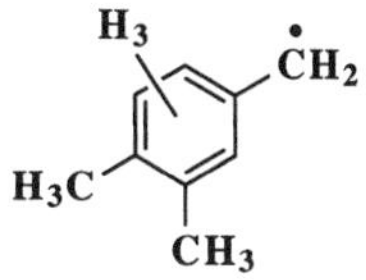 H_3 — ring — $\overset{\bullet}{C}H_2$; H_3C, CH_3	476	----	hexane	90
122	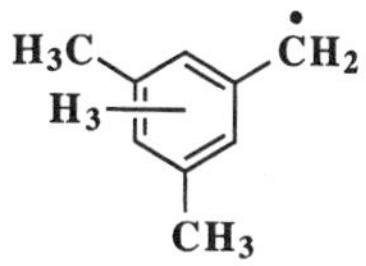 H_3C — ring — $\overset{\bullet}{C}H_2$; H_3, CH_3	490	0.62	ethanol	86,87
123	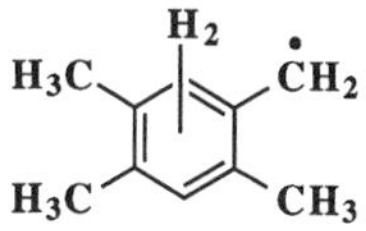 H_3C — ring (H_2) — $\overset{\bullet}{C}H_2$; H_3C, CH_3	451	0.48	3-methylpentane	84,85
124	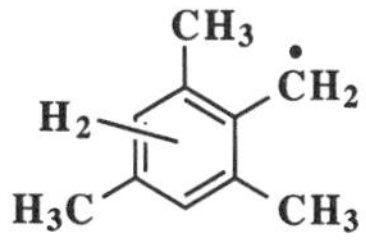 CH_3 — ring — $\overset{\bullet}{C}H_2$; H_2, H_3C, CH_3	477	----	hexane	90
125	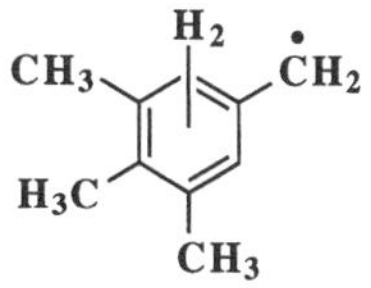 CH_3 — ring (H_2) — $\overset{\bullet}{C}H_2$; H_3C, CH_3	489	----	hexane	90

TABLE 6   Continued

	Benzyl Radical	λ_{max} (emission) (nm)	τ (μs)	Solvent	Ref.
126	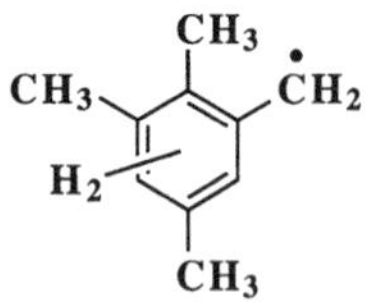 CH₃, CH₃, H₂, CH₃ with •CH₂	513	----	hexane	90
127	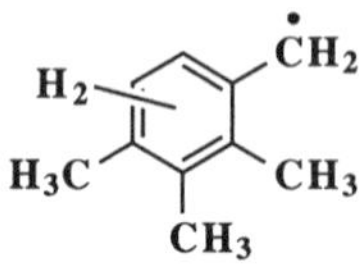 H₂, H₃C, CH₃ with •CH₂	489	----	hexane	90
128	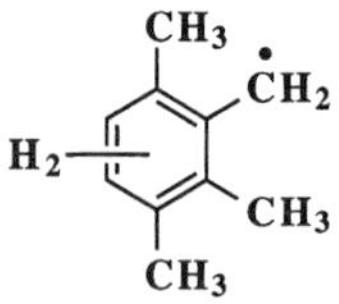 CH₃, H₂, CH₃, CH₃ with •CH₂	515.5	----	hexane	90
129	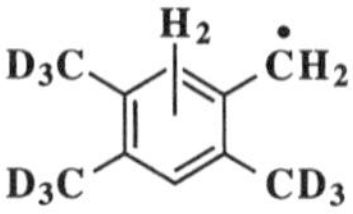 H₂, D₃C, D₃C, CD₃ with •CH₂	broad	0.59	3-methylpentane	85
130	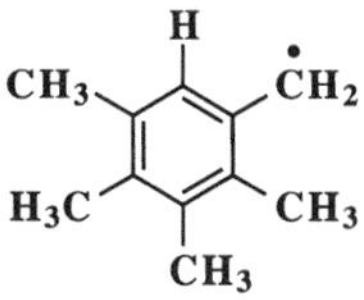 H, CH₃, H₃C, CH₃ with •CH₂	504 (hexane)	0.16 (ethanol)		87,90
131	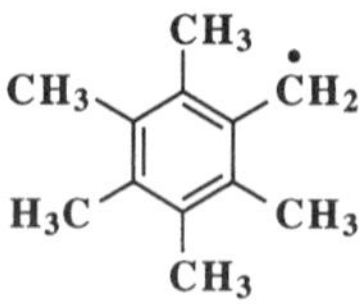 CH₃, CH₃, H₃C, CH₃ with •CH₂	533	0.24	3-methylpentane	85

TABLE 7 Spectral and Lifetime Data for Fluorescence from Excited Arylmethyl Radicals in Low-Temperature (77 K) Matrices

	Arylmethyl Radical	λ_{max} (emission) nm	τ (ns)	Φ_{fluor}	Solvent	Ref.
132	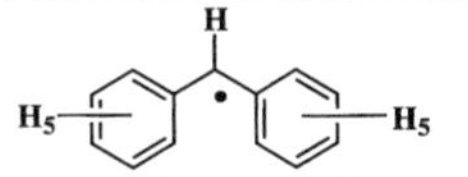	523	430	0.29	methyl-cyclohexane/methyl-pentane	91
133	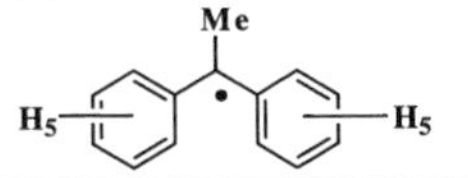	518	220	0.72	methyl-cyclohexane/methyl-pentane	91
134	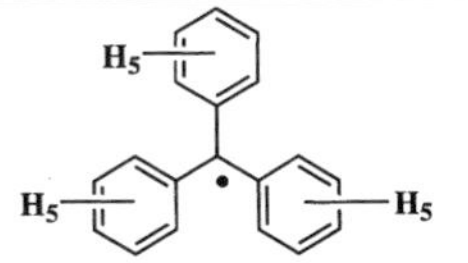	516	235	0.73	methyl-cyclohexane/methyl-pentane	91-93
135	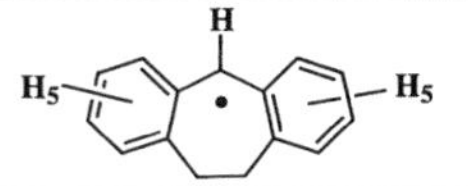	534	340	0.22	methyl-cyclohexane/methyl-pentane	91

however, and reflects the importance of the O−H (O−D) bond in both chemical and physical deactivation of the excited radical (vide infra) [109–113].

V. EXCITED RADICALS: PHOTOCHEMISTRY

A. Bond Cleavage

Unlike vicinal dibromides such as 1,2-dibromoethane [124] and *trans*-10,11-dibromodibenzosuberone [125] which undergo photochemical loss of bromine followed by facile thermal cleavage of the second C−Br bond, 2-(1,2-dibromo-ethyl)naphthalene (**211**) and 2-bromo-9-(1,2-dibromoethyl)anthracene (**212**) [126] require two photons for complete debromination (Scheme 9).

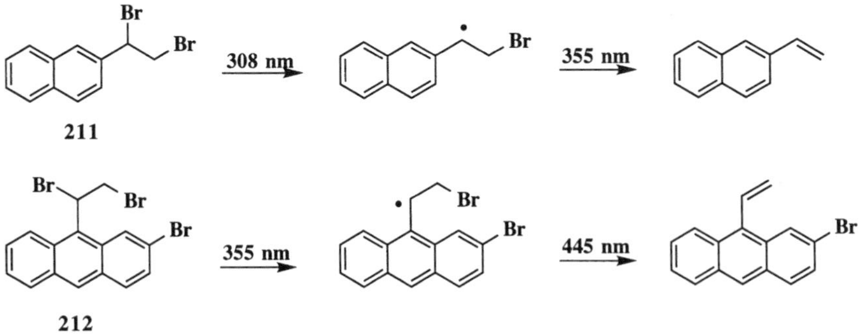

SCHEME 9

TABLE 8 Spectral and Lifetime Data for Fluorescene from Excited Arylmethyl Radicals in Room-Temperature Solution

	Arylmethyl Radical	λ_{max} (emission) nm	$\lambda_{0,0}$ nm	λ_{max} (absorption) nm	τ (ns)	Φ_{fluor}	Solvent	Ref.
104	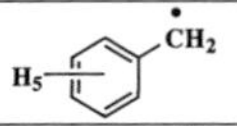 H_5–ring–$\overset{\bullet}{C}H_2$		475		<1		hexane	94
114	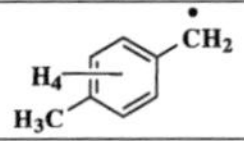 H_4/H_3C–ring–$\overset{\bullet}{C}H_2$	485			14		hexane	94
123	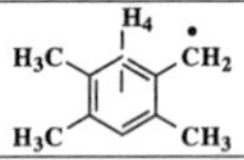 H_3C, H_4, H_3C, CH_3 ring–$\overset{\bullet}{C}H_2$	508			5		toluene	6
131	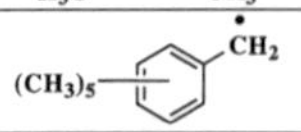 $(CH_3)_5$–ring–$\overset{\bullet}{C}H_2$		503		5.5		toluene	6
136	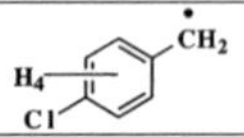 H_4/Cl–ring–$\overset{\bullet}{C}H_2$	470			81		hexane	94
137	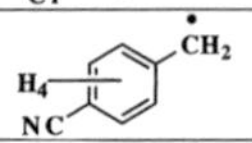 H_4/NC–ring–$\overset{\bullet}{C}H_2$			295	58		hexane	95
138	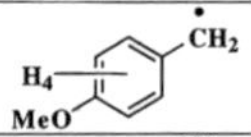 H_4/MeO–ring–$\overset{\bullet}{C}H_2$	490			120		hexane	96
139	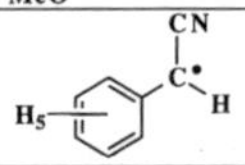 H_5–ring–$\overset{\bullet}{C}(CN)H$		477		<4		hexane	97
140	H_4/H_5 biphenyl–$\overset{\bullet}{C}H_2$	576	385		14		benzene	98,99
132	H_5–ring–$\overset{\bullet}{C}H$–ring–H_5	525	355		280 / 255	0.29	acetonitrile / cyclohexane	100,101 / 102
141	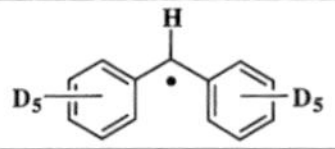 D_5–ring–$\overset{\bullet}{C}H$–ring–D_5	527			390		cyclohexane	102
142	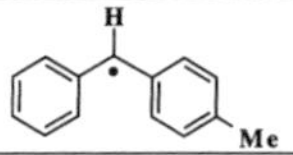 ring–$\overset{\bullet}{C}H$–ring–Me	535			218		cyclohexane	102
143	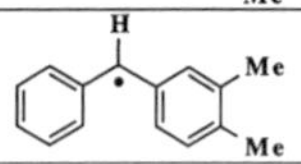 ring–$\overset{\bullet}{C}H$–ring–Me,Me	535			212		cyclohexane	102
144	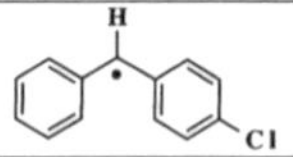 ring–$\overset{\bullet}{C}H$–ring–Cl	537			197		cyclohexane	102
145	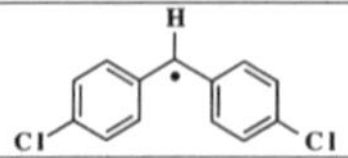 Cl–ring–$\overset{\bullet}{C}H$–ring–Cl	546			178		cyclohexane	102
146	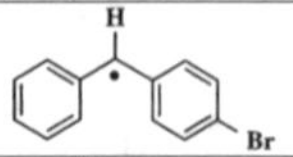 ring–$\overset{\bullet}{C}H$–ring–Br	538			142		cyclohexane	102
147	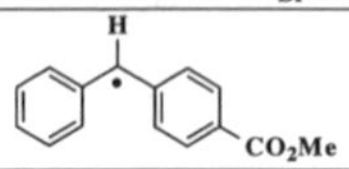 ring–$\overset{\bullet}{C}H$–ring–CO_2Me	552			130		cyclohexane	102

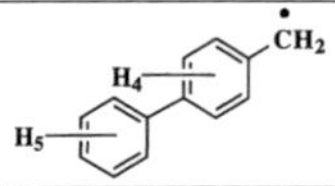

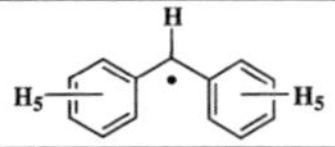

	Arylmethyl Radical	λ_{max} (emission) nm	$\lambda_{0,0}$ nm	λ_{max} (absorption) nm	τ (ns)	Φ_{fluor}	Solvent	Ref.
148	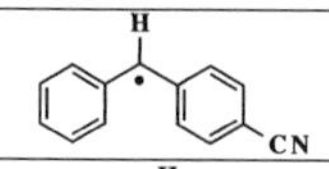		560		80		cyclohexane	102
149	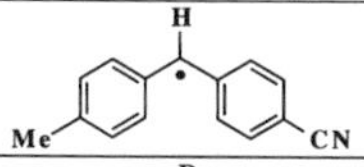		566		83		cyclohexane	102
150	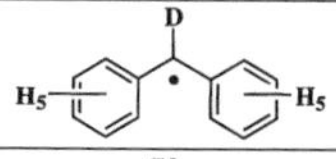				250		cyclohexane	6
151	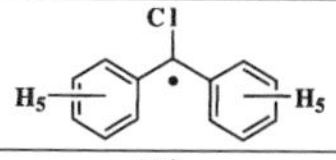		560-600		2-4		various	4
152	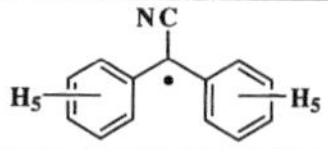		522	358	110 41		hexane methanol	97
134	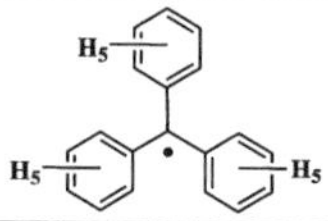		520	440			cyclohexane	103
153	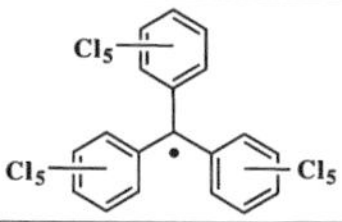			440	8		carbon tetrachloride	104
135	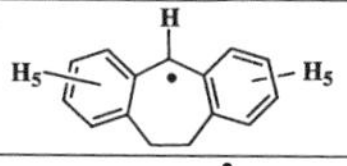			360	260		cyclohexane	105
154	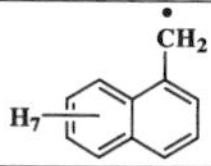		586	430	35		cyclohexane	106
155	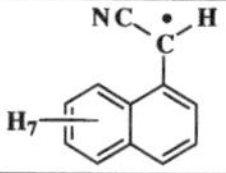		618	415	29 18.1		hexane methanol	97
156	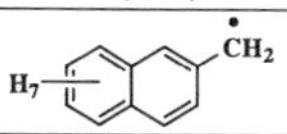		607		27		methanol	99
157	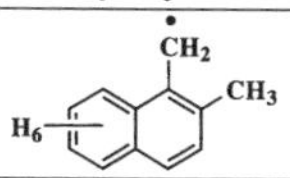		594				hexane	67,107
158	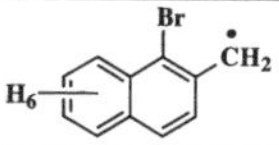		624		24		cyclohexane	99
159	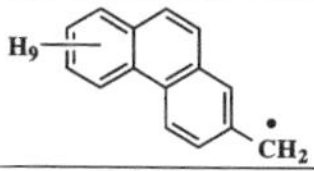		593	400	79 75		toluene	99
160	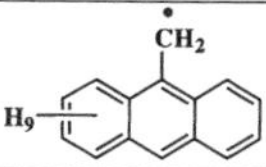		545		21		hexane	107,108

TABLE 9 Spectral and Lifetime Data for Fluorescence from Excited Ketyl Radicals in Low-Temperature (77 K) Matrices

	Ketyl Radical	λ_{max} (emission) nm	τ (ns)	Φ_{fluor}	Solvent	Ref.
161		575	21		poly(vinyl alcohol) film (PVA)	109
			17	0.16	ethanol	110
			21		ether-isopentane-alcohol (EPA)	110
162		578	14	0.06	PVA	109
163		581	15	0.06	PVA	109
164		564	25	0.12	PVA	109
165		578	12	0.04	PVA	109
166		564	15	0.13	PVA	109
167		595		<0.03	PVA	109
168		581	14	0.08	PVA	109
169		595	12	0.07	PVA	109
170		584	13	0.07	PVA	109
171		578	10	0.07	PVA	109
172		575	26	0.10	PVA	109
173		543	48	0.35	PVA	109

In these compounds, the intermediate arylmethyl radical is long-lived. However, further excitation of the radicals in benzene results in production of the bromine atom–benzene charge transfer complex ($\lambda_{max} \approx 550$ nm), an indication of photo-induced cleavage of the remaining C−Br bond. This is further supported by observation of the corresponding vinyl products.

Similar transient behavior—formation of the charge transfer complex—was observed following irradiation of the radical derived from 1-bromo-(2-bromomethyl)naphthalene, i.e., the 1-bromo-2-naphthylmethyl radical (**158**) [99]. However, product evidence was ambiguous with respect to the production of the 1,3-biradical intermediate (Scheme 10).

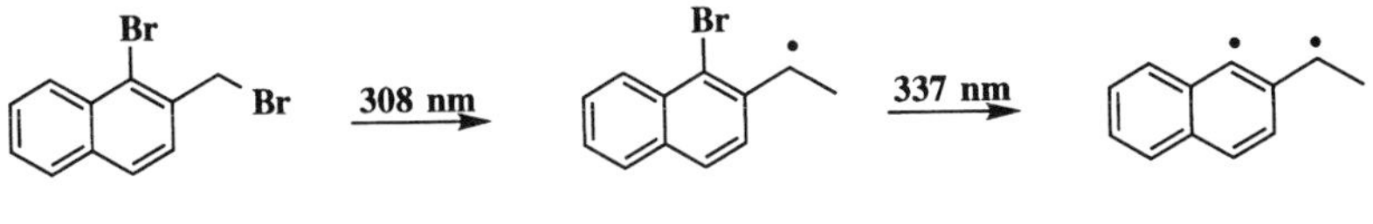

158

SCHEME 10

Laser irradiation of 1,8-bis-(substituted)naphthalenes [chloro- (**213**), bromo- (**214**), phenoxy (**215**), thiophenoxy (**216**), selenophenoxy (**217**)] is believed to follow a similar mechanism [127–130]. Some time ago, biphotonic production of acenaphthene had been suggested to proceed via the photoexcited intermediate monohalogenated naphthylmethyl radical produced from **213**, **214** [127]. More recently, Ouchi and Koga reported a two-color laser flash photolysis (308 + 351 nm) and laser jet study of **215–217**, again showing acenaphthene formation via the intermediate monosubstituted radical [129]. A more detailed study of **213** and **214**, combining laser flash and laser jet techniques, gave definitive evidence for the proposed mechanism and provided more mechanistic details [130] (Scheme 11). The preradical photochemistry itself is quite complex, with C–X

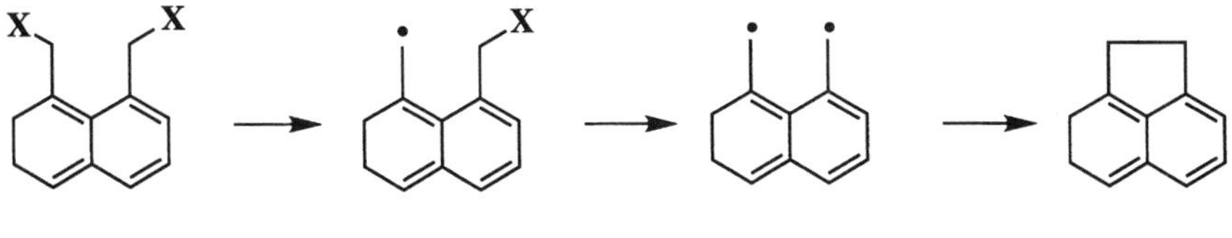

213 - 217

SCHEME 11

cleavage proceeding directly via the precursor S_3 (n, σ^* or π, π^*), S_1, and T_1 (derived by ISC from S_1) states and via an upper (σ^*) triplet produced by ISC from S_2 (as has been suggested for mono-halosubstituted compounds). Photolysis of the monohalo-radical at 351 nm then leads to biradical formation and subsequently to acenaphthene. Again, this study shows that the combination of time-resolved transient analysis and laser jet product analysis is a powerful photochemical mechanistic tool.

TABLE 10 Spectral and Lifetime Data for Fluorescence from Excited Ketyl Radicals in Room-Temperature Solution

	Ketyl Radical	λ_{max} (emission) nm	τ (ns)	Φ_{fluor}	Solvent	Ref.
161		574	3.9	0.11	toluene	111-113
174		570	8.7		toluene-d$_8$	112,113
175			4.4		toluene	112
176		570	10.5		toluene-d$_8$	112,113
177		585	3.5	0.09	toluene	114
178		575			toluene	7
179		588	3.1		toluene	113
180		577			toluene	7
181		610	<1	0.02	toluene	114
182		623	<1	0.003	toluene	113,114
168		586			toluene	7
183		574			toluene	7
169		596	5.9	0.12	toluene	113,114

	Ketyl Radical	λ_{max} (emission) nm	τ (ns)	Φ_{fluor}	Solvent	Ref.
184	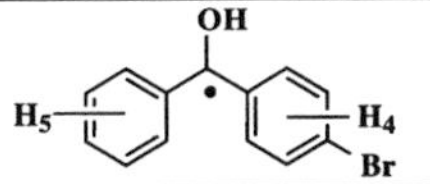 H_5–C$_6$H$_4$–Ċ(OH)–C$_6$H$_4$–H_4 (4-Br)	585			toluene	7
185	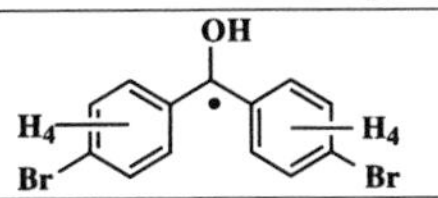 (4-Br)H_4–C$_6$H$_4$–Ċ(OH)–C$_6$H$_4$–H_4 (4-Br)	605	<1	0.02	toluene	114
186	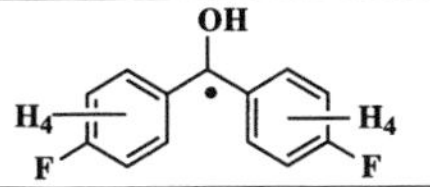 (F)H_4–C$_6$H$_4$–Ċ(OH)–C$_6$H$_4$–H_4 (F)	586			toluene	7
187	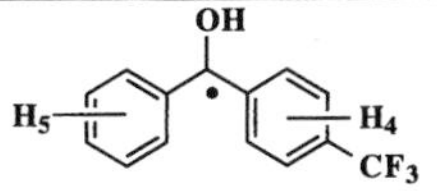 H_5–C$_6$H$_4$–Ċ(OH)–C$_6$H$_4$–H_4 (4-CF$_3$)	590	6.0	0.15	toluene	114
188	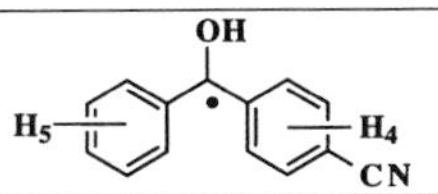 H_5–C$_6$H$_4$–Ċ(OH)–C$_6$H$_4$–H_4 (4-CN)	613	7.8		toluene	113
189	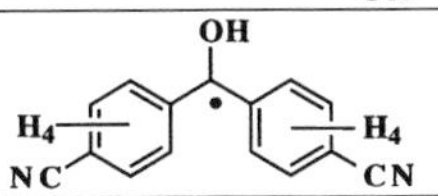 (NC)H_4–C$_6$H$_4$–Ċ(OH)–C$_6$H$_4$–H_4 (CN)	586			toluene	7
190	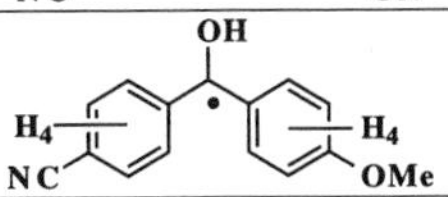 (NC)H_4–C$_6$H$_4$–Ċ(OH)–C$_6$H$_4$–H_4 (OMe)	639			toluene	7
191	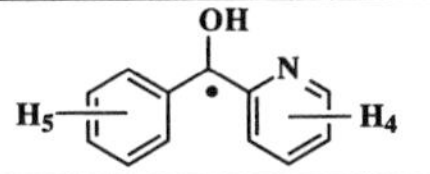 H_5–C$_6$H$_4$–Ċ(OH)–(pyridin-2-yl)H_4	567	5.9		toluene	6
192	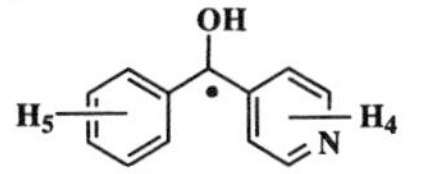 H_5–C$_6$H$_4$–Ċ(OH)–(pyridin-4-yl)H_4	559	7.4		toluene	6
193	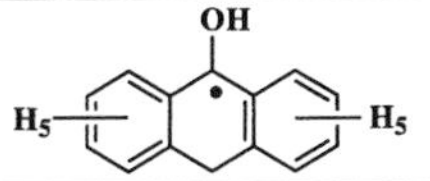 dibenzo ketyl radical, OH, H_5 / H_5	580	18		benzene	115
194	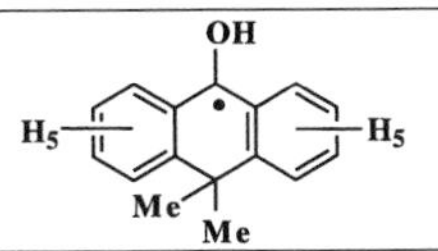 dibenzo ketyl radical, OH, H_5 / H_5, Me Me	560	33		toluene	116
195	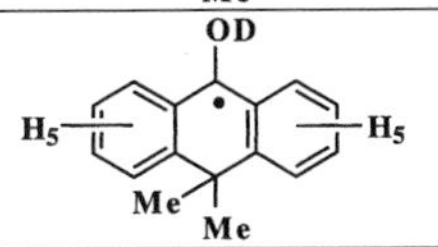 dibenzo ketyl radical, OD, H_5 / H_5, Me Me	560	44		toluene-d_8	115,116

TABLE 10 Continued

	Ketyl Radical	λ_{max} (emission) nm	τ (ns)	Φ_{fluor}	Solvent	Ref.
196	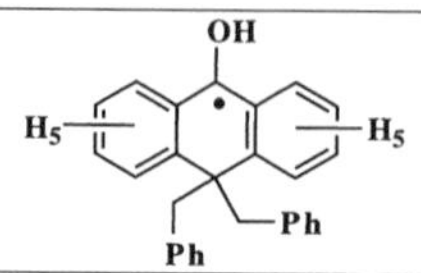	564	7.9		toluene	115
197	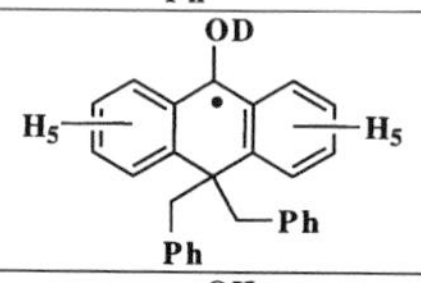	564	9.7		toluene-d$_8$	115
198	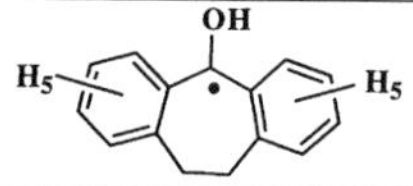	575	5.5		toluene	2

Biradical formation via photolysis of an intermediate mono-halo radical has also been suggested for 1,3-bis-(chloromethyl)benzene (**218**) with spectral evidence for both the biradical and mono-halo radical obtained at room temperature [131] (Scheme 12). However, it was not possible to determine whether biradical

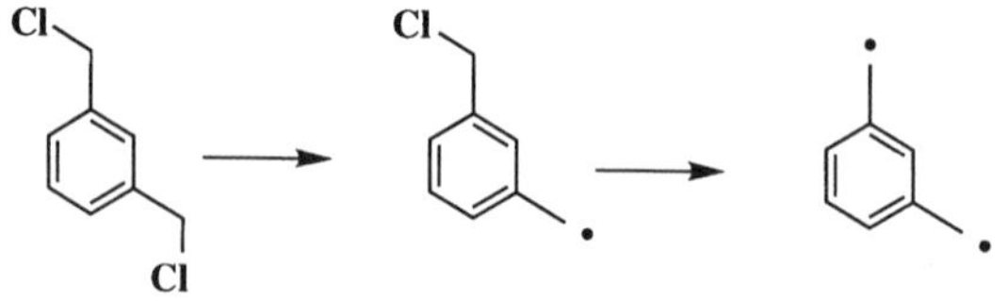

218

SCHEME 12

formation was due to a biphotonic process involving the radical or an upper triplet state, or a monophotonic cleavage of both C—Cl bonds (tandem excited state process; see Wilson and Schnapp [3]).

Bond cleavage from excited radicals has also been observed in a laser jet study of the *bis*-ether (**219**). Thus, the excited 9-anthrylmethyl radical is suggested to undergo loss of phenoxy radical to yield the 9,10-bis-anthrylmethyl biradical (Scheme 13). This is a three-photon process, with two photons required to produce the monoradical and an additional photon needed to generate the biradical. The authors note that the biradical itself may have photochemistry, forming the 9,10-bis-methylether (**220**) upon further photolysis in methanol [132].

A family of α,α-dialkoxy substituted benzyl radicals (**221**) has been shown to undergo β-cleavage both from their ground and excited states. This is in contrast to the stability of benzyl-type radicals toward permanent photochemical

TABLE 11 Spectral and Lifetime Data for Fluorescence from Excited Heteroatom-Centered Radicals at Room Temperature and in Low-Temperature (77 K) Matrices

	Hetero Radical	Temperature	$\lambda_{0,0}$ (emission) nm	τ (ns)	Solvent	Ref.
199	^tBu, ^tBu, ^tBu substituted phenoxyl radical (H_2)	room	400		benzene	117
200	furanyl radical with two phenyl groups (H_5, H_5)	room	617	10.7	benzene/di-$tert$-butyl peroxide	118,119
201	furanyl radical with phenyl (H_5) and Me-phenyl (H_4) groups	room	627	12.5	benzene/di-$tert$-butyl peroxide	118,119
202	furanyl radical with phenyl (H_5) and MeO-phenyl (H_4) groups	room	640	11.6	benzene/di-$tert$-butyl peroxide	118,119
203	thiyl radical (H_5), S	77 K	638 400	29,000 ($D_1 \to D_0$) 0.073 ($D_2 \to D_0$)	3-methylpentane	120
204	H_2N-phenyl thiyl radical (H_4), S	room	~580	< 100	2,2,4-trimethylpentane	121
205	H_2N-phenyl thiyl radical (H_4), S	77 K	410	12	dimethyl ether	122
206	Me-phenyl NH radical (H_4)	77 K	400		dimethyl ether	122
207	Me-phenyl NH radical (H_4)	77 K	405		dimethyl ether	122
208	Me-phenyl NH radical (H_4)	77 K	412		dimethyl ether	122
209	phenyl NEt radical (H_5)	77 K	417	12	dimethyl ether	122
210	phenyl SiMe$_2$ radical (H_5)	77 K	504	1000	3-methylpentane	123

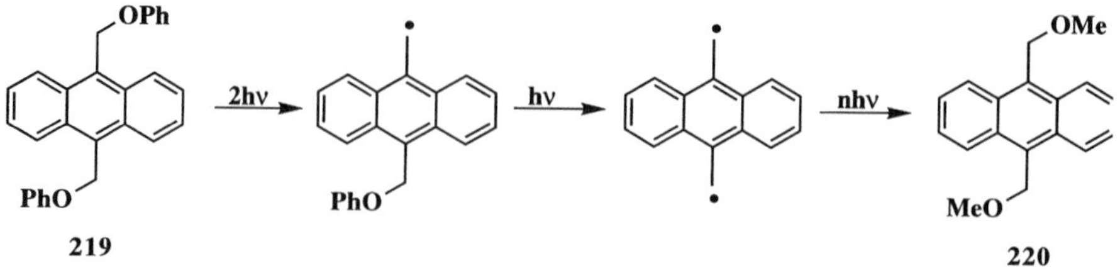

SCHEME 13

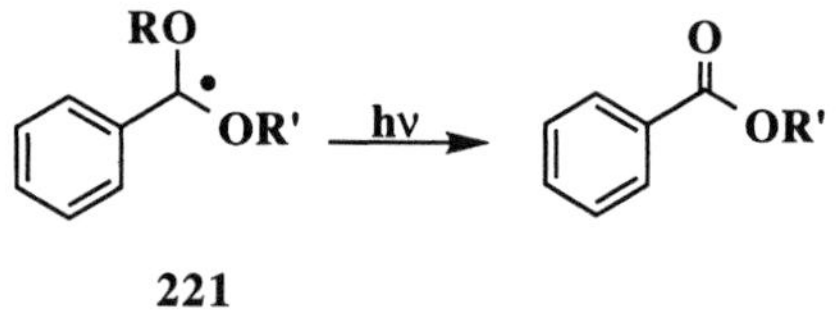

221

SCHEME 14

change (Scheme 14). An initial study involving the symmetrical substituted derivatives (R = R' = Me) was later expanded to include asymmetrically substituted derivatives (R, R' = Me, Et, i-Pr; R ≠ R'). In the absence of light these asymmetric radicals cleave according to the stability of the radical fragment. In the excited radical, however, no such selectivity was observed, indicating that the energy of the excited state is more than sufficient to compensate for the instability of the fragment [133,134].

The 4-methoxybenzoyloxy radical (222) also undergoes cleavage from its excited state (Scheme 15). Irradiation of the radical at 700 nm leads to photobleaching that has been assigned to decarboxylation yielding the 4-methoxyphenyl radical [135].

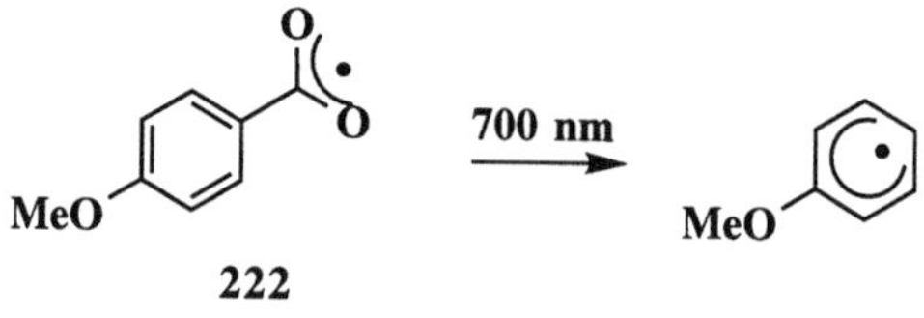

222

SCHEME 15

Bond cleavage in excited radicals has also been suggested to lead to carbene formation (Scheme 16). For example, in a low-temperature matrix, photolysis of dichlorodiphenylmethane (223) leads to production of dipenylcarbene via the chlorodiphenylmethyl radical [136]. In an analogous fashion, the 10,11-dihydro-5H-dibenzocycloheptenyl radical (135) undergoes photolysis in polar solvents to yield a product identified as having an o-xylylene structure (224), which is presumably formed by rearrangement from an intermediate carbene. Interestingly,

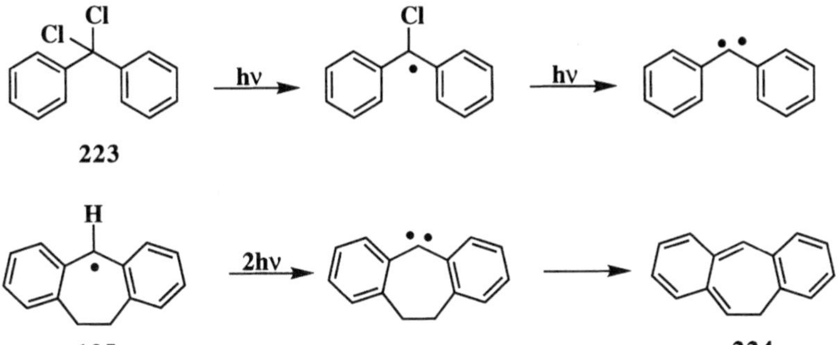

SCHEME 16

the lowest excited state of (135) is inert toward permanent chemical change, and thus it is an upper excited state of the radical that undergoes carbene formation.

The photochemistry of benzophenone-derived ketyl radicals [unsubstituted (161); 4-Me (177); 4,4'-Me$_2$ (179); 4-MeO (181); 4,4'-MeO$_2$ (182); 4,4'-Cl$_2$ (169); 4,4'-Br$_2$ (185); 4-CF$_3$ (187)], as well as ketyl radicals derived from closely related compounds [anthrone (193); 10,10-dimethylanthrone (194)], has been studied extensively by the two-color technique and has been found to involve, among other processes, cleavage of the O—H bond, to yield the corresponding benzophenone [113–115,137]. For 161 in cyclohexane solvent, this process contributes along with fluorescence and other nonradiative and chemical processes (vide infra) to a large quantum yield for disappearance of the radical ($\Phi = 0.27$). Confirmation of cleavage was provided by a decrease in ketyl radical derived products, an increase in the benzophenone yield and the detection by Raman spectroscopy of molecular hydrogen. (The H atom produced by cleavage from the excited ketyl radical abstracts hydrogen from the solvent.) Further support came from the significant deuterium isotope effect observed (O—H vs. O—D), $k_H/k_D = 5.8$.

A much larger quantum yield of disappearance ($\Phi = 0.75$) was observed for the 10,10-dibenzylanthrone ketyl radical (196) upon irradiation [115] (Scheme 17).

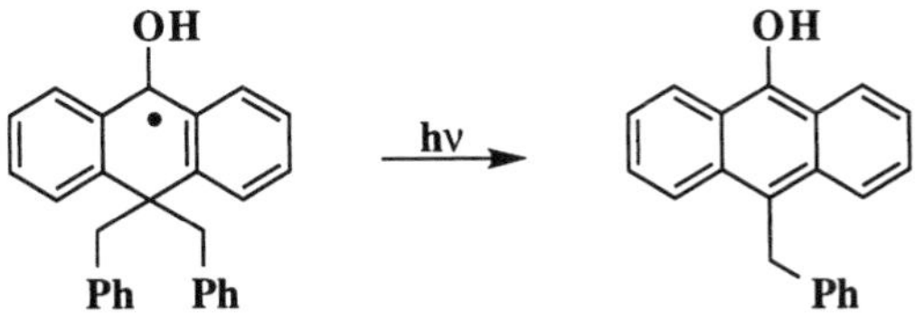

SCHEME 17

However, the dramatic increase in this case was attributed primarily to cleavage remote from the radical center to yield a benzyl radical.

B. Bond Formation

The photochemistry of triphenylmethyl radicals in solution has been of interest for some time because the relatively long lifetimes of these species (some are essentially stable) have made their study accessible with low-intensity light sources. Relatively recently, laser flash photolysis has been used to confirm the tendency of the excited radicals to undergo ring closure to form the 9-phenyldihydrofluorenyl radicals.

In the case of the excited triphenylmethyl radical itself (**134**), ring closure occurs with a 10-ns lifetime (Scheme 18). Confirmation of this process was

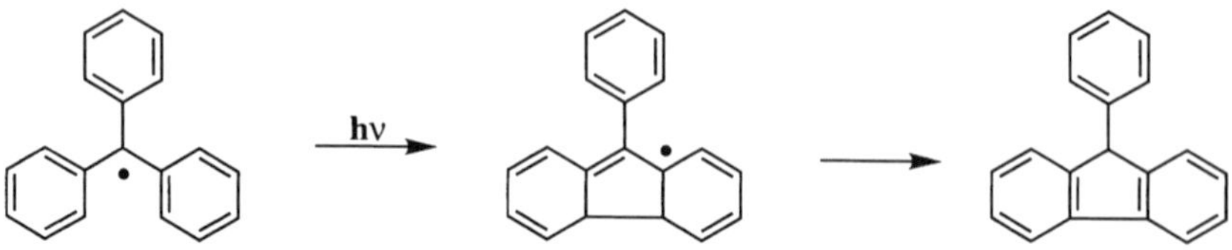

134

SCHEME 18

provided by the observation of a 490-nm transient, similar to the 9-phenylfluorenyl radical absorption, which formed concurrently with the photoinduced disappearance of the triphenylmethyl radical [103,105]. Earlier steady-state experiments also indicated the formation of products consistent with production of the fluorenyl radical [138].

Excitation of the perchlorotriphenylmethyl radical leads to analogous product formation, although in this case the final perchloro-9-phenylfluorenyl radical is believed to form initially via a charge-separated intermediate and then a ring-closed species that subsequently loses two chlorine atoms [104].

Different behavior is observed for the stable allyl radical, bis-(9-fluorenyl)-phenylmethyl radical (**225**) [139] (Scheme 19). Photolysis of this radical in poor H atom donating solvents is suggested to lead to the intermediate (**226**) which subsequently reacts with molecular oxygen yielding a variety of cleavage and oxygenation products. The excited radical lifetime in this case is less than 30 ps.

Cyclization to form the ring-closed dihydrofluorenyl radical is a process that appears to be limited to diarylmethyl radicals that are substituted at the central carbon atom. Thus the excited triphenylmethyl, diphenylethyl, and diphenylcyclopropylmethyl radicals all form ring-closed radicals whereas the parent diphenylmethyl radical does not. The reason for this different behavior lies in the steric crowding produced by introducing a substituent at the central carbon atom. Whereas in the unsubstituted diphenylmethyl radical the angle between the phenyl

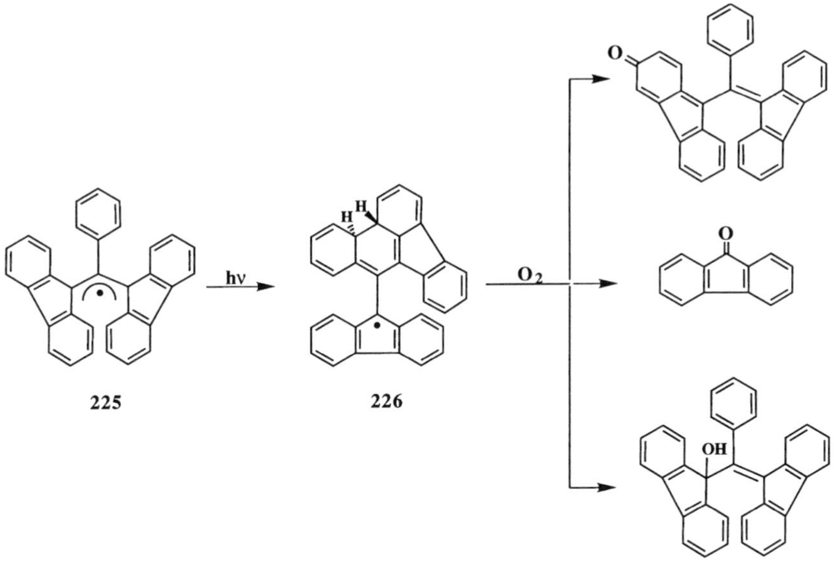

225 226

SCHEME 19

rings is sufficiently large to allow the rings to take up a co-planar geometry, in the substituted radicals the rings are forced out-of-plane, producing good overlap at the ortho positions and thereby facilitating the cyclization [3,105,140].

The excited states of ketyl radicals also undergo bond formation reactions (Scheme 20). Laser jet investigations of **161** and 9-hydroxyxanthenyl radical

227

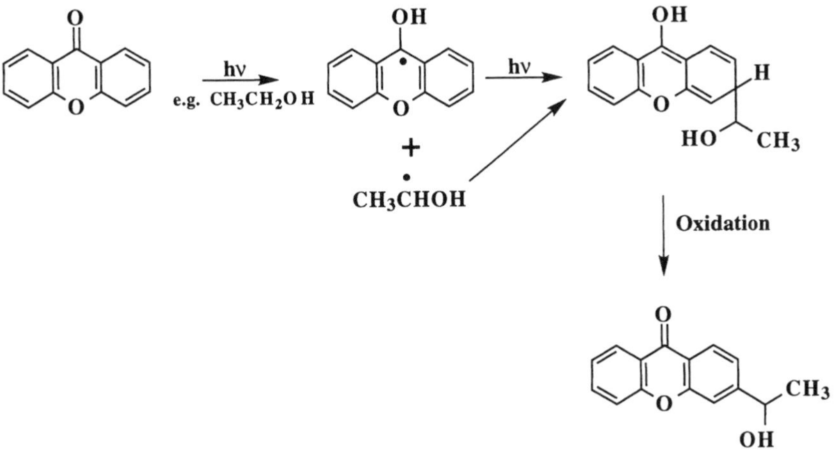

SCHEME 20

(**227**) in alcohol solvents has shown that while head-to-head radical-radical coupling occurs for the ground-state radicals, the excited radicals prefer a head-to-tail pathway involving attack at a ring position [141–143].

C. Photoionization

In addition to the decay pathways already described, excited arylmethyl radicals are also known to undergo photoionization in polar solvents. The triphenylmethyl radical (**134**) undergoes monophotonic ionization with an efficiency of 75–85% when irradiated at 308 nm or 248 nm in aqueous solution as evidenced by the spectroscopic detection of e_{aq}^- and the triphenylmethyl carbocation [144]. In contrast, diphenylmethyl radical (**132**) ionization is a two-photon process (from the ground-state radical), requiring the formation of an upper doublet state of the radical [101,145,146]. That photoionization occurs in this system was originally indicated by the lack of reactivity when the radical was irradiated in various solvents at 337 nm. However, when irradiation was carried out at 347 nm in ethanol or water, efficient photobleaching was observed. While this bleaching was originally believed to be due to carbene extrusion, subsequent laser jet studies in methanol and carbon tetrachloride (CCl_4) and laser flash studies in acetonitrile and water:alcohol mixtures have revealed a more complex excited state chemistry. In the laser jet experiments, at low laser intensities in CCl_4, charge transfer quenching and subsequent chlorodiphenylmethane formation occur. However at high intensity in methanol, methoxydiphenylmethane is formed as a result of photoionization and trapping of the resultant carbocation by solvent. It was concluded that the CCl_4 trapping reaction occurs from the lowest excited state, whereas photoionization takes place from an upper state. Given this mechanism, it is possible to rationalize the original laser flash data. Apparently, the absorption coefficient of the ground state radical which is close to a maximum at 337 nm, is sufficiently large to prevent the excited radical from absorbing a second photon when irradiation takes place at this wavelength. At 347 nm, on the other hand, the excited radical competes more efficiently for the laser photons and the upper excited state can be produced efficiently.

More recently, photoionization following the same mechanism as described for **132** has been reported for 1-naphthylmethyl radical (**154**) [147] and 4-biphenylmethyl radical (**140**) [148] (Scheme 21). Thus, the lowest excited state of the radical forms a chloro-adduct in carbon tetrachloride, presumably as a result of charge transfer followed by trapping, while an upper excited state produced by biphotonic excitation photoionizes and the carbocation product is quenched by alcohol.

It is notable that the study of **154** involved a combination of laser flash and laser jet techniques which again points to the utility of the latter in obtaining product information about transients that are produced on too small a scale to be studied by laser flash photolysis.

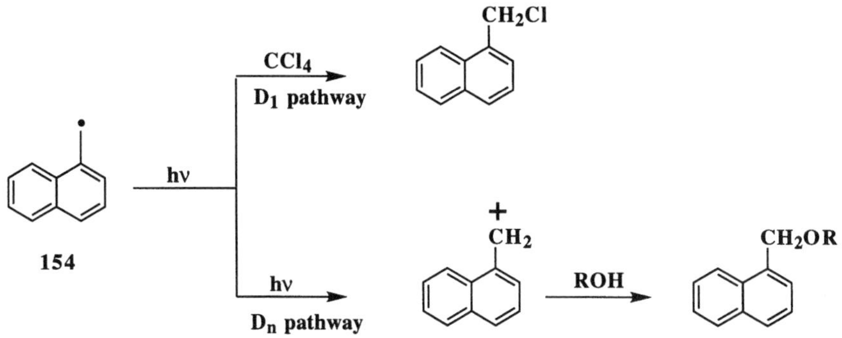

SCHEME 21

[While Scheme 21 shows the pathways specifically for **154**, it is also applicable to **132** and **140**.]

Photoionization is also an important (although not the major) decay pathway for the excited diphenylketyl radical (**161**) in polar solvents such as acetonitrile [137] (Scheme 22). This was confirmed with the observation of benzophenone radical anion in laser flash experiments following 515-nm photolysis of ground-state **161**. The anion radical is apparently produced as a result of electron ejection followed by trapping of the electron by ground-state benzophenone.

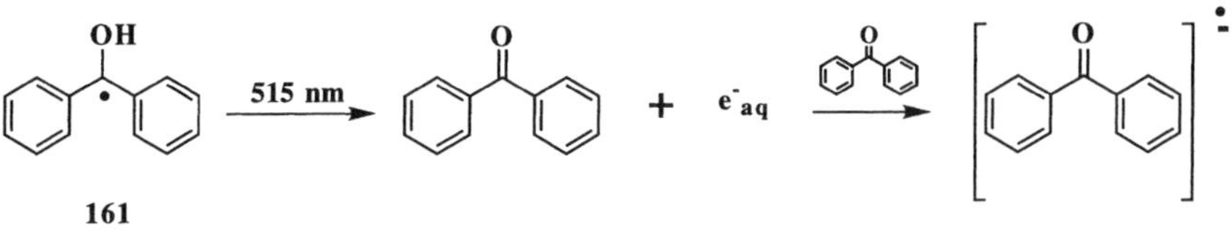

SCHEME 22

D. Intermolecular Reactions

Tables 12 to 14 list information about the known intermolecular reactions of excited radicals. The reactions are listed for each radical and separate tables are used for excited benzyl, arylmethyl (other than benzyl), and ketyl radicals.

In general, the excited radicals shown have been found to possess enhanced donor/acceptor properties, especially in interactions with such quenchers as oxygen, dienes, amines, and halides. To date, the most well-characterized intermolecular processes are charge (electron) transfer reactions between excited diphenylmethyl radicals and electron donors and acceptors. Thus, excited **132** reacts with methyl benzoates and benzyl bromides (Schemes 23 and 24) with rates that increase with the increasing electron withdrawing ability of substituents on the

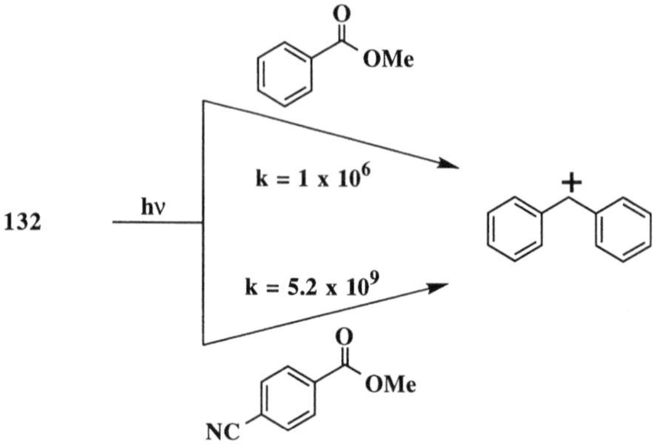

SCHEME 23

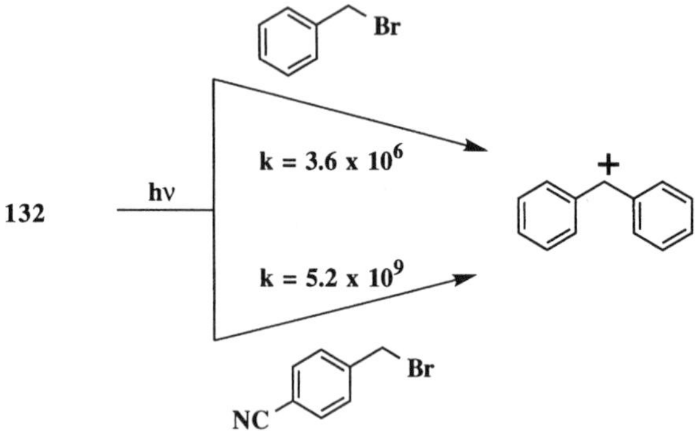

SCHEME 24

quenchers. Transient conductivity and absorption measurements confirmed the formation of the diphenylmethyl cation in these quenching reactions [149].

Charge transfer reactions of **132**, cyano-substituted diphenylmethyl radicals (**148, 152**), and 2-phenanthrylmethyl radical (**159**) with CCl_4 and CBr_4 have also been reported [97,99,101]. A decreased rate constant for CBr_4 quenching of the α-cyano-substituted diphenylmethyl radical (**152**) relative to the parent radical provided indirect evidence of charge transfer [97,145]. More recently, the diphenylmethyl cation was observed directly in acetonitrile solvent with added CCl_4 [145,149] (Scheme 25).

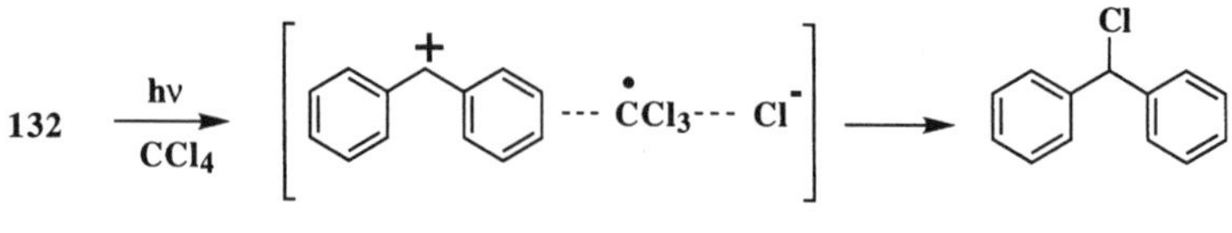

SCHEME 25

TABLE 12 Rate Constants for the Reactions of Excited Benzyl and Substituted Benzyl Radicals with Added Quenchers in Room-Temperature Solution

	Radical	Quencher	k $M^{-1}s^{-1}$	Solvent	Ref.
228	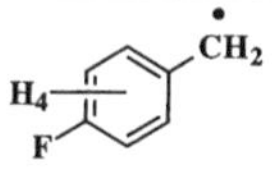 H4—⟨⟩—CH2 F	O_2	1.9×10^{10}	hexane	98
		1,4-cyclohexadiene	6.3×10^{6}	hexane	95
136	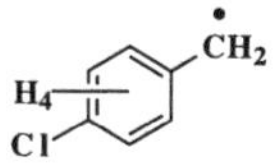 H4—⟨⟩—CH2 Cl	O_2	1.8×10^{10}	hexane	98
		1,4-cyclohexadiene	2.0×10^{8}	hexane	95
137	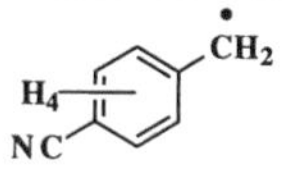 H4—⟨⟩—CH2 NC	O_2	6.7×10^{9}	hexane	98
		1,4-cyclohexadiene	3.2×10^{9}	hexane	95
138	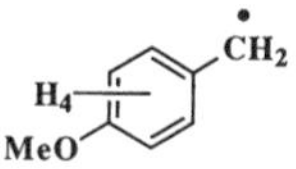 H4—⟨⟩—CH2 MeO	O_2	2.2×10^{10}	hexane	98
		1,4-cyclohexadiene	9.4×10^{5}	hexane	95

Direct spectroscopic evidence for charge transfer was also obtained using methyl viologen as an acceptor. Formation of the reduced species (methyl viologen radical cation) was observed upon irradiation of **132** and **154** [101,106] (Scheme 26).

132 $\xrightarrow{h\nu}$

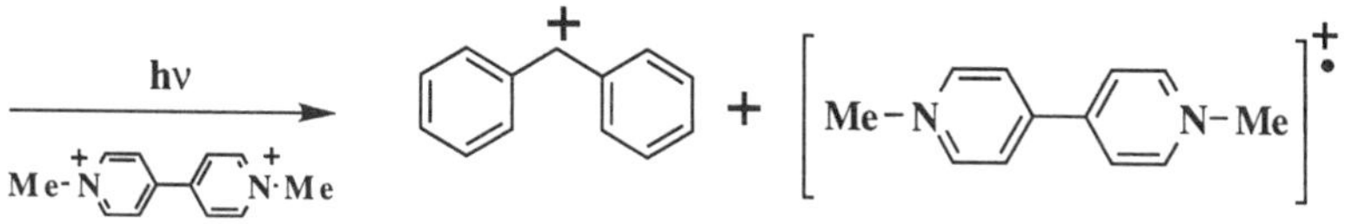

λ_{max}= 390, 600 nm

SCHEME 26

TABLE 13 Rate Constants for the Reactions of Excited Arylmethyl and Substituted Arylmethyl Radicals wth Added Quenchers in Room-Temperature Solution

	Arylmethyl Radical	Quencher	k $M^{-1}s^{-1}$	Solvent	Ref.
132	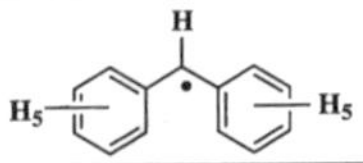	O_2	8.7×10^9	cyclohexane	101
		1,4-cyclohexadiene	1.1×10^6	cyclohexane	101
		1,3-cyclooctadiene	3.4×10^8	hexane	97
		2,5-dimethyl-2,4-hexadiene	3.2×10^8	hexane	97
		methyl methacrylate	4.0×10^6	cyclohexane	101
		triethylamine	2.1×10^8	methanol	101
			4.2×10^8	hexane	97
		n-butylamine	1.3×10^7	hexane	97
		1,3,3,5,7,7-hexamethyl-1,5-diazacyclooctane	4.5×10^9	cyclohexane	101
		1-methylnaphthalene	2.9×10^7	hexane	97
		carbon tetrachloride	3.3×10^8	acetonitrile	149
		carbon tetrabromide	3.8×10^{10}	acetonitrile	97
		m-dicyanobenzene	4.3×10^9	acetonitrile	149
		p-dicyanobenzene	1.6×10^{10}	acetonitrile	149
		methyl benzoate	1×10^6	acetonitrile	149
		methyl p-chlorobenzoate	3×10^6	acetonitrile	149
		methyl m-chlorobenzoate	1.1×10^8	acetonitrile	149

Arylmethyl Radical	Quencher	k $M^{-1}s^{-1}$	Solvent	Ref.
	methyl p-bromobenzoate	1.3×10^8	acetonitrile	149
	methyl m-chlorobenzoate	2.3×10^8	acetonitrile	149
	methyl p-cyanobenzoate	5.2×10^9	acetonitrile	149
	methyl p-(trifluoromethyl)benzoate	4.2×10^9	acetonitrile	149
	methyl p-nitrobenzoate	1.3×10^{10}	acetonitrile	149
	dimethyl terephthalate	2.0×10^8	acetonitrile	149
	benzyl chloride	1.3×10^6	acetonitrile	149
	benzyl bromide	3.6×10^6	acetonitrile	149
	p-bromobenzyl bromide	8.6×10^7	acetonitrile	149
	p-fluorobenzyl bromide	9.0×10^6	acetonitrile	149
	m-fluorobenzyl bromide	1.1×10^8	acetonitrile	149
	p-cyanobenzyl bromide	5.2×10^9	acetonitrile	149
	p-nitrobenzyl bromide	1.7×10^{10}	acetonitrile	149
	methylviologen	1.3×10^{10}	4:1 acetonitrile-water	101
	trifluoroacetic anhydride	1.4×10^8	acetonitrile	149
	di-tert-butyl nitroxide	5.3×10^9	acetonitrile	151

TABLE 13   Continued

	Arylmethyl Radical	Quencher	k M⁻¹s⁻¹	Solvent	Ref.
		2,2,6,6-tetramethyl-piperidinyl-1-oxy radical	5.0×10^9	acetonitrile	151
		4-hydroxy-2,2,6,6-tetramethyl-piperidinyl-1-oxy radical	4.8×10^9	acetonitrile	151
		galvinoxyl radical	7.8×10^7	benzene	151
148	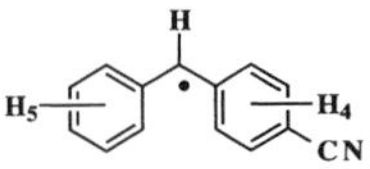	1,3-cyclooctadiene	2.9×10^8	hexane	97
		2,5-dimethyl-2,4-hexadiene	1.4×10^{10}	hexane	97
		triethylamine	1.5×10^{10}	hexane	97
		n-butylamine	6.7×10^7	hexane	97
		1-methylnaphthalene	6.4×10^7	hexane	97
		carbon tetrabromide	4.3×10^9	hexane	97
152	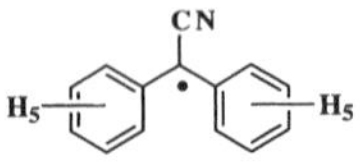	1,3-cyclooctadiene	1.2×10^9	hexane	97
		2,5-dimethyl-2,4-hexadiene	1.6×10^{10}	hexane	97
		triethylamine	2.3×10^{10}	hexane	97
		n-butylamine	9.7×10^9	hexane	97
		1-methylnaphthalene	1.7×10^{10}	hexane	97
		carbon tetrabromide	$<1 \times 10^6$	hexane	97

	Arylmethyl Radical	Quencher	k $M^{-1}s^{-1}$	Solvent	Ref.
154	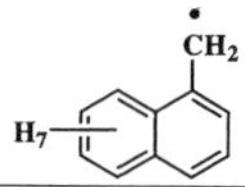	O_2	4.7×10^9	cyclohexane	106
		1,4-cyclohexadiene	$<5 \times 10^5$	1,4-cyclohexadiene	106
		1,3-cyclooctadiene	5×10^4	hexane	97
		triethylamine	$<2 \times 10^6$	triethylamine	2
		1,3,3,5,7,7-hexamethyl-1,5-diazacyclooctane	2.6×10^9	cyclohexane	106
		carbon tetrachloride	4×10^6	hexane	97
		methylviologen	3.8×10^{10} 2.6×10^9	methanol cyclohexane	106 106
		tri-n-butylstannane	$\sim 4 \times 10^5$	tri-n-butylstannane	106
155	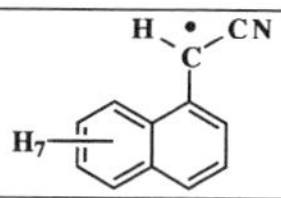	1,3-cyclooctadiene	3.8×10^8	hexane	97
		2,5-dimethyl-2,4-hexadiene	1.6×10^{10}	hexane	97
		triethylamine	1.4×10^9	hexane	97
159	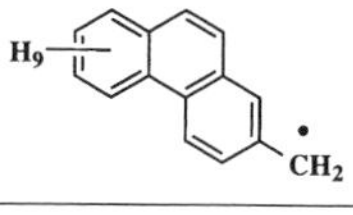	O_2	5.7×10^9	benzene	99
		carbon tetrachloride	1×10^6	benzene	99
153	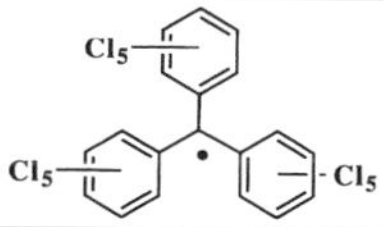	N,N-diethylaniline	1.6×10^{10}	carbon tetrachloride	104
		triphenylamine	$1. \times 10^{10}$	carbon tetrachloride	104
		thianthrene	1.5×10^{10}	carbon tetrachloride	104

TABLE 14 Rate Constants for the Reactions of Excited Diphenylketyl and Substituted Diphenylketyl Radicals with Added Quenchers in Solution at Room Temperature

	Ketyl Radical	Quencher	k M^{-1}s^{-1}	Solvent	
161		carbon tetrabromide	3.8 x 10^{10}	acetonitrile	150
		cyclohexane	4 x 10^7	cyclohexane	152
		2-propanol	1.1 x 10^6	cyclohexane	152
		di-*tert*-butyl peroxide	1.9 x 10^9	benzene	113
		p-chlorophenyldiazonium tetrafluoroborate (4-C$_6$H$_5$N$_2$$^+$)(BF$_4$)$^-$	2.1 x 10^{10}	acetonitrile	150
		diphenyliodonium tetrafluoroborate ((C$_6$H$_5$)$_2$I)$^+$(BF$_4$)$^-$	1.2 x 10^{10}	acetonitrile	150
		triphenylsulfonium tetrafluoroborate ((C$_6$H$_5$)$_3$S)$^+$(BF$_4$)$^-$	1.5 x 10^{10}	acetonitrile	150
		dimethylphenylsulfonium tetrafluoroborate ((CH$_3$)$_2$C$_6$H$_5$S)$^+$(BF$_4$)$^-$	9.7 x 10^9	acetonitrile	150
		trimethylsulfonium iodide ((CH$_3$)$_3$S)$^+$(I)$^-$	1.2 x 10^9	acetonitrile	150
		triphenylbenzylphosphonium chloride ((C$_6$H$_5$)$_3$C$_6$H$_5$CH$_2$P)$^+$(Cl)$^-$	1 x 10^8	acetonitrile	150
174		methyl methacrylate	3.9 x 10^9	toluene-d$_8$	112
194		1,4-cyclohexadiene	8.4 x 10^5	toluene	116
		methyl methacrylate	3.9 x 10^8	toluene	116
		triethylamine	1.4 x 10^9	toluene	116
		2-propanol	9.3 x 10^8	toluene	116

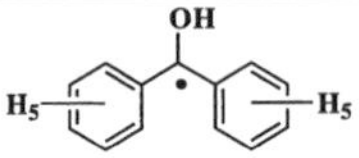

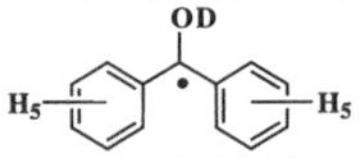

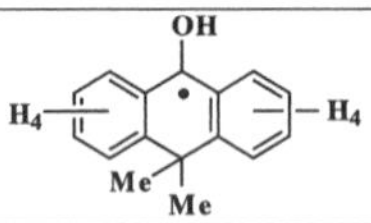

Charge transfer reactions involving excited arylmethyl and ketyl radicals have also been observed for a series of amines [97,101,104,106,116], 1-methyl-naphthalene [97], a variety of onium salt acceptors (diazonium, iodonium and sulfonium) [150], and with methylmethacrylate [101,112,116] as donors.

Charge transfer has also been demonstrated in the reactions of excited cyano-substituted radicals (**152, 155**) with a series of dienes [97]. Whereas reaction with the unsubstituted excited parent radicals is believed to involve H-atom transfer [95,97,101,106,116] (vide infra), correlation between the free energies of electron transfer and the observed quenching rates for excited **152** and **155** with 2,5-dimethyl-2,4-hexadiene and 1,3-cyclooctadiene indicates a charge transfer process.

Hydrogen atom abstraction from dienes is a recognized decay pathway for excited benzyl radicals and is particularly efficient for the 4-substituted radicals [4-CN (**137**); 4-F (**228**); 4-Cl (**136**)] [95]. The reason given for the enhancement of rate by substitution, which is observed particularly for the 4-cyanobenzyl radical, is the switch of the lowest doublet excited state from $1A_2$ to $2B_2$ configuration [98]. In the latter configuration, more electron density is to be found at the benzylic carbon, thus facilitating H-atom abstraction. Excited **132** and **154** are also believed to undergo abstraction from dienes, although the rates are relatively slow [4]. As noted above, upon cyano-substitution the rate increases but this is associated with a change of mechanism to charge transfer. Excited ketyl radicals also exhibit enhanced reactivity toward H-atom donors. In addition to reaction with dienes, laser jet studies have demonstrated the reaction of excited **161** with alcohols [141,143] (Scheme 27).

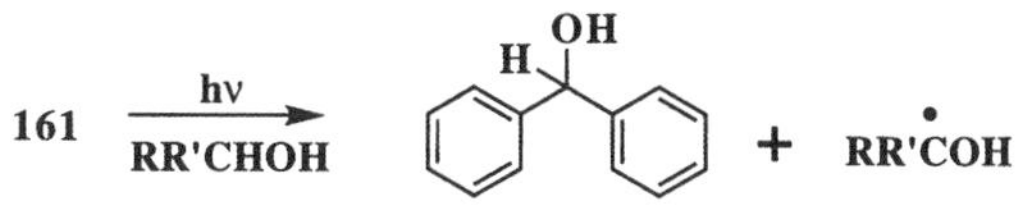

SCHEME 27

The rate of reaction of excited radicals with oxygen is also enhanced in comparison with the corresponding ground-state radical [98,99,101,106]. In most cases the interaction is believed to be a physical one since no permanent depletion of the ground state radical is observed. Again, in the case of the 4-cyanobenzyl radical (**137**), different behavior is observed, in this case lower reactivity than for the excited parent radical. This has been attributed to a change in quenching mechanism to one that involves peroxy radical formation and permanent depletion of the ground-state radical [98] (Scheme 28). The localization of electron density

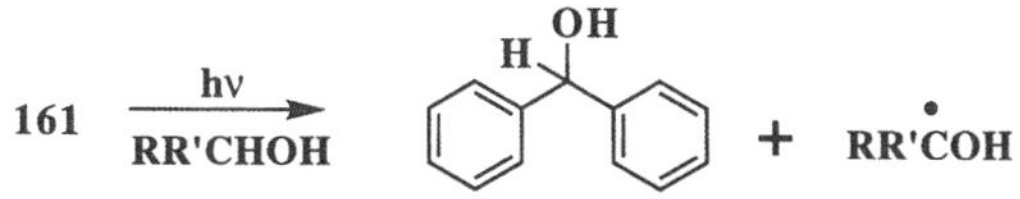

SCHEME 28

at the benzylic carbon is again believed responsible for promoting this chemical reaction.

VI. SUMMARY

The recent development of the two-color and laser jet techniques has facilitated the exploration of the chemistry and photophysics of upper excited state and excited reaction intermediates, including neutral radicals. Upper states and excited radicals participate in many of the same reactions as lower states (but often with different efficiency) as well as in many reactions that are not accessible to lower states because of energy considerations. The use of two-color and laser jet techniques in several recent studies is a powerful combination that is just now beginning to be utilized in the study of multiphoton transients.

REFERENCES AND NOTES

1. Turro, N. J.; Ramamurthy, V.; Cherry, W.; Farneth, W. *Chem. Rev.* 1978, *78*, 127.
2. Scaiano, J. C.; Johnston, L. J. in *Organic Photochemistry*; Padwa, A., ed., Marcel Dekker: New York, 1989; vol. 10, p. 309.
3. Wilson, R. M.; Schnapp, K. A. *Chem. Rev.* 1993, *93*, 223.
4. Johnston, L. J. *Chem. Rev.* 1993, *93*, 251.
5. Maciejewski, A.; Steer, R. P. *Chem. Rev.* 1993, *93*, 67.
6. Scaiano, J. C.; Johnston, L. J; McGimpsey, W. G.; Weir, D. *Acc. Chem. Res.* 1988, *21*, 22.
7. McGimpsey, W. G. in *Handbook of Organic Photochemistry*; Scaiano, J. C., ed., CRC Press: Boca Raton, 1989; vol. 1, p. 419.
8. Kasha, M. *Discuss. Faraday Soc.* 1950, *9*, 14.
9. For a description of these methods see Refs. 6 and 3.
10. Beer, M.; Longuet-Higgins, H. C. *J. Chem. Phys.* 1955, *23*, 1390.
11. Murata, S.; Iwanaga, C.; Toda, T.; Kokubun, H. *Ber. Bunsenges. Phys. Chem.* 1972, *76*, 1176.
12. Murata, S.; Iwanaga, C.; Toda, T.; Kokubun, H. *Chem. Phys. Lett.* 1972, *13*, 101.
13. Binsch, G.; Heilbronner, E.; Kankow, R.; Schmidt, D. *Chem. Phys. Lett.* 1967, *1*, 135.
14. Dhingra, R. C.; Poole, J. A. *J. Chem. Phys.* 1968, *48*, 4829.
15. Falk, K. J.; Knight, A. R.; Maciejewski, A.; Steer, R. P. *J. Am. Chem. Soc.* 1984, *106*, 8292.
16. Falk, K. J.; Steer, R. P. *J. Am. Chem. Soc.* 1989, *111*, 6518.
17. Maciejewski. A.; Steer, R. P. *J. Am. Chem. Soc.* 1983, *105*, 6738.
18. Maciejewski, A.; Safarzadeh-Amiri, A.; Verral, R. E.; Steer, R. P. *Chem. Phys.* 1984, *87*, 295.
19. Maciejewski, A.; Demmer, D. R.; James, D. R.; Safarzadeh-Amiri, A.; Verral, R. E.; Steer, R. P. *J. Am. Chem. Soc.* 1985, *107*, 2831.
20. Ho, C.-J.; Motyka, A. L.; Topp, M. R. *Chem. Phys. Lett.* 1989, *158*, 51.
21. Maciejewski, A.; Steer, R. P. *J. Photochem.* 1984, *24*, 303.

22. Maciejewski, A.; Steer, R. P. *Chem. Phys. Lett.* 1983, *100*, 540.
23. Rao, V. P.; Steer, R. P. *J. Photochem. Photobiol. A: Chem.* 1989, *47*, 277.
24. Hui, M. H.; de Mayo, P.; Suau, R.; Ware, W. R. *Chem. Phys. Lett.* 1975, *31*, 257.
25. Gillispie, G. D.; Lim, E. C. *J. Chem. Phys.* 1976, *65*, 2022.
26. Gillispie, G. D.; Lim, E. C. *Chem. Phys. Lett.* 1979, *63*, 355.
27. Liu, R. S. H.; Edman, J. R. *J. Am. Chem. Soc.* 1969, *91*, 1492.
28. Campbell, R. O.; Liu, R. S. H. *J. Am. Chem. Soc.* 1973, *95*, 6560.
29. Ladwig, C. C.; Liu, R. S. H. *J. Am. Chem. Soc.* 1974, *96*, 6210.
30. Gillispie, G. D.; Lim, E. C. *Chem. Phys. Lett.* 1979, *63*, 193.
31. Bohne, C.; Kennedy, S. R.; Boch, R.; Negri, F.; Orlandi, G.; Siebrand, W.; Scaiano, J. C. *J. Phys. Chem.* 1991, *95*, 10300.
32. Kobayashi, S.; Kikuchi, K.; Kokubun, H. *Chem. Phys.* 1978, *27*, 399.
33. Wilson, T., Halpern, A. M. *J. Am. Chem. Soc.* 1980, *102*, 7272.
34. Fukumura, H.; Kikuchi, K.; Koike, K.; Kokubun, H. *Chem. Phys. Lett.* 1986, *123*, 226.
35. Catalani, L. H.; Wilson, T. *J. Am. Chem. Soc.* 1987, *109*, 7458.
36. Fukumura, H.; Kikuchi, K.; Koike, K.; Kokubun, H. *J. Photochem. Photobiol. A: Chem.* 1988, *42*, 283.
37. McGimpsey, W. G.; Scaiano, J. C. *J. Am. Chem. Soc.* 1989, *111*, 335.
38. Redmond, R. W.; Kochevar, I. E.; Kreig, M.; Smith, G. A.; McGimpsey, W. G. *J. Phys. Chem.* 1997, *101*, 2773.
39. Richter, C.; Hub, W.; Traber, R.; Schneider, S. *Photochem. Photobiol.* 1987, *45*, 671.
40. Brewer, W. E.; Studer, S. L.; Standiford, M.; Chou, P.-T. *J. Phys. Chem.* 1989, *93*, 6088.
41. Martinez, M. L.; Studer, S. L.; Chou, P.-T. *J. Am. Chem. Soc.* 1990, *112*, 2427.
42. Sepiol, J.; Kolos, R. *Chem. Phys. Lett.* 1990, *167*, 445.
43. Chou, P.-T.; Martinez, M. L.; Studer, S. J. *J. Phys. Chem.* 1991, *95*, 10306.
44. Tokumura, K.; Yagata, N.; Fujiwara, Y.; Itoh, M. *J. Phys. Chem.* 1993, *97*, 6656.
45. Chou, P.T.; Martinez, M. L.; Studer, S. L. *Chem. Phys. Lett.* 1992, *195*, 586.
46. Tokumura, K.; Kurauchi, M.; Oyama, O. *J. Photochem. Photobiol. A: Chem.* 1994, *81*, 151.
47. Liu, R. S. H.; Edman, J. R. *J. Am. Chem. Soc.* 1968, *90*, 213.
48. Liu, R. S. H.; *J. Am. Chem. Soc.* 1968, *90*, 1899.
49. Liu, R. S. H.; Kellogg, R. E. *J. Am. Chem. Soc.* 1969, *91*, 250.
50. McGimpsey, W. G.; Scaiano, J. C. *Chem. Phys. Lett.* 1987, *138*, 13.
51. McGimpsey, W. G.; Scaiano, J. C. *J. Am. Chem. Soc.* 1988, *110*, 2299.
52. Wang, Z.; Weininger, S. J.; McGimpsey, W. G. *J. Phys. Chem.* 1993, *97*, 374.
53. Gannon, T.; McGimpsey, W. G. *J. Org. Chem.* 1993, *58*, 5639.
54. Wang, Z.; Ren, Y.; Zhu, H.; Weininger, S. J.; McGimpsey, W. G. *J. Am. Chem. Soc.* 1995, *117*, 4367.
55. For example, Kroon, J.; Oliver, A. M.; Paddon-Row, M. N.; Verhoeven, J. *J. Am. Chem. Soc.* 1990, *112*, 4868.
56. Kolc, J.; Michl, J. *J. Am. Chem. Soc.* 1973, *95*, 7391.
57. Meinwald, J.; Samuelson, C. E.; Okeda, M. *J. Am. Chem. Soc.* 1970, *92*, 7604.
58. Turro, N. J.; Ramamurthy, V.; Pagni, R. M.; Botcher, J. A. *J. Org. Chem.* 1977, *42*, 92.
59. Murai, H.; Obi, K. *J. Phys. Chem.* 1975, *79*, 2246.

60. Berger, M.; Steel, C. *J. Am. Chem. Soc.* 1975, *97*, 4817.

61. For example: Kawata, S.; Schaffner, K. *Helv. Chim. Acta* 1969, *52*, 173.

62. For example: de Mayo, P.; Shizuka, J. *J. Am. Chem. Soc.* 1973, *95*, 3942.

63. Kito, N.; Ohno, A. *Bull. Chem. Soc. Jpn.* 1973, *46*, 2487.

64. Couture, A.; Ho, K.; Hoshino, M.; de Mayo, P.; Suau, R.; Ware, W. R. *J. Am. Chem. Soc.* 1976, *98*, 6218.

65. Dzvonik, M.; Yang, S. C.; Bersohn, R. *J. Phys. Chem.* 1974, *61*, 4408.

66. Freedham, A.; Yang, S. C.; Kawasaki, M.; Bersohn, R. *J. Phys. Chem.* 1980, *72*, 1028.

67. Hilinski, E. F.; Huppert, D.; Kelley, D. F.; Milton, S. V.; Rentzepis, P. M. *J. Am. Chem. Soc.* 1984, *106*, 1951.

68. Scaiano, J. C.; Arnold, B. R.; McGimpsey, W. G. *J. Phys. Chem.* 1994, *98*, 5431.

69. Hall, M.; Pandit, R.; Chen, L.; McGimpsey, W. G. *J. Photochem. Photobiol. A: Chem.* 1997, *111*, 27.

70. Elisei, F.; Latterini, L.; Aloisi, G. G.; D'Auria, M. *J. Phys. Chem.* 1995, *99*, 5365.

71. McGimpsey, W. G.; Scaiano, J. C. *J. Am. Chem. Soc.* 1987, *109*, 2179.

72. Johnston, L. J.; Scaiano, J. C. *J. Am. Chem. Soc.* 1987, *109*, 5487.

73. Guerin, G.; Johnston, L. J.; Quach, T. *J. Org. Chem.* 1988, *53*, 2826.

74. Wilson, R. M.; Schnapp, K. A.; Glos, M.; Bohne, C.; Dixon, A. C. *J. Chem. Soc. Chem. Comm.* 1997, 149.

75. Wang, Z.; McGimpsey, W. G. *J. Phys. Chem.* 1993, *97*, 9668.

76. Martin, M.; Breheret, F. T.; Lacourbes, B. *J. Phys. Chem.* 1980, *84*, 70.

77. Smith, G. A.; McGimpsey, W. G. *J. Phys. Chem.* 1994, *98*, 2923.

78. Johnston, L. J.; Redmond, R. W. *J. Phys. Chem. A.* 1997, *101*, 4660.

79. Garcia, C.; Smith, G. A.; McGimpsey, W. G.; Kochevar, I. E.; Redmond, R. W. *J. Am. Chem. Soc.* 1995, *117*, 10871.

80. Hashimoto, S.; Thomas, J. K. *J. Photochem. Photobiol. A: Chem.* 1991, *55*, 377.

81. Lamotte, M.; Pereyre, J.; Lapouyade, R.; Joussot-Dubien, J. *J. Photochem. Photobiol. A: Chem* 1991, *58*, 225.

82. Ogino, T.; Takahashi, Y.; Kobayashi, Y.; Awano, K.; Fukazawa, Y. *J. Chem. Soc. Chem. Comm.* 1992, 103.

83. Ripoche, J. *Spectrochim. Acta* 1967, *23A*, 1003.

84. Laposa, J. D.; Morrison, V. *Chem. Phys. Lett.* 1974, *28*, 270.

85. Bromberg, A.; Friedrich, D. M.; Albrecht, A. C. *Chem. Phys.* 1974, *6*, 353.

86. Izumida, T.; Ichikawa, T.; Yoshida, J. *J. Chem. Phys.* 1980, *84*, 60.

87. Okamura, T.; Obi, K.; Tanaka, I. *Chem. Phys. Lett.* 1974, *26*, 218.

88. Johnson, P. M.; Albrecht, A. C. *J. Chem. Phys.* 1968, *48*, 851.

89. Okamura, T.; Tanaka, I. *J. Phys. Chem.* 1975, *79*, 2728.

90. Branchard-Larcher, C.; Migirdicyan, E.; Baudet, J. *Chem. Phys.* 1973, *2*, 95.

91. Bromberg, A.; Meisel, D. *J. Phys. Chem.* 1985, *83*, 2507.

92. Chu, T. L.; Weissman, S. I. *J. Chem. Phys.* 1954, *22*, 21.

93. Lewis, G. N.; Lipkin, D.; Magel, T. T. *J. Am. Chem. Soc.* 1944, *66*, 1579.

94. Tokumura, K.; Udagawa, M.; Ozaki, T.; Itoh, M. *Chem. Phys. Lett.* 1987, *141*, 558.

95. Tokumura, K.; Ozaki, T.; Itoh, M. *J. Am. Chem. Soc.* 1989, *111*, 5999.

96. Tokumura, K.; Ozaki, T.; Udagawa, M.; Itoh, M. *J. Phys. Chem.* 1989, *93*, 161.

97. Weir, D. *J. Phys. Chem.* 1990, *94*, 5870.

98. Tokumura, K.; Ozaki, T.; Nosaka, H.; Saigusa, Y.; Itoh, M. *J. Am. Chem. Soc.* 1991, *113*, 4974.

99. Weir, D.; Johnston, L. J.; Scaiano, J. C. *J. Phys. Chem.* 1988, *92*, 1742.

100. Bromberg, A.; Schmidt, K. H.; Meisel, D. *J. Am. Chem. Soc.* 1984, *106*, 3056.

101. Scaiano, J. C.; Tanner, M.; Weir, D. *J. Am. Chem. Soc.* 1985, *107*, 4396.

102. Weir, D.; Scaiano, J. C. *Chem. Phys. Lett.* 1986, *128*, 156.

103. Schmidt, J. A.; Hilinski, E. F. *J. Am. Chem. Soc.* 1988, *110*, 4036.

104. Fox, M. A.; Gaillard, E.; Chen, C.-C. *J. Am. Chem. Soc.* 1987, *109*, 7088.

105. Bromberg, A.; Schmidt, K. H.; Meisel, D. *J. Am. Chem. Soc.* 1985, *107*, 83.

106. Johnston, L. J.; Scaiano, J. C. *J. Am. Chem. Soc.* 1985, *107*, 6368.

107. Tokumura, K.; Udagawa, M.; Itoh, M. *Stud. Phys. Org. Chem.* 1987, *31*, 79.

108. Tokumura, K.; Mizukami, N.; Udagawa, M.; Itoh, M. *J. Phys. Chem.* 1986, *90*, 3873.

109. Hiratsuka, N.; Yamazaki, T.; Maekawa, Y.; Hikida, T.; Mori, Y. *J. Phys. Chem.* 1986, *90*, 774.

110. Obi, K.; Yamaguchi, H. *Chem. Phys. Lett.* 1978, *54*, 448.

111. Razi Naqvi, K.; Wild, U. P. *Chem. Phys. Lett.* 1976, *41*, 570.

112. Johnston, L. J.; Lougnot, D. J.; Scaiano, J. C. *Chem. Phys. Lett.* 1986, *129*, 205.

113. Johnston, L. J.; Lougnot, D. J.; Wintgens, V.; Scaiano, J. C. *J. Am. Chem. Soc.* 1988, *110*, 518.

114. Redmond, R. W.; Scaiano, J. C.; Johnston, L. J. *J. Am. Chem. Soc.* 1982, *114*, 9768.

115. Netto-Ferreira, J. C.; Murphy, W. F.; Redmond, R. W.; Scaiano, J. C. *J. Am. Chem. Soc.* 1990, *112*, 4472.

116. Netto-Ferreira, J. C.; Scaiano, J. C. *J. Chem. Soc. Chem. Comm.* 1989, 435.

117. Okamura, T.; Yip, R. W. *Bull. Chem. Soc. Jpn.* 1978, *51*, 937.

118. Bhattacharyya, K.; Das, P. K.; Fessenden, R. W.; George, M. V.; Gopidas, K. R.; Hug, G. L. *J. Phys. Chem.* 1985, *89*, 4164.

119. Bhuttacharyya, K.; Das, P. K.; Fessenden, R. W.; George, M. V.; Gopidas, K. R.; Hiratsuka, H.; Hug, G. L.; Rajadurai, S.; Samanta, A. *J. Am. Chem. Soc.* 1989, *111*, 3542.

120. Jinguji, M.; Imamura, T.; Obi, K.; Tanaka, I. *Chem. Phys. Lett.* 1984, *109*, 31.

121. Morine, G. A.; Kuntz, R. R. *Chem. Phys. Lett.* 1979, *67*, 552.

122. Smirnov, V. A.; Plotnikov, V. G. *Russ. Chem. Rev.* 1986, *55*, 929.

123. Hiratsuka, H.; Masatomi, T.; Tonokura, K.; Shizuka, H. *Chem. Phys. Lett.* 1990, *169*, 317.

124. Scaiano, J. C.; Barra, M.; Calabrese, G.; Sinta, R. *J. Chem. Soc. Chem. Commun.* 1992, 1419.

125. Gannon, T.; McGimpsey, W. G. *J. Org. Chem.* 1993, *58*, 913.

126. Zhang, B.; Pandit, C. R.; McGimpsey, W. G. *J. Phys. Chem.* 1994, *98*, 7022.

127. Ouchi, A.; Yabe, A. *Tetrahedron Lett.* 1990, *31*, 1727.

128. Adam, W.; Denninger, U.; Finzel, R.; Kita, F.; Platsch, H.; Walker, H.; Zang, G. *J. Am. Chem. Soc.* 1992, *114*, 5027.

129. Ouchi, A.; Koga, Y. *Tetrahedron Lett.* 1995, *36*, 8999.

130. Ouchi, A.; Koga, Y., Adam, W. *J. Am. Chem. Soc.* 1997, *119*, 592.

131. Haider, K.; Platz, M. S.; Despres, A.; Lejeune, V.; Migirdicyen, E.; Bally, T.; Haselbach, E. *J. Am. Chem. Soc.* 1988, *110*, 2318.

132. Adam, W.; Schneider, K.; Stapper, M.; Steenken, S. *J. Am. Chem. Soc.* 1997, *119*, 3285.

133. Jent, F.; Paul, H.; Fischer, H. *Chem. Phys. Lett.* 1988, *146*, 315.
134. Banks, J. T.; Scaiano, J. C.; Adam, W.; Schulte Oestrich, R. *J. Am. Chem. Soc.* 1993, *115*, 2473.
135. Chateauneuf, J.; Lusztyk, J.; Ingold, K. U. *J. Am. Chem. Soc.* 1988, *110*, 2877.
136. Haider, K.; Platz, M. S. *J. Phys. Org. Chem.* 1989, *2*, 623.
137. Redmond, R. W.; Scaiano, J. C.; Johnston, L. J. *J. Am. Chem. Soc.* 1990, *112*, 398.
138. Letsinger, R. L.; Collat, R.; Magnusson, M. *J. Am. Chem. Soc.* 1954, *76*, 4185.
139. Breslin, D. T.; Fox, M. A. *J. Phys. Chem.* 1993, *97*, 13341.
140. Ruberu, S. R.; Fox, M. A. *J. Phys. Chem.* 1993, *97*, 143.
141. Adam, W.; Kita, F.; Schulte Oestrich, R. *J. Photochem. Photobiol. A: Chem.* 1994, *80*, 187.
142. Adam, W.; Kita, F. *J. Am. Chem. Soc.* 1994, *116*, 3680.
143. Adam, W.; Walther, B. *Tetrahedron* 1996, *52*, 10399.
144. Faria, J. L.; Steenken, S. *J. Am. Chem. Soc.* 1990, *112*, 1277.
145. Adam, W.; Schulte Oestrich, R. *J. Am. Chem. Soc.* 1992, *114*, 6031.
146. Faria, J. L.; Steenken, S. *J. Phys. Chem.* 1993, *97*, 1924.
147. Adam, W.; Schneider, K.; Steenken, S. *J. Org. Chem.* 1997, *62*, 3727.
148. Adam, W.; Schneider, K. *J. Chem. Soc. Perkin 2* 1997, 441.
149. Arnold, B. R.; Scaiano, J. C.; McGimpsey, W. G. *J. Am. Chem. Soc.* 1992, *114*, 9978.
150. Baumann, H.; Merckel, C.; Timpe, H. J.; Graness, A.; Kleinschmidt, J.; Gould, I. R.; Turro, N. J. *Chem. Phys. Lett.* 1984, *103*, 497.
151. Samanta, A.; Bhattacharyya, K.; Das, P. K.; Kamat, P. V.; Weir, D.; Hug, G. L. *J. Phys. Chem.* 1989, *93*, 3651.
152. Nagarajan, V.; Fessenden, R. W. *Chem. Phys. Lett.* 1984, *112*, 207.

7

Environmental Photochemistry with Semiconductor Nanoparticles

Prashant V. Kamat
University of Notre Dame, Notre Dame, Indiana

K. Vinodgopal
Indiana University Northwest, Gary, Indiana

I. INTRODUCTION

A. General Aspects and Problems

As we step into the twenty-first century, we are faced with the challenge of purification of our water and air resources. While we enjoy the comforts and benefits that chemistry has provided us, from drugs to dyes, from composites to computer chips, we are faced with the task of treating wastes generated during manufacturing processes and the proper disposal of various products and by-products. Consider for example, the problems faced with two classes of compounds, halogenated organics, and commercial colorants. The ubiquitous nature of halogenated organics, from polyvinyl chloride to the wood preservative penta-chlorophenol, has spawned considerable discussion about the desirability of such halogenated organics in the environment. Chlorophenols are a serious health concern in drinking water treatment because they are either present in the source water or are a byproduct of chlorine-based disinfecting treatment methods. The chemistry of chlorophenol is representative of the chemistry of a class of much more toxic pollutants such as chlorinated dioxins, chlorinated dibenzofurans, and PCBs.

Another class of compounds that has emerged as a focus of attention vis-à-vis their impact on the environment is dyes and pigments, the single largest group of industrial chemicals produced [1–3]. A substantial percentage of the colorant is lost during the dyeing process and therefore remediation efforts have largely been focused on removing these dyes from the wastewater effluents of textile mills and other colorant manufacturers. Incomplete decolorization of the effluent prior to discharge shifts the burden of treatment downstream to publicly owned water treatment facilities (POWTs). In the POWT, these dyes often end up as sludges, which are then dewatered and eventually deposited in landfills [4]. There is a substantial economic impetus therefore to develop a flow reactor, which could be used onstream by the mill to treat its colorant effluent and recycle the water.

Within the overall category of dyestuffs, azo dyes constitute a significant portion and probably have the least desirable consequences in terms of the surrounding ecosystem. Some of the more common commercially used azo dyes are summarized below.

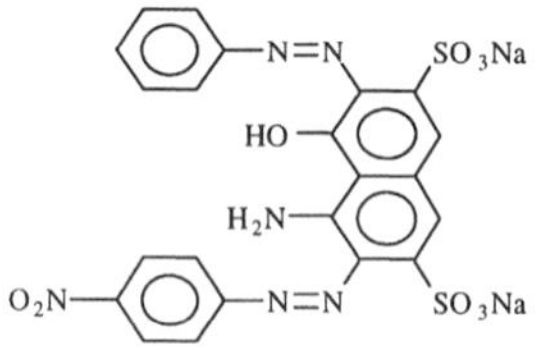

Naphthol Blue Black 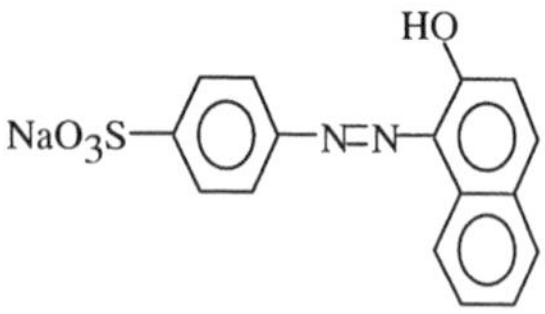   **Acid Orange 7**

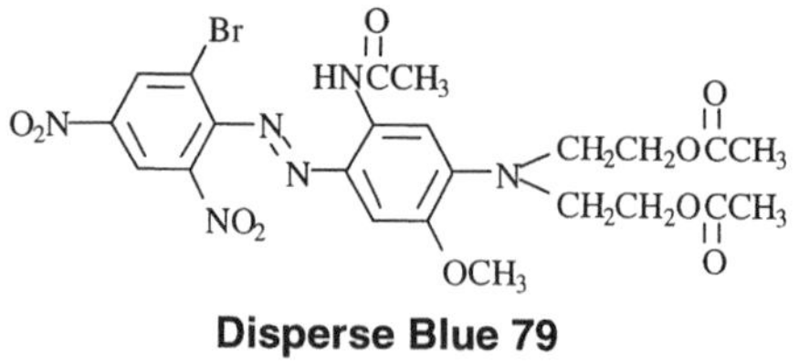

Disperse Blue 79

Conventional treatment processes for these dyes prior to discharge involve ozonation. While a simple water-soluble monoazo dye such as AO7 is easily oxidized by ozone, the larger diazo dyes such as NBB are difficult to destroy by ozonation. While oxidative ozonation is a popular treatment process, the sodium borohydride catalyzed reductive decolorization of these dyes is also used. The other example shown above, DB79, is water insoluble and is finding increased application in the dyeing of polyester and other synthetic fabrics. Traces of the disperse dye have been reported in river bed sediments in Quebec, Canada, receiving dye waste from upstream manufacturers.

Our review here will highlight the role of semiconductor-initiated photo-chemistry as an environmental remediation method for the treatment of organic chemicals. While the role of sunlight-induced photochemistry in creating environmental problems such as urban photochemical smog and the polar ozone holes has been well documented, the potential applications of photochemical methods in resolving environmental problems are less obvious.

B. Advanced Oxidation Process (AOP) in the Treatment of Chemical Contaminants

Environmental photochemistry using semiconductor nanoclusters is part of a general group of chemical remediation methods known as Advanced Oxidation Processes (AOP). These methods are based on one distinguishing feature—the generation and use of hydroxide radicals as the primary oxidant for the degradation of organic pollutants. AOPs such as UV-peroxide, ozonation, and Photo-Fenton process have already been proved useful to carry out the oxidation of organic compounds. Three other AOPs, viz., semiconductor-based photocatalysis, sonolysis, and *gamma*-radiolysis have also emerged as viable processes in recent years. It is only in recent years that serious consideration has been given to these methods for treating aqueous organic pollutants.

Scheme I illustrates the common experimental approach used to investigate salient features of these three advanced oxidation processes.

Free radicals are formed when water is irradiated with ionizing radiation

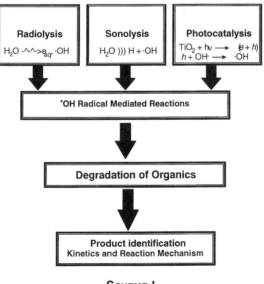

SCHEME I

such as γ-rays or a high-energy electron beam. These radiolytically generated radicals are very effective in degrading organic compounds [5–16]. In the absence of specific scavengers, hydroxyl radicals ($^{\cdot}OH$), and hydrated electrons (e_{aq}) are the major reactive species produced in a neutral or alkaline aqueous solution. By scavenging aqueous electrons with suitable scavengers such as N_2O, one can induce oxidative degradation of the organic substrates with $^{\cdot}OH$ radicals.

Sonochemical degradation methods are relatively new and involve sonication of the aqueous solutions containing the organic pollutant [17–21]. Propagation of an ultrasound wave in aqueous solution leads to the formation of cavitation bubbles; a prerequisite for these bubbles is the presence of a dissolved gas [22]. The collapse of these bubbles spawns extreme conditions such as very high temperatures and pressures, which in turn lead to the dissociation of H_2O and the production of radical species such as $^{\cdot}OH$, $HOO^{\cdot}$, etc. In recent years, evidence has accumulated indicating that higher ultrasound frequencies at ~400 kHz are more favorable for the production of $^{\cdot}OH$ radicals [21,23]. Several recent studies have also focused on the aspect of understanding chemical reactivity of these radicals in a sonolytic reaction [24–28]. Our initial results on degradation of azo dyes suggest substantially faster rates as well as better mineralization as compared with photocatalytic schemes.

Photocatalytic methods involve illumination of a large band gap semiconductor particle such as TiO_2 either dispersed as a slurry in the contaminated aqueous solutions or as immobilized films. While a substantial body of literature exists on photocatalysis [29–40], barriers to successful commercialization still prevail. The major problems with this technique include deactivation of the photocatalyst surface, and recovery of the photocatalyst in slurry systems.

C. Photocatalysis with Semiconductor Nanoclusters

During the past decade, semiconductor nanoclusters have been employed for carrying out photochemical transformations of organic and inorganic compounds (see, for example, Refs. 41–44). By making use of the principles of photoelectrochemistry, semiconductor nanoclusters have been employed in the conversion of light energy [45,46] and photocatalytic detoxification of air and water [33,38–40,47,48]. They also have potential applications in the area of microelectronics, photovoltaics, imaging and display technologies, sensing devices, and thin film coatings. Several terms are used to describe the state of matter that is labeled collectively as nanoparticles. Clusters, quantum size particles (Q particles), quantum dots, and nanoparticles all describe assemblies of atoms or molecules whose size is a paramount factor in determining the properties of the material.

The participation of a semiconductor nanoparticle in a photocatalytic process can be either direct or indirect as illustrated in Fig. 1a and b, respectively.

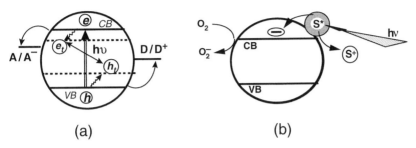

(a)   (b)

FIGURE 1 Photoinduced charge transfer processes in semiconductor nanoclusters. (a) Under bandgap excitation and (b) sensitized charge injection by exciting adsorbed sensitizer (S). CB and VB refer to conduction and valence bands of the semiconductor and e_t and h_t refer to trapped electrons and holes, respectively.

Charge separation in semiconductor particles occurs when subjected to bandgap excitation. The photogenerated electrons and holes are capable of oxidizing or reducing the adsorbed substrates (Fig. 1a). Alternatively the semiconductor nanoclusters also promote a photocatalytic reaction by acting as a mediator for the charge transfer between two adsorbed molecules (Fig. 1b). This process which is commonly referred as photosensitization is extensively used in photoelectrochemistry and imaging science. In the first case, the bandgap excitation of a semiconductor particle is followed by the charge transfer at the semiconductor/electrolyte interface. However, in the second case, the semiconductor particle quenches the excited state by accepting an electron and then transfers the charge to another substrate or generates photocurrent. The energy of the conduction and valence bands of the semiconductor and redox potential of the adsorbed molecule control the reaction course of the photochemical reaction. In this chapter we will discuss how these two processes can be beneficially utilized to induce chemical transformations of the organic substrates.

II. PHOTOCATALYSIS USING DIRECT BANDGAP EXCITATION

A. Photochemical Properties of Semiconductor Nanoclusters

Free charge carriers generated upon optical excitation either get trapped at the surface vacancies or undergo charge recombination (1) to (3).

$$TiO_2 + hv \rightarrow TiO_2 \ (e + h) \tag{1}$$

$$TiO_2(e) \rightarrow e_t \tag{2}$$

$$TiO_2(h) \rightarrow h_t \tag{3}$$

where e and h are the free charge carriers in conduction and valence bands and e_t and h_t are the trapped electrons and holes, respectively. The trapped charge carriers are sufficiently long-lived (μs-s) and can be characterized using absorption, emission, EPR, and microwave absorption spectroscopy methods. Bard and his co-workers [49–51] were the first to demonstrate the effect of charge trapping on the photoelectrochemical properties of TiO_2 particles under illumination.

Luminescence measurement is another useful technique to probe charge trapping as well as recombination between trapped charge carriers. Based on the analysis of luminescence decay Brus has estimated that hole localization on the CdSe surface occurs within 150 fs [52]. The chemical changes associated with trapping of charge carriers in CdS, PbS, CdSe have also been time-resolved [53–57]. Dynamics of charge carrier trapping, recombination, and photophysical processes have been studied by several researchers [53,54,57–72].

EPR studies of photoexcited anatase TiO_2 particles have shown that the photogenerated electrons are trapped at Ti^{4+} sites within the bulk [73–75]. One of the early efforts to understand the dynamics of charge separation in colloidal TiO_2 particles came from Rothenberger et al. [59] who interpreted the charge-trapping process based on a stochastic kinetic model. Recently, picosecond and femtosecond transient absorption spectroscopy has also been carried out to study the dynamics of charge trapping processes in TiO_2 and ZnO nanoparticles [76–78]. Similarly, electron trapping in ZnO [79–81]. WO_3 [82], and SnO_2 [83] colloids have been investigated by laser flash photolysis. Often such trapping processes are completed within 180 fs [77]. The mean lifetime of the electron-hole pair was also determined to be 30 ps. The effect of hole scavengers in extending the lifetime of the trapped electron in colloidal TiO_2 has also been demonstrated recently [84]. The broad absorption of these trapped electrons in the visible and red region results in the blue coloration of semiconductor colloids. Thus, these processes have applications in developing photochromic materials. Similarly, trapped electrons in colloidal metal chalcogenides such as CdS undergo chemical changes by reacting with Cd^{2+} sites. Recently, Zhang et al. [72,85] have concluded that the electron trapping in CdS colloids occurs in less than 100 fs.

The basic hydroxide groups on the surface of TiO_2 can be considered deep traps for valence band holes. While the majority of these trapped holes recombine with conduction band and trapped electrons, a small fraction of $O^{-\cdot}$ undergoes dimerization at the surface to form stable titanium peroxide. An indirect method of hole trapping was achieved in this study by reacting TiO_2 colloids with pulse radiolytically generated $\cdot$OH radicals. The transient spectrum recorded after irradiation of aqueous colloidal TiO_2 suspension with an electron pulse showed an absorption band with onset at 470 nm and a maximum around 350 nm. A slightly red-shifted band (λ_{max} = 420 nm) has also been observed in UV-irradiated TiO_2 colloids [58,71]. Recent EPR studies have identified these trapped holes as oxygen anion radical covalently bound to titanium ions, Ti^{IV}-O-Ti^{IV}-O$\cdot$ [75]. Based on their EPR study, Micic et al. [86,87] have found the nature of the trapped holes in alcoholic medium to be slightly different. The adsorbed alcoholic groups were found to be associated with the trapped holes. The quintet EPR signal of the trapped holes was ascribed to Ti-O-Ti-O˙CHCH$_3$ on the TiO$_2$ surface. Warman and his co-workers have carried out time-resolved microwave conductivity measurements to probe the electronic processes in semiconductor materials [88–92].

B. Interfacial Reduction and Oxidation Processes

If the photogenerated electrons and holes survive long enough to react with adsorbed substrate, these interfacial electron transfer processes can be used for photocatalytic reactions. (See for example, Refs. 93–97.) When the acceptor molecules are adsorbed on the surface, the interfacial electron transfer competes directly with the charge recombination process [98]. The rate constant for the interfacial electron transfer in this case is greater than 5×10^{10} s^{-1}. However, dynamic electron transfer can also be observed when the electron trapping occurs with a high efficiency and these trapped electrons survive for a long time. For example, heterogeneous rate constants of the order of 10^8–10^9 M^{-1} s^{-1} have been observed in the reduction of oxazine dyes in WO$_3$ colloidal suspensions [82].

When a deaerated suspension of colloidal WO$_3$ or TiO$_2$ containing NBB or Direct Blue 1 (DB1) was irradiated with UV light, the characteristic blue color of both dyes quickly disappeared. Figure 2 shows the absorption spectra recorded during the UV photolysis of WO$_3$ colloids containing NBB. As we continue the photolysis, the absorption band of the dye decreased. The failure to see IR absorption band during this irradiation period indicated that electrons did not accumulate when dye molecules were present in the colloidal suspension (spectra a to c in Fig. 2A). As indicated in the previous section, the photolysis of WO$_3$ colloids results in electron-hole separation followed by the trapping of electrons. These trapped electrons are then scavenged by the dye (D) to form a colorless reduced product (reaction 4).

$$WO_3(e_t) + D \rightarrow D^{˙-} \rightarrow \text{colorless product} \qquad (4)$$

Once the dye molecules are stoichiometrically reduced, the photogenerated electrons accumulate within the colloidal particles. This is evident from the broad absorption band in the near-IR region of the spectrum d (Fig. 2A) which was recorded 20 min after photolysis.

The charge transfer kinetics in the above example is a diffusion-controlled process and is completed within 100 μs following the laser pulse excitation. The two traces shown in Fig. 2B can thus be utilized to monitor the rate of reaction between trapped electrons and dye molecules in solution. The quantum efficiency for the photocatalytic reduction of these two dyes as measured from the magnitude of irreversible bleaching was ~4.5%.

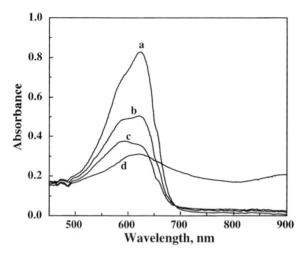

FIGURE 2A   Photocatalytic reduction of DB79 in a WO$_3$ colloidal suspension: Absorption spectra of a deaerated suspension of WO$_3$ colloids (0.4 M) and DB79 (47 μM) following UV photolysis in a 2-mm pathlength cell. The spectra were recorded at (a) 0, (b) 4, (c) 7, and (d) 20 min after UV illumination. (From Ref. 99.)

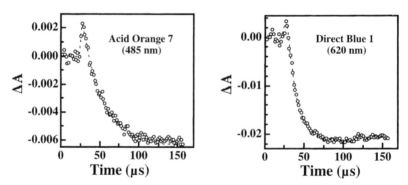

FIGURE 2B   Absorption-time profile showing the irreversible bleaching of the azo dyes. *Left*: Acid Orange 7. *Right*: Direct Blue 1 recorded at monitoring wavelengths 485 and 620 nm, respectively. The traces were recorded following 308-nm excitation of WO$_3$ colloids in aqueous dye solution. (From Ref. 99.)

Similarly, one-electron reduction of organic dyes such as thiazine, oxazine, phenazine, squaraine, and methyl orange as well as viologen and fullerenes have also been carried out in various colloidal semiconductors [44]. The values of quantum yield of reduction vary from 0.01 to 1.0. The quantum yield of reduction is dependent on both the association constant and the energetics of the conduction band and the redox couple. The energy difference between the two acts as a driving force for heterogeneous electron transfer at the semiconductor colloid.

Photogenerated holes can also be simultaneously utilized to oxidize substrates at the semiconductor surface [43,100–102]. The holes trapped at the TiO_2 surface can survive for a duration of microseconds to milliseconds and can react with OH^- to generate ·OH radicals in aqueous medium. These ·OH radicals can further induce secondary oxidation. Oxidation of halide ions (X^-) has been investigated in various colloidal semiconductor suspensions by laser flash photolysis [103,104].

$$TiO_2(h_{VB} \text{ or } h_t) + X^- \rightarrow TiO_2 + X_2^{-\cdot} \tag{5}$$

The transient product formed in this heterogeneous process can easily be followed from the characteristic absorption in the 300–400 nm region. The efficiency of halide oxidation at TiO_2 follows the sequence $Cl^- < Br^- < I^-$, correlating with the decrease in the oxidation potential. Scavenging of holes by alcohol or SCN^- had a marked effect on the reduction of tetranitromethane [58] and oxazine dyes [105] at the TiO_2 particle. Simultaneous capture of holes and electrons at the TiO_2 surface with suitable scavengers enhanced the efficiency of interfacial charge transfer by a factor of 2.

By decreasing the particle size, it is possible to shift the conduction band to more negative potentials and the valence band to more positive potentials as a result of quantization effects [106]. Because of the increase in the effective bandgap, the band edges of the quantized semiconductor particle attain new positions relative to the band edges of the bulk material. Hence, redox processes that cannot occur in bulk materials can be energetically favored in quantized small particles as the conduction and the valence bands become stronger reductant and oxidants, respectively.

C. Environmental Applications of Photocatalysis

One of the significant applications of photocatalysis is in the area of environmental chemistry. Research efforts are focused on developing photocatalytic systems to destroy organic compounds in both water and gaseous systems. TiO_2 is one of the most extensively investigated photocatalysts because it is harmless to the environment, is stable in aqueous environments, and is relatively inexpensive. Its ability to photocatalyze the oxidation of organic materials has been known for a long time in the paint industry, which makes extensive use of TiO_2 as a white paint pigment [107]. Attempts have been made earlier to exploit the photocatalytic ability of semiconductor particles such as TiO_2 to split the water molecule into hydrogen and oxygen as an alternative energy source (see reviews and collections [108,109]).

As indicated in the previous section, photoinduced charge carriers in semiconductor nanoparticles induce redox reactions. In aqueous solutions the holes are scavenged by surface hydroxyl groups to generate ˙OH radicals which then promote the oxidation of organics [11,14,110]. Often this oxidation mediated by ˙OH leads to the mineralization of hazardous chemicals. Heterogeneous electron transfer at the semiconductor electrolyte interface is greatly influenced by the parameters such as the surface charge, site specific interactions and adsorption, and energetics of the semiconductor and substrate. These parameters can be controlled by varying the composition and reaction parameters of the medium in which the TiO_2 particles are suspended. Varying pH not only shifts the band energies of the metal oxide semiconductor (0.059 mV/pH unit) but also varies the surface charge. Most conventional slurry based photocatalytic reactions are usually carried out under oxygen saturated conditions. The oxygen acts as an electron scavenger, thereby facilitating charge separation. This electron scavenging is important since it becomes a rate-limiting factor in such oxidation processes [49, 111–113]. Photocatalytic oxidation carried out in slurries usually indicates that no degradation occurred without oxygen or some other oxidant.

4-Chlorophenol is a good model for an environmental pollutant since it is present in the waste streams from a wide variety of industrial activities where chlorine and phenol may be used, as in pesticide manufacture. Therefore, the mineralization of chlorophenols is of direct interest to many industries, municipalities, and federal regulatory agencies, while at the same time providing insight into the chemistry of more complicated and dangerous chemicals. For these reasons chloro and other halogenated phenols are regarded as standard model compounds for environmental research and have been studied extensively.

The photocatalytic degradation of 4-chlorophenol (4-CP) [114–121], and other chlorophenols [114,115,122–126], has been demonstrated in TiO_2 studies. Various techniques have been employed to monitor the degradation of these compounds. Figure 3 shows one such example in which the changes observed in the absorption spectra of 4-CP solution were recorded during the photolysis of an immobilized TiO_2 particulate film. The absorption peak corresponding to the 4-CP disappear completely following the photolysis. The increased absorption at early irradiation times is due to the reaction intermediates such as hydroquinone.

The photocatalytic degradation of chlorophenols on ZnO has also been demonstrated [127]. The photocatalytic degradation of other chlorinated aromatic compounds [127], phenol [128–134], fluorinated aromatic compounds [135], and other substituted phenols and aromatic compounds [Izumi 1981, #738; Matthews 1984, #2386; Abdullah 1990, #2099; [136–141] have been demonstrated. The degradation of halogenated aromatic pollutants such as polychlorinated biphenyls (PCBs) [142] and polybrominated dibenzofurans [143] has also been attempted.

The degradation of many other halogenated compounds with TiO_2 photocatalysis has been studied [112,122,144–148], as has the degradation of a number of other model compounds including carboxylic acids [110,130,149–151]. Other

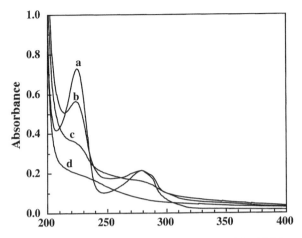

FIGURE 3   Degradation of 4-CP at an immobilized TiO_2 particulate film maintained at 0.6 V SCE. Absorption spectra of 4-CP (initial concentration of 0.8 mM) were recorded following the UV photolysis at intervals (a) 0, (b) 25, (c) 60, and (d) 155 min. (From Ref. 116.)

pollutants investigated include herbicides such as atrazine [129,152], organophosphorus insecticides [153], and permetherins [154]. Dyes [155,156], and nitroaromatics such as TNT [157–159], nitrophenols [141,160,161], and nitrobenzene and nitrosobenzene [162] have also been degraded with TiO_2 photocatalysis. The photocatalyzed degradation of model organic compounds can provide additional insight into the degradation of halogenated aromatic compounds. While there is no doubt that many of these aromatic molecules can be completely mineralized using TiO_2 photocatalysis, the detection of the intermediates and trace amounts of byproducts often pose analytical challenges.

In the case of semiconductor assisted photocatalysis organic compounds are eventually mineralized to carbon dioxide, water, and in the case of chlorinated compounds, chloride ions. It is not unusual to encounter reports with detection of different intermediates in different laboratories have been observed. For example, in the degradation of 4-CP the most abundant intermediate detected in some reports was hydroquinone (HQ) [114,115,123], while in other studies 4-chlorocatechol, 4-CC (3,4-dihydroxychlorobenzene) was most abundant [14,116–118, 121,163]. The controversy in the reaction intermediate identification stems mainly from the surface and hydroxyl radical mediated oxidation processes. Moreover, experimental parameters such as concentration of the photocatalyst, light intensity, and concentration of oxygen also contribute in guiding the course of reaction pathway. The photocatalytic degradation of 4-CP in TiO_2 slurries and thin films

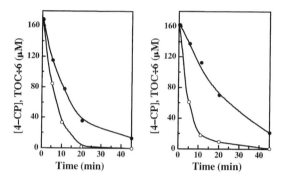

FIGURE 4   The photocatalytic degradation of ~170 μm 4-CP solution using 450-W Hg lamp, in 0.125 g/L (left) and 1.0 g/L (right) slurries of TiO$_2$ compared with the decline in total organic carbon TOC. The TOC is presented in terms of mM of carbon divided by 6, for easy comparison with 4-CP. (From Ref. 164.)

was studied to esstablish the degradation pathways [14,117,164,165]. Figure 4 shows the comparison between the rate of consumption of 4-CP and the rate of decrease in TOC for the degradation of 4-CP at two different catalyst loadings. While the degradation rate of 4-CP was similar for these two catalyst loadings, the rate of overall degradation of organic species is significantly faster for higher concentration of the photocatalyst.

III. PHOTOCHEMISTRY ON SEMICONDUCTOR SURFACES

A. Principle of Photosensitization

The process of charge injection from the excited state into the conduction band of the semiconductor has been studied with respect to extending the photoresponse of large bandgap semiconductor materials and has applications in imaging science and solar energy conversion [43,166–171]. This is often achieved by attaching sensitizer molecules by direct adsorption or covalent linkage. These dyes are directly adsorbed on the semiconductor surface either by electrostatic interaction or by transfer interaction. Functional groups such as carboxylate and phosphonates are useful for binding the dyes to oxide surfaces. The interaction between the semiconductor surface and the dye molecules often results in spectral changes which include displacement or broadening of the absorption bands, appearance of new charge transfer bands, and changes in the extinction coefficient of absorption. Ruthenium(II) polypyridyl complexes have so far been proved to be most efficient in sensitizing nanocrystalline semiconductor films. Much interest has recently been directed toward the synthesis and sensitizing properties of Ru(bpy)$_2$(dcbpy)$^{2+}$

and related ruthenium complexes because of their strong visible absorption and resistance to ligand substitution, as well as their ability to interact with semiconductor surfaces.

Our earlier surface photochemical studies with organic dyes have indicated that the intrinsic semiconductor property of the oxide support plays an important role in controlling the photochemistry of adsorbed molecules. Semiconducting oxides such as TiO_2 directly participate in the surface photochemical reaction while nonreactive oxides such as SiO_2 ZrO_2 or Al_2O_3 do not influence the excited behavior of adsorbed substrate. The sensitizing dyes bound to the semiconductor surface exhibit relatively low emission yield. For example, comparison of the relative quantum yields of $Ru(bpy)_2(dcbpy)^{2+*}$ on these oxide surfaces indicated that the fluorescence yields are significantly lower (~5%) on a semiconductor (TiO_2, $E_g \approx 3$ eV) surface than on an insulator surface (alumina, $E_g \approx 9$ eV). In fact, luminescence decay of the sensitizer on the semiconductor surface is a convenient method to probe the kinetics of charge injection process. Other spectroscopic techniques such as emission, resonance Raman, diffuse reflectance, microwave absorption, and nanosecond and picosecond laser flash photolysis are also useful to probe the interfacial charge transfer between an excited sensitizer and the semiconductor particle. Figure 5 illustrates the principle of the charge injection from excited singlet (S_1) and triplet (T_1) states into a semiconductor nanocrystallite.

Dyes such as erythrosin B [172], eosin [173–177], rose bengal [178,179], rhodamines [180–185], cresyl violet [186–191], thionine [192], chlorophyll *a* and *b* [193–198], chlorophyllin [197,199], anthracene-9-carboxylate [200,201], perylene [202,203] 8-hydroxyquinoline [204], porphyrins [205], phthalocyanines [206,207], transition metal cyanides [208,209], $Ru(bpy)_3^{2+}$ and its analogs [83,170,210–218], cyanines [169,219–226], squaraines [55,227–230], and phenylfluorone [231] which have high extinction coefficients in the visible, are often employed to extend the photoresponse of the semiconductor in photoelectrochemical systems. Visible light sensitization of platinized TiO_2 photocatalyst by surface-coated polymers derivatized with ruthenium tris(bipyridyl) complex has also been attempted [232,233]. Because the singlet excited state of these dyes is short lived it becomes essential to adsorb them on the semiconductor surface with

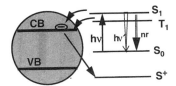

FIGURE 5 Excited state processes of a dye molecule adsorbed on a semiconductor nanocluster.

an electrostatic, hydrophobic, or chemical interaction so that electron transfer can occur with the lifetime of the singlet excited state. Spectroscopic techniques such as emission, resonance Raman, diffuse reflectance, microwave absorption, and nanosecond and picosecond laser flash photolysis are often employed to probe the interfacial charge transfer between an excited sensitizer and the semiconductor particle.

According to the model proposed by Gerischer [234] and Spitler [235] the probability of an electron transfer (j_c) from conduction band with an energy level E_c is proportional to an exponential factor described by the expression (6).

$$j_c \propto \exp \frac{\{-(E_c - {}^\circ E_{D/D+})^2}{4L^*kT} \qquad (6)$$

where L^* represents the reorganization energy in the excited state and is in the range of 0.5–1.5 eV. The energy difference ($E_c - {}^\circ E_{D^*/D+}$) acts as a driving force for the charge injection process and governs the efficiency of sensitization.

B. Mechanistic and Kinetic Aspects

Kinetics

The sensitizer molecules adsorbed on TiO_2 surface have a significantly shorter fluorescence lifetime than in the homogeneous solution and this decrease in lifetime has been attributed to the charge injection process [83,181–183,186–188, 197,218,225,236–239]. Heterogeneous electron transfer rate constants in the range of 10^7–10^{11} have been reported in these studies.

The charge injection from singlet excited sensitizer into the conduction band of a large bandgap semiconductor is usually considered to be an ultrafast process occurring in the picosecond time domain. Charge injection process in the case of organic dyes such as anthracene carboxylate [201,240], squaraines [241], cresyl violet [68,188], coumarin [242], and $Ru(H_2O)_2^{2-}$ [68,188] has been shown to occur in the subpicosecond timescale. A convenient way to probe the charge injection process is to monitor the formation of dye cation radical. Figure 6 compares the transient absorption and decay arising from singlet excited 9AC and anthracene carboxylate cation following 390 nm laser pulse excitation of 9AC in the absence and presence of TiO_2. The cation radical is formed as a result of charge injection from excited 9AC into TiO_2 nanoclusters:

$$9AC^* + TiO_2 \rightarrow 9AC^{+\cdot} + TiO_2(e) \qquad (7)$$

The timescale for the forward electron transfer reaction as determined from these ultrafast investigations was ≤ 1 ps.

On the contrary, relatively smaller charge injection rate constants (10^8–10^9 s^{-1}) have been reported by several research groups investigating the photophysical behavior of ruthenium complexes adsorbed on various semiconductor surfaces

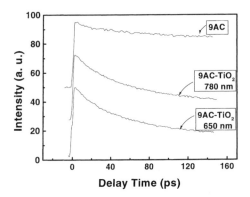

FIGURE 6 Transient absorption data for free 9AC and 9AC-TiO$_2$ recorded at magic angle polarization. The pump wavelength was 390 nm and the probe laser wavelength was 780 nm for 9AC and either 780 nm or 650 nm for 9AC-TiO$_2$. (From Ref. 201.)

[243–245]. Similarly, the charge injection from the triplet excited dyes into TiO$_2$ and ZnO colloids has also been shown to occur on a slower timescale [175,192, 230]. Our kinetic measurements suggest that the electron transfer from the excited Ru(bpy)$_2$(dcbpy)$^{2+}$ occurs with a relatively slower rate than the singlet excited organic dyes but is comparable to the triplet excited dyes. It should be noted that the excited state of Ru(bpy)$_2$(dcbpy)$^{2+}$ involves metal to ligand charge transfer state and implications are that such an electronic configuration of the excited state plays an important role in controlling the electron injection rates. Trapping and detrapping processes can also play a major role in influencing the kinetics of the charge injection process.

Microwave absorption and luminescence decay measurements have been independently carried out to monitor the charge injection from excited Ru(bpy)$_2$-(dcbpy)$^{2+}$ into SnO$_2$, ZnO, and TiO$_2$ nanocrystallites [245]. Since microwave conductivity arises as a result of mobile charge carriers within semiconductor particles it is possible to probe the charge injection process by monitoring the growth in the microwave absorption [178].

C. Use of Humic Acid as Sensitizer

Humic substances (HS) are polymeric oxidation products that result from the decomposition of plant and animal residues. As a consequence of their colloidal state in natural waters, they play an important role in the transport of organic pollutants. Thus hydrophobic organic pollutants such as polycyclic aromatic hydrocarbons, DDT, and PCBs are known to bind well to humic substances, thereby enhancing the former's water solubility. One important characteristic of humic substances lies in their ability to initiate photochemical transformation of organic compounds in natural water and their eventual degradation. Upon irradiation humic substances (e.g. humic and fulvic acids) produces hydrated electrons as phototransients [246]. This characteristic has rendered feasible the use of fulvic acid and humic acid as sensitizers for extending the photoresponse of large bandgap semiconductor such as TiO_2 or ZnO [247–249]. By using emission and transient spectroscopy it was possible to demonstrate the photosensitization properties of humic substances. The net charge transfer efficiency as determined by the emission quenching of Suwanee River fulvic acid by ZnO was 73% [247]. It has also been shown that following visible light excitation of the humic acid, this trapped electron on the semiconductor surface can be utilized to reduce another substrate, such as oxazine dyes [248]:

$$HS + h\text{V} \rightarrow HS^* - TiO_2 \rightarrow HS^{+\cdot} + TiO_2(e) \tag{8}$$

$$TiO_2(e) + \text{oxazine} \rightarrow TiO_2 + \text{oxazine}^{-\cdot} \tag{9}$$

The transient absorption spectrum in Fig. 7 shows the formation of a radical anion of oxazine 725 following the excitation of humic acid adsorbed on TiO_2 particles. A similar approach has also been employed recently to reduce Cr(VI) ions using humic acid/ZnO system [250]. The use of humic acid in such a semiconductor mediated reduction process has significant environmental implications. The presence of naturally occurring metal oxide semiconductors in soil systems along with humic substances presents a possible pathway for natural reductive processes to clean up the environment.

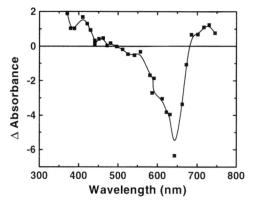

FIGURE 7   Photosensitized reduction of oxazine dye using humic acid as sensitizer. The transient absorption spectrum was recorded following 532-nm laser pulse excitation of humic acid containing the dye and TiO_2.

D. Degradation of Textile Dyes and Other Colored Compounds

In a sensitizer based photoelectrochemical cell, a redox couple (e.g., I_3^-/I^-) is employed to regenerate the sensitizer [170]. In the absence of a regenerative system, the oxidized sensitizer readily undergoes further degradation. The principle of this approach is illustrated in Fig. 1b. The photosensitized oxidation of a variety of colored compounds such as methylene blue [180,251], phenosafranin [252], fullerenes (C_{60} and C_{70}) [253,254] rose bengal [179], diphenylisobenzofuran [143], nitrophenol [141,255], 4-chlorophenol [256], azo dyes [257,258], Solvent Red 1 [259], and ruthenium trisbipyridyl complex [244] has been carried out on semiconductor surfaces. In all of the cases mentioned above, the colored compound adsorbed on the semiconductor surface is completely or partially bleached under steady illumination. The advantages of this process are (1) the utilization of visible light for degrading colored compounds and (2) the ability to degrade hazardous colored organics in potentially difficult matrices such as sludge cakes containing dyestuffs from treatment plants.

Steady-State Photolysis

When air equilibrated samples of the dye Acid Orange 7 adsorbed on TiO_2 were irradiated with visible light, ($\lambda > 380$ nm) they readily underwent degradation and the colored titania powders were completely bleached. Figure 8A shows the

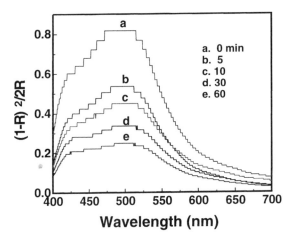

a. 0 min
b. 5
c. 10
d. 30
e. 60

FIGURE 8A Diffuse reflectance spectra recorded following the steady-state photolysis (visible light) of Acid Orange 7 adsorbed on TiO_2 nanoparticles (0.02 mmol AO7/g of TiO_2). The ordinate scale is expressed in Kubelka-Munk units, where R is the reflectivity measured at the corresponding wavelength. (From Ref. 258.)

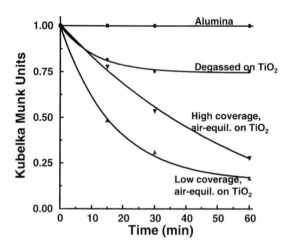

Time (min)

FIGURE 8B   Normalized decay traces representing (a) the degradation of AO7 on alumina [air-equilibrated; 0.02 mmole of AO7/g of Al$_2$O$_3$)], (b) AO7 on TiO$_2$ (0.02 mmole of AO7/g of TiO$_2$), (c) AO7 on TiO$_2$ at high coverage (0.10 mmole of AO7/g of TiO$_2$), and (d) AO7 on TiO$_2$ at low coverage (0.02 mmole of AO7/g of TiO$_2$). Sample (b) was degassed while the others were equilibrated in air before the photolysis. (From Ref. 258.)

diffuse reflectance spectra of aerated samples of AO7 on TiO$_2$ nanoparticles. The spectra were recorded at various time intervals following photolysis with visible light. Almost complete photobleaching of the absorption band at 480 nm is achieved for these low coverage samples in ~30 min.

Normalized decay traces shown in Fig. 8B indicate the importance of O$_2$ and the semiconductor support for inducing the dye degradation. The behavior of degassed samples of AO7 on TiO$_2$ is quite different. The degradation following steady-state photolysis of these degassed samples did not proceed to completion. This is good evidence for the necessity of oxygen for scavenging photoinjected electrons during the photosensitization process. In the absence of oxygen, the recombination between injected electrons and the cation radical of the dye results in the regeneration of the sensitizer. The photochemical degradation did not proceed when dyes were adsorbed on neutral surfaces such as alumina. Similar dye degradation have also been confirmed for dyes such as naphthol blue black adsorbed on TiO$_2$ particles [257].

As indicated in the previous section the charge injection from the dye into the semiconductor is the primary event responsible for the surface promoted photooxidation process. The electron from the excited dye molecule is injected into the conduction band of the TiO$_2$ and the cation radical formed at the surface quickly undergoes degradation to yield stable products:

$$Dye + h\nu \rightarrow {}^1Dye* \text{ or } {}^3Dye* \qquad (10)$$

$${}^1Dye* \text{ or } {}^3Dye* + TiO_2 \rightarrow Dye^{+\cdot} + TiO_2(e) \qquad (11)$$

$$TiO_2(e) + O_2 \rightarrow O_2^{-\cdot} + TiO_2 \qquad (12)$$

$$Dye^{+\cdot} \rightarrow products \qquad (13)$$

$$Dye^{+\cdot} + O_2^{-\cdot} \rightarrow products \qquad (14)$$

Independent diffuse reflectance laser flash measurements have been performed to characterize formation of the cation radical and the chemical events leading to permanent changes.

Diffuse Reflectance FTIR

The IR spectra of AO7 on TiO_2 were recorded following steady-state photolysis at various time intervals. The AO7 adsorbed on TiO_2 prior to photolysis shows a characteristic absorption peak at 1500 cm^{-1} which is an azo-bond sensitive vibration and at least five other peaks observed at 1620, 1596, 1568, 1555 and 1450 cm^{-1} that can be attributed to aromatic skeletal vibrations. Substantial changes are observed upon photolysis of the TiO_2 sample. No spectral changes were observed when the experiments were carried out with AO7 adsorbed on insulator surface such as alumina. This confirms the importance of semiconducting TiO_2 surface in promoting dye degradation. The FTIR spectrum of AO7 on TiO_2 is shown in Fig. 9.

The biggest decrease is observed in the azo-bond sensitive vibration at 1500 cm^{-1}, suggesting that the molecule is being cleaved at the azo bond. Nearly all of the aromatic skeletal vibrations also disappear, suggesting that the molecule is undergoing irreversible chemical changes. The spectra in Fig. 11 also show a new

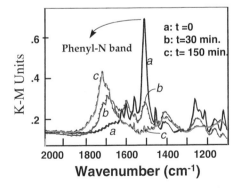

FIGURE 9   In situ diffuse reflectance FTIR spectra of AO7 on TiO_2 surface (0.10 mmoles of AO7/g of TiO_2) during steady-state photolysis. The spectra were recorded at time intervals of (a) 0, (b) 60 min, and (c) 24 h of irradiation with visible light. (From Ref. 258.)

band appearing at 1700 cm^{-1} that is characteristic of a carboxylic acid functional group. Spectrum c was obtained after 24 h of irradiation and shows only two peaks at ~1700 and 1400 cm^{-1}.

Literature references on the ozonation of AO7 indicate that the molecule in such an oxidative degradation is cleaved at the point where the azo bond is attached to the naphthalene moiety giving rise to the intermediate, 1,2 naphthoquinone (NQ) [260,261]. This intermediate is oxidized subsequently to a phthalic acid derivative. The spectrum of the photolyzed AO7 was found to be similar, especially in the area of 1400 to 1800 cm^{-1}, to the spectra of the photolyzed naphthoquinone samples. It is reasonable to conclude therefore that naphthoquinone is an intermediate from the photosensitized degradation of AO7 on TiO$_2$. The naphthoquinone then undergoes further degradation to form a phthalic acid derivative. The end product of oxidation of AO7 is an aromatic carboxylic acid as characterized from the peak at 1700 cm^{-1}.

Resolving the decay mechanism of the oxidized dye radical generated at the surface subsequent to charge injection is a little more challenging. The experimental evidence that the decay occurs only in the presence of oxygen suggests molecular oxygen's capability to scavenge electrons and thus suppress the recombination between AO7$^{+\cdot}$ and the trapped electron on the TiO$_2$ surface. It is likely that the reduced oxygen species such as O$_2^{-\cdot}$ produced at the surface is also likely to participate in the oxidation.

A possible mechanism for the decay of the AO7$^{+\cdot}$ to produce NQ and benzene sulfonic acid is shown in Scheme II.

The FTIR results lend credence to this proposed mechanism. More intriguing however is the degradation pathway for naphthoquinone to phthalic acid. NQ

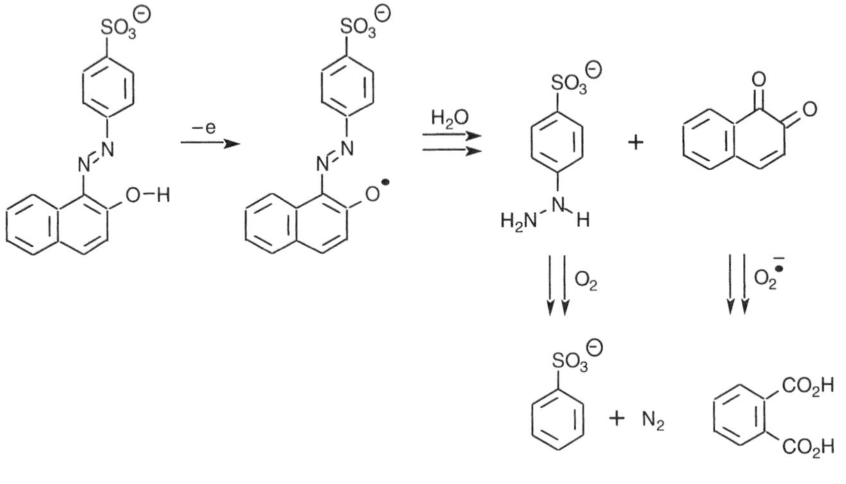

SCHEME II

when adsorbed on TiO_2 forms a grayish colored powder so that self-sensitized decay of the NQ by charge injection into the semiconductor is quite possible and our FTIR studies of NQ on TiO_2 provide evidence for such decay. References are available in organic literature of the oxidation of naphthoquinone to phthalic acid by superoxide [262]. Within the limitations of the experimental data it is hard to distinguish between the two oxidation processes, especially since the end products are likely to be the same. Our FTIR studies give us some insight into the nature of the products insofar as we are able to pinpoint the ultimate product to be a carboxylic acid. It is difficult to believe that the decay on the surface could proceed to such an end product without the participation of oxygen in some form.

IV. NOVEL APPROACHES TO IMPROVE THE EFFICIENCY OF PHOTOCATALYTIC PROCESSES

A. Electrochemically Assisted Photocatalysis

In a single-crystal semiconductor (*n*-type) based photoelectrochemical cell, the problem of achieving charge separation is easily overcome by applying an anodic bias as was first demonstrated by Honda and Fujishima [263]. Using a single crystal TiO_2, they were able to carry out the photoelectrolysis of water under the influence of an anodic bias. This concept to manipulate the photocatalytic reaction by electrochemical method can be extended to nanostructured semiconductor thin films [39,116]. The principle of electrochemically assisted photocatalysis is illustrated in Fig. 10.

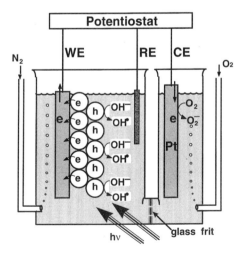

FIGURE 10   Principle of electrochemically assisted photocatalysis.

The thin semiconductor particulate film prepared by immobilizing semiconductor nanoclusters on a conducting glass surface acts as a photosensitive electrode in an electrochemical cell. An externally applied anodic bias not only improves the efficiency of charge separation by driving the photogenerated electrons via the external circuit to the counter electrode compartment but also provides a means to carry out selective oxidation and reduction in two separate compartments. This technique has been shown to be very effective for the degradation of 4-chlorophenol [116,117], formic acid [149], and surfactants [150] and textile azo dyes [264,265].

The absorption spectra of an aqueous solution of NBB recorded following excitation of an OTE/TiO$_2$ electrode is shown in Fig. 11. The anodic bias of 0.83 V vs. SCE applied in these experiments is less than the oxidation potential of AO7. The solution is constantly bubbled with a stream of nitrogen. The absorption peaks corresponding to the orange dye are reduced and decolorization of a 42-ppm solution is achieved in about 2 h. No such decolorization occurs when a blank OTE electrode (i.e., without TiO$_2$) is used with a bias of 0.83 V. The effectiveness of Electrochemically Assisted Photocatalysis was further confirmed from the comparatively slow rate of degradation that is observed when no potential is applied to the OTE/TiO$_2$ electrode and oxygen gas is bubbled through the solution.

B. Composite Semiconductors

The feasibility of synthesizing composite semiconductor nanoclusters by chemical precipitation or electrochemical deposition opens up a wide array of possi-

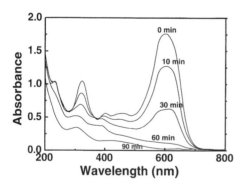

FIGURE 11   The absorption spectrum of a 0.2-mM NBB dye solution recorded at different time intervals following electrochemically assisted photocatalysis. An immobilized TiO$_2$ particulate film cast on a conducting glass electrode was maintained at an electrochemical bias of 0.8 V SCE during the photocatalytic degradation experiment. (From Ref. 265.)

bilities in utilizing these materials for chemical sensors, electrooptics, microelectronics, imaging technology, and photovoltaics. A variety of interesting properties of capped and coupled semiconductor systems have recently been reported by several researchers (see for example, Refs. 265–271). Such composite semiconductor systems not only extend the photoresponse of large bandgap semiconductors but also rectify the flow of photogenerated charge carriers [268,272] and improve the efficiency of dye sensitization [193,273] and interfacial charge transfer processes.

Composite semiconductor nanoclusters can be classified into two categories, namely, capped- and coupled-type heterostructures. The capped nanoclusters essentially have a core-shell geometry while in a coupled system two semiconductor nanoclusters are in contact with each other. The principle of charge separation in capped and coupled semiconductor systems is illustrated in Fig. 12.

While the mechanism of charge separation in a capped semiconductor system is similar to that in a coupled semiconductor system, the interfacial charge transfer or charge collection at this multicomponent semiconductor system is significantly different. Only one of the charge carriers is accessible at the surface in a capped semiconductor system, thus making selective charge transfer possible at the semiconductor/electrolyte interface. The other charge carrier (e.g., the electron in example of Fig. 12) gets trapped within the inner semiconductor particle and is not readily accessible. In a coupled semiconductor system both holes and electrons are accessible for selective oxidation and reduction processes on different particle surfaces.

An interesting aspect of composite semiconductor nanoclusters is their ability to rectify the charge carrier flow following the bandgap excitation of the

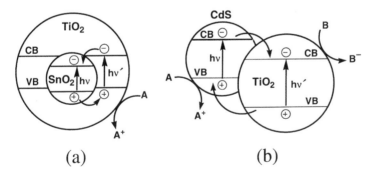

FIGURE 12 Principle of charge separation in semiconductor heterostructures: (a) capped (or Core-Shell) geometry and (b) coupled geometry. Electrons accumulate at the conduction band (CB) of SnO_2 while holes accumulate at the valence band (VB) of TiO_2.

semiconductor. For example, it is possible to rectify the flow of electrons in nanostructured TiO_2 film by coating a thin CdSe film [268]. Uekawa et al. [274] have also succeeded in developing a *p-n* junction diode. Such a capping process is not only convenient for building tandem semiconductor structures but also provides an economical way to deposit expensive semiconductor material on an inexpensive support. In order to make these systems practically viable, it is necessary to understand the photoeffects of nanostructured semiconductor composites. Photoinduced charge transfer processes in single and multicomponent semiconductor nanostructures are discussed in this chapter.

HgS@CdS and Other Composite Sulfide Systems

Materials with nearly similar lattice constants provided an interesting combination to develop composite semiconductor particles. Weller and his co-workers have investigated optical properties of HgS and CdS in a variety of core-shell geometries [275,276]. Detailed image processing analysis reveals that in the case of HgS@CdS the CdS shell grows epitaxially on HgS, thus yielding a nanometer-sized quantum well. These capped colloids exhibit electronic properties that differ from simple superposition of the electronic properties of two semiconductor particles. The photogenerated exciton in such a composite particle is not confined strictly in the core or shell but feels a potential which is formed by the entire composite particle [275,277]. One can expect such a mixing of the electronic levels since the bulk exciton diameter in CdS and HgS is larger than the diameter of the whole particle. Theoretical models for quantum confinement in semiconductor heterostructures have been developed recently [278].

Deposition of HgS on the CdS core on the other hand was shown to yield a variety of interesting fluorescence properties [275,279]. The charge recombination in such a composite system is modulated by controlling the deposition of HgS. The energy-level diagram describing these possibilities is illustrated in Fig. 13.

When HgS formed on CdS is quantized (as in Fig. 13b) the observed fluorescence is same as that of native CdS. In the case of excessive deposition of HgS, the photogenerated electrons and holes are quickly transferred to HgS and emission from HgS is observed. The energetic situation more closely resembles Fig. 13c. Similarly, in the case of CdS@PbS colloids the excited energy gap and photoluminescence energy were shown to be dependent on the core/shell ratio of the composite [280]. These studies have further led to the development of novel quantum dot quantum well structures such as CdS@HgS@CdS [276]. The picosecond electronic relaxation processes in these systems have been time resolved recently by Kamalov et al. [281]. Efforts are also being made in our laboratory to investigate the interparticle electron transfer between gold and CdS composite nanoclusters [282]. Excitation of outer CdS layer with 355-nm laser pulse caused the bleaching of plasmon band of the gold core.

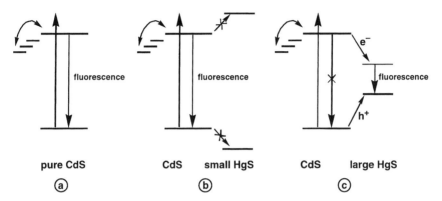

FIGURE 13 Energy schemes illustrating the photophysical processes that occur in (a) native CdS, (b) CdS with small deposition of HgS, and (c) CdS with larger deposition of HgS. (From Ref. 275.)

SnO$_2$/TiO$_2$ Systems

Significant enhancement in the photocatalytic degradation rates have been achieved by employing two or more semiconductor nanoclusters in the coupled configuration [264,265,283,284]. By coupling two large-bandgap metal oxide semiconductors with suitable energetics (e.g., SnO$_2$ and TiO$_2$) these researchers succeeded in achieving efficient charge separation. Since the conduction band (CB) of SnO$_2$ (E_{CB} for SnO$_2$ = 0 V vs. NHE at pH 7) is lower than that of the TiO$_2$ (E_{CB} = −0.5 V vs. NHE at pH 7) the former acts as a sink for the photogenerated electrons. The holes move in the opposite direction from the electrons, and accumulate on the TiO$_2$ particle thereby making charge separation more efficient. Mixing of two different colloid suspensions at the desired ratio and applying it to the conducting surface of OTE produces composite semiconductor films. These films are considered to be in a coupled geometry since the two particles are physically in contact with each other.

In order to make the SnO$_2$/TiO$_2$ system useful for photocatalysis it is advantageous to carry out these experiments under the influence of an externally applied electrochemical bias. It provides the necessary potential gradient within the nanostructured semiconductor film to drive away the accumulated electrons via the external circuit and thus promote oxidation at the semiconductor/electrolyte interface. The photocatalytic degradation of textile azo dyes such as Acid Orange 7 (AO7) and Naphthol Blue Black (NBB) have been investigated using a two-compartment electrochemical cell [39,264,265]. Unlike in slurry systems, the photocatalytic degradation experiments can be carried out in the absence of electron scavengers such as O_2. The principle and methodology of electro-chemically assisted photocatalysis is reported elsewhere [116,117].

The major photoelectrochemical reactions that initiate redox processes in the electrode compartments can be summarized as follows:

At photoanode: $SnO_2/TiO_2 + h\nu \rightarrow \{SnO_2(e \ldots .h)\}\{TiO_2(e\ldots .h)\}$

$$SnO_2(e) + TiO_2(h) \qquad (8)$$
$$TiO_2(h) + OH^- \rightarrow TiO_2 + OH^\cdot \qquad (9)$$
$$TiO_2(h) \text{ or } OH^\cdot + \text{azo dye} \rightarrow \text{products} \qquad (10)$$

At dark cathode: $Pt(e) + O_2 \rightarrow O_2^- \qquad (11)$
$$O_2^- + e + 2H^+ \rightarrow H_2O_2 \qquad (12)$$

The role of coupled semiconductor films in enhancing the photocatalytic degradation rates of azo dyes has been probed by varying the composition of the two components in the composite film. Such an experiment indicates the optimum ratio of the two semiconductors that is necessary to obtain the fastest degradation of azo dye. An example of the electrochemically assisted photocatalytic degrada-tion rates of Acid Orange 7 (AO7) as a function of the total mass of the semicon-ductor when both TiO_2 and SnO_2 are present in equal amounts is shown as a bar diagram in Fig. 14.

The rates obtained when either SnO_2 or TiO_2 is used separately (but at the same mass) is also compared in the bar diagram. It is clear from Fig. 14 that

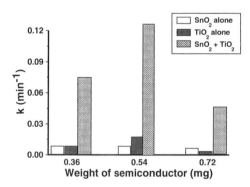

FIGURE 14   Comparison of photocatalytic degradation rates of AO7 with SnO_2, TiO_2, and SNO_2/TiO_2 particulate films coated on OTE electrodes. In each set of experiments, the total weight of the semiconductor catalyst was kept constant at the indicated value. The bias potential was 0.83 V vs. SCE and the electrolyte was 42 ppm AO7 in water (pH ~6). (From Ref. 265.)

degradation is rapid with the coupled systems, but the fastest rates are obtained when the mass ratio of SnO_2 to TiO_2 is at least 2:1 or higher.

V. CONCLUDING REMARKS

While most of the published reports focus on using a single method for oxidative degradation of organic compounds, combining two different AOPs to improve the selectivity and efficiency remains to be a challenge of the future. Therefore, an obvious question is whether one can combine photocatalysis with other advanced oxidation processes. One such possibility is combining the techniques of photocatalysis and sonolysis. Both mass transfer in the heterogeneous medium and formation of oxidizing radicals are expected to facilitate the overall degradation process. For example, Rajeshwar et al. [285] have recently reported a similar approach to enhance the bactericidal activity of irradiated TiO_2 suspensions in water. The success of this experiment suggests a potential new direction in improving the economic viability of these AOPs.

ACKNOWLEDGMENTS

PVK acknowledges the support of the Office of Basic Energy Sciences, U.S. Department of Energy. KV acknowledges the support of Indiana University Northwest through a Grant-in-Aid. This is Contribution No. NDRL-4016 from the Notre Dame Radiation Laboratory.

REFERENCES

1. Vaidya, A. A.; Datye, K. V. Environmental pollution during chemical processing of synthetic fibres., *Colourage* 1982, *14*, 3.
2. Zollinger, H. *Color Chemistry: Synthesis, Properties and Applications of Organic Dyes and Pigments*; VCH Publishers: New York, 1987.
3. Tincher, W. C. Processing waste water from carpet mills, *Textile Chemist and Colorist*, 1989, *21*, 33.
4. Ganesh, R.; Boardman, G. D.; Michelson, D. Fate of azo dyes in sludges. *Water Research* 1994, *28*, 1367.
5. Bahnemann, D.; Asmus, K. D.; Willson, R. L. Free radical induced one-electron oxidation of the chlorpromazine and promethazine, *J. Chem. Soc., Perkin Trans.* 1983, 2, 1661.
6. Moenig, J.; Bahnemann, D.; Asmus, K. D. One electron reduction of carbon tetrachloride in solutions: a trichloromethyldioxy-free radical mediated, *Chem. Biol. Interact.* 1983, *47*, 15.
7. Asmus, K. D.; Bahnemann, D.; Krischer, K.; Lal, M. One-electron induced degradation of halogenated methanes in oxygenated and anoxic aqueous solutions, *Life Chem. Rep.* 1985, *3*, 1.

8. Getoff, N.; Solar, S. Radiation induced decomposition of chlorinated phenols in water, *Radiat. Phys. Chem.* 1988, *31*, 121.
9. Getoff, N.; Solar, S. Radiolysis and pulse radiolysis of chlorinated phenols in aqueous solutions, *Radiat. Phys. Chem.* 1986, *28*, 443.
10. Mao, Y.; Schoeneich, C.; Asmus, K. D. Radical mediated degradation mechanisms of halogenated organic compounds as studied by photocatalysis at TiO_2 and by radiation chemistry, in Photocatalytic Purification and Treatment of Water and Air; D. F. Ollis; H. Al-Ekabi, eds., Elsevier Science Publishers B.V.: Amsterdam, 1993, p. 49.
11. Goldstein, S.; Czapski, G.; Rabani, J. Oxidation of phenol by radiolytically generated ˙OH and chemically generated $SO_4^{˙-}$. A distinction between ˙OH transfer and hole oxidation in the photolysis of TiO_2 colloid solution, *J. Phys. Chem.* 1994, *98*, 6586.
12. Hilarides, R. J.; Gray, K. A.; Guzzetta, J.; Cortellucci, N.; Sommer, C. Radiolytic degradation of 2,3,7,8-TCDD in artificially contaminated soils, *Environ. Sci. Technol.* 1994, *28*, 2249.
13. Kuruc, J.; Sahoo, M. K.; Locaj, J.; Hutta, M. Radiation degradation of waste waters. I. Reverse phase-high performance liquid chromatography and multicomponent UV-VIS analysis of gamma-irradiated aqueous solutions of nitrobenzene, *J. Radioanal. Nucl. Chem.* 1994, *183*, 99.
14. Stafford, U.; Gray, K. A.; Kamat, P. V. Radiolytic and TiO_2 assisted photocatalytic degradation of 4-chlorophenol. A comparative study, *J. Phys. Chem.* 1994, *98*, 6343.
15. Aguila, A.; O'Shea, E.; Kamat, P. V. Radiolytic reduction and oxidation of diethyl benzylphosphonate. A pulse radiolysis study, *Advance Oxidation Technology* 1998, in press.
16. Nasr, C.; Vinodgopal, K.; Hotchandani, S.; Kamat, P. V. Excited state and reduced forms of a textile diazo dye, Naphthol blue black. Spectral characterization using laser flash photolysis and puise radiolysis. *Radiat. Phys. Chem.* 1997, *49*, 159.
17. Hart, E. J.; Fischer, C. H.; Henglein, A. Sonolysis of hydrocarbons in aqueous solution, *Radiat. Phys. Chem.* 1990, *36*, 511.
18. Hart, E. J.; Henglein, A. Sonolysis of formic acid-water mixtures, *Radiat. Phys. Chem.* 1988, *32*, 11.
19. Serpone, N.; Terzian, R.; Colarusso, P.; Minero, C.; Pelizzetti, E.; Hidaka, H. Sonochemical oxidation of phenol and three of its intermediate products in aqueous media: Catechol, hydroquinone, and benzoquinone. Kinetic and mechanistic aspects, *Res. Chem. Intermed.* 1992, *18*, 183.
20. Serpone, N.; Colarusso, P. Sonochemistry I. Effects of ultrasounds on heterogeneous chemical reactions. A useful tool to generate radicals and to examine reaction mechanisms. *Res. Chem. Intermed.* 1994, *20*, 635.
21. Petrier, C.; Lamy, M. F.; Francony, A.; Benahcene, A.; David, B.; Renaudin, V.; Gondrexon, N. Sonochemical degradation of phenol in dilute aqueous solutions: Comparison of the reaction rates at 20 and 487 kHz. *J. Phys. Chem.* 1994, *98*, 10514.
22. Suslick, K. S. Sonochemistry, *Science* 1990, *247*, 1439.
23. Colarusso, P.; Serpone, N. Sonochemistry II. Effects of ultrasounds on homogeneous chemical reactions and in environmental detoxification, *Res. Chem. Intermed.* 1996, *22*, 61.

24. Gutierrez, M.; Henglein, A.; Dohrmann, J. K. H atom reactions in the sonolysis of aqueous solutions, *J. Phys. Chem.* 1987, *91*, 6687.

25. Gutierrez, M.; Henglein, A. Chemical action of pulsed ultrasound: Observation of an unprecedented intensity effect, *J. Phys. Chem.* 1990, *94*, 3625.

26. Hart, E. J.; Henglein, A. Free radical and free atom reactions in the sonolysis of aqueous iodide and formate solutions, *J. Phys. Chem.* 1985, *89*, 4342.

27. Henglein, A.; Gutierrez, M. Chemical reactions by pulsed ultrasound: Memory effects in the formation of NO_3^- and NO_2^- in aerated water, *Int. J. Radiat. Biol. Relat. Stud. Phys., Chem. Med.* 1986, *50*, 527.

28. Henglein, A.; Gutierrez, M. Chemical effects of continuous and pulsed ultrasound: A comparative study of polymer degradation and iodide oxidation, *J. Phys. Chem.* 1990, *94*, 5169.

29. Serpone, N.; Pelizzetti, E., eds., *Photocatalysis. Fundamentals and Applications*, John Wiley & Sons: New York, 1989.

30. Ollis, D. F.; Pelizzetti, E.; Serpone, N. Heterogeneous photocatalysis in the environment: Application to water purification. In *Photocatalysis. Fundamentals and Applications*, N. Serpone and E. Pelizzetti, eds., Wiley, New York, 1989, p. 603.

31. Anpo, M. Photocatalysis on small particle TiO_2 catalysts. Reaction intermediates and reaction mechanisms, *Res. Chem. Intermed.* 1989, *11*, 67.

32. Ollis, D. F.; Pelizzetti, E.; Serpone, N. Destruction of water contaminants, *Environ. Sci. Technol.* 1991, *25*, 1523.

33. Fox, M. A. Photocatalysis: Decontamination with sunlight, *Chemtech* 1992, *22*, 680.

34. Herrmann, J. M.; Guillard, C.; Pichat, P. Heterogeneous photocatalysis: An emerging technology for water treatment, *Catal. Today* 1993, *17*, 7.

35. Zeltner, W. A.; Hill, Jr., C. G.; Anderson, M. A. Supported titania for photodegradation, *Chem. Tech.* 1994, 21.

36. Bahnemann, D.; Cunningham, J.; Fox, M. A.; Pelizzetti, E.; Pichat, P.; Serpone, N. Photocatalytic treatment of waters. In *Aquatic and Surface Chemistry*, G. R. Helz, R. G. Zepp, D. G. Crosby, eds., Lewis Publishers, Boca Raton, Florida, 1994, p. 261.

37. Hoffmann, M. R.; Martin, S. T.; Choi, W.; Bahnemann, D. W. Environmental applications of semiconductor photocatalysis, *Chem. Rev.* 1995, *95*, 69.

38. Rajeshwar, K. Photoelectrochemistry and the environment, *J. Appl. Electrochem.* 1995, *25*, 1067.

39. Vinodgopal, K.; Kamat, P. V. Combine electrochemistry with photocatalysis, *CHEMTECH April* 1996, 18.

40. Stafford, U.; Gray, K. A.; Kamat, P. V. Photocatalytic degradation of organic contaminants. Halophenols and related model compounds, *Heterogeneous Chemistry Reviews* 1996, *3*, 77.

41. Henglein, A. Small-particle research: Physicochemical properties of extremely small colloidal metal and semiconductor particles, *Chem. Rev.* 1989, *89*, 1861.

42. Henglein, A.; Weller, H. Colloidal semiconductors: Size quantization, sandwich structures, photo-electron emission, and related chemical effects, *Photochemical Energy Conversion*, J. R. Norris, Jr. and D. Meisel, eds., Elsevier: New York, 1989.

43. Kamat, P. V. Photochemistry on nonreactive and reactive (semiconductor) surfaces, *Chem. Rev.* 1993, *93*, 267.

44. Kamat, P. V. Interfacial charge transfer processes in colloidal semiconductor systems, *Progr. React. Kinetics* 1994, *19*, 277.

45. Hagfeldt, A.; Graetzel, M. Light-induced redox reactions in nanocrystalline systems, *Chem. Rev.* 1995, *95*, 49.

46. Kamat, P. V. Native and surface modified semiconductor nanoclusters. In *Molecular Level Artificial Photosynthetic Materials.* Progress in Inorganic Chemistry Series, vol. 44, J. Meyer, ed., Wiley: New York, 1997, p. 273.

47. Ollis, D. F. Contaminant degradation in water. Heterogeneous photocatalysis degrades halogenated hydrocarbon contaminants, *Environ. Sci. Technol.* 1985, *19*, 480.

48. Serpone, N.; Lawless, D.; Terzian, R.; Minero, C.; Pelizzetti, E. Heterogeneous photocatalysis: Photochemical conversion of inorganic substances in the environment: Hydrogen sulfide, cyanides, and metals. In *Photochemical Conversion and Storage of Solar Energy,* E. Pelizzetti, M. Schiavello, eds., Kluwer Acad. Publ.: Dordrecht, The Netherlands, 1991.

49. Dunn, W. W.; Aikawa, Y.; Bard, A. J. Characterization of particulate titanium dioxide photocatalysts by photoelectrophoretic and electrochemical measurements, *J. Am. Chem. Soc.* 1981, *103*, 3456.

50. Ward, M. D.; Bard, A. J. Photocurrent enhancement via trapping of photogenerated electrons of TiO$_2$ particles, *J. Phys. Chem.* 1982, *86*, 3599.

51. Ward, M. D.; White, J. R.; Bard, A. J. Electrochemical investigation of the energetics of particulate titanium dioxide photocatalysts. The methyl viologen-acetate system, *J. Am. Chem. Soc.* 1983, *105*, 27.

52. Brus, L. Radiationless transitions in CdSe quantum crystallites, *Isr. J. Chem.* 1993, *33*, 9.

53. Albery, W. J.; Brown, G. T.; Darwent, J. R.; Saievar, I. E. Time-resolved photoredox reactions of colloidal CdS, *J. Chem Soc., Faraday Trans.,* 1985, *181*, 1999.

54. Kamat, P. V.; Ebbesen, T. W.; Dimitrijevic, N. M.; Nozik, A. J. Photoelectrochemistry in semiconductor particulate systems. Part 12. Primary photochemical events in CdS semiconductor colloids as probed by picosecond laser flash photolysis, transient bleaching, *Chem. Phys. Lett.* 1989, *157*, 384.

55. Kamat, P. V.; Gopidas, K. R.; Dimitrijevic, N. M. Picosecond charge transfer processes in ultrasmall CdS and CdSe semiconductor particles, *Mol. Cryst. Liq. Cryst.* 1990, *183*, 439.

56. Nenadovic, M. T.; Comor, M. I.; Vasic, V.; Micic, O. I. Transient bleaching of small PbS colloids. Influence of surface properties, *J. Phys. Chem.* 1990, *94*, 6390.

57. Rajh, T.; Micic, O. I.; Lawless, D.; Serpone, N. Semiconductor photophysics. 7. Photoluminescence and picosecondcharge carrier dynamics in CdS quantum dots confined in a silicate glass, *J. Phys. Chem.* 1992, *96*, 4633.

58. Bahnemann, D.; Henglein, A.; Lilie, J.; Spanhel, L. Flash photolysis observation of the absorption spectra of trapped positive holes and electrons in colloidal TiO$_2$, *J. Phys. Chem.* 1984, *88*, 709.

59. Rothenberger, G.; Moser, J.; Graetzel, M.; Serpone, N.; Sharma, D. K. Charge carrier trapping and recombination dynamics in small semiconductor particles, *J. Am. Chem. Soc.* 1985, *107*, 8054.

60. Arbour, C.; Sharma, D. K.; Langford, C. H. Electron trapping in colloidal TiO$_2$

photocatalysts: 20 ps to 10 ns kinetics. In *Photochemistry and Photophysics of Coordination Compounds*, Springer Verlag: Berlin, 1987.

61. Hilinski, E. F.; Lucas, P. A.; Wang, Y. A picosecond bleaching study of quantum-confined cadmium sulfides microcrystallites in a polymer film, *J. Phys. Chem.* 1988, *89*, 3435.

62. Benjamin, D.; Huppert, D. Surface recombination velocity measurements of CdS single crystals immersed in electrolytes. A picosecond luminescence study, *J. Phys. Chem.* 1988, *92*, 4676.

63. Kasinski, J. J.; Gomez, J. L. A.; Min, L.; Bao, Q.; Miller, R. J. D. Picosecond dynamics of electron transfer at semiconductor liquid junctions, *J. Lumin.* 1988, *41*, 555.

64. Morgan, J. R.; Natarajan, L. V. Picosecond transient grating study of charge carrier dynamics in colloidal cadmium sulfide, *J. Phys. Chem.* 1989, *93*, 5.

65. Meyer, G. J.; Leung, L. K.; Yu, J. C.; Lisensky, G. C.; Ellis, A. B. Semiconductor-olefin adducts. Photoluminescence properties of CdS and CdSe in the presence of butenes, *J. Am. Chem. Soc.* 1989, *111*, 5146.

66. O'Neil, M.; Marohn, J.; McLendon, G. Dynamics of electron-hole pair recombination in semiconductor clusters, *J. Phys. Chem.* 1990, *94*, 4356.

67. O'Neil, M.; Marohn, J.; McLendon, G. Picosecond measurements of exciton trapping in semiconductor clusters, *Chem. Phys. Lett.* 1990, *168*, 208.

68. Kietzmann, R.; Willig, F.; Weller, H.; Vogel, R.; Nath, D. N.; Eichberger, R.; Liska, P.; Lehnert, J. Picosecond time resolved electron injection from excited cresyl violet monomers and Cd_3P_2 quantum dots into TiO_2, *Mol. Cryst. Liq. Cryst.* 1991, *194*, 169.

69. Rosenwaks, Y.; Thacker, B. R.; Nozik, A. J.; Shapira, Y.; Huppert, D. Recombination dynamics at InP/liquid interfaces, *J. Phys. Chem.* 1993, *97*, 10421.

70. Rosenwaks, Y.; Thacker, B. R.; Ahrenkiel, R. K.; Nozik, A. J. Electron transfer dynamics at p-GaAs/liquid interfaces, *J. Phys. Chem.* 1992, *96*, 10096.

71. Lepore, G. P.; Langford, C. H.; Vichova, J.; Vlcek, A. J. Photochemistry and picosecond absorption spectra of aqueous suspensions visible spectrum, *J. Photochem. Photobiol.* 1993, *A75*, 67.

72. Zhang, J. G.; O'Neil, R. H.; Roberti, T. W. Femtosecond studies of photoinduced electron dynamics at the liquid-solid interface of aqueous CdS colloids, *J. Phys. Chem.* 1994, *98*, 3859.

73. Serwicka, E.; Schlierkamp, M. W.; Schindler, R. N. Localization of conduction band electrons in polycrystalline TiO_2 studies by ESR, *Z. Naturforsch* 1981, *32*.

74. Howe, R. F., Graetzel, M. EPR study of hydrated anatase under UV irradiation, *J. Phys. Chem.* 1987, *91*, 3906.

75. Howe, R. F., Graetzel, M. EPR observation of trapped electrons in colloidal TiO_2, *J. Phys. Chem.* 1985, *89*, 4495.

76. Colombo, D. P. J.; Bowman, R. M. Femtosecond diffuse reflectance spectroscopy of TiO_2 powders, *J. Phys. Chem.* 1995, *99*, 11752.

77. Skinner, D. E.; Colombo, D. P. J.; Cavaleri, J. J.; Bowman, R. M. Femtosecond investigation of electron trapping in semiconductor nanoclusters, *J. Phys. Chem.* 1995, *99*, 7853.

78. Serpone, N.; Lawless, D.; Khairutdinov, R.; Pelizzetti, E. Subnanosecond relaxation dynamics in TiO$_2$ colloidal sols (Particle sizes Rp = 1–13.4 nm). Relevance to heterogeneous photocatalysis, *J. Phys. Chem.* 1995, *99*, 16655.

79. Kamat, P. V.; Patrick, B. Photophysics and photochemistry of quantized ZnO colloids, *J. Phys. Chem.* 1992, *96*, 6829.

80. Cavaleri, J. J.; Skinner, D. E.; Colombo, D. P. J.; Bowman, R. M. Femtosecond study of the size-dependent charge carrier dynamics in ZnO nanocluster solutions, *J. Phys. Chem.* 1995, *103*, 5378.

81. Colombo, D. P. J.; Rousal, K. A.; Saeh, J.; Skinner, D. E.; Bowman, R. M. Femtosecond study of the size-dependent charge carrier dynamics in ZnO nanocluster solutions, *Chem. Phys. Lett.* 1995, *232*, 207.

82. Bedja, I.; Hotchandani, S.; Kamat, P. V. Photoelectrochemistry of quantized WO$_3$ colloids. Electron storage, electrochromic, and photoelectrochromic effects, *J. Phys. Chem.* 1993, *97*, 11064.

83. Bedja, I.; Hotchandani, S.; Kamat, P. V. Preparation and characterization of thin SnO2 nanocrystalline semiconductor films and their sensitization with bis(2,2′-bipyridine)(2,2′-bipyridine-4-4′-dicarboxylic acid)ruthenium complex, *J. Phys. Chem.* 1994, *98*, 4133.

84. Kamat, P. V.; Gopidas, K. R. Charge transfer processes in semiconductor colloids. In *Picosecond and Fentosecond Spectroscopy from Laboratory to Real World*, vol. 22, SPIE-Int. Soc. Opt. Eng., Los Angeles, 1990, p. 115.

85. Zhang, J. Z.; O'Neil, R. H.; Roberti, T. W.; McGowen, J. L.; Evans, J. E. Femtosecond studies of trapped electrons at the liquid-solid interface of aqueous CdS colloids. *Chem. Phys. Lett.* 1994, *218*, 479.

86. Micic, O. I.; Zhang, Y.; Cromack, K. R.; Trifunac, A. D.; Thurnauer, M. C. Photoinduced hole transfer from TiO$_2$ to methanol molecules in aqueous solution studied by electron paramagnetic resonance, *J. Phys. Chem.* 1993, *97*, 13284.

87. Micic, O. I.; Zhang, Y.; Cromack, K. R.; Trifunac, A. D.; Thurnauer, M. C. Trapped holes on TiO$_2$ colloids studied by electron paramagnetic resonance, *J. Phys. Chem.* 1993, *97*, 7277.

88. Warman, J. M.; de Haas, M. P.; Graetzel, M.; Infelta, P. P. Microwave probing of electronic processes in small particle suspensions, *Nature* 1984, *310*, 306.

89. Warman, J. M.; de Haas, M. P.; Wentinck, H. M. The study of radiation induced conductivity changes in microheterogeneous materials using microwaves, *Radiat. Phys. Chem.* 1989, *34*, 581.

90. Warman, J. M.; de Haas, M. P.; van Hovell tot Westerflier, S. W. F. M.; Binsma, J. J. M.; Kolar, Z. I. Electronic processes in semiconductor materials studied by nanosecond time-resolved microwave conductivity. I. Cadmium sulfide macroscopic crystal, *J. Phys. Chem.* 1989, *93*, 5895.

91. Warman, J. M.; de Haas, M. P.; Pichat, P.; Koster, T. P. M.; van der Zouwen, A. E. A.; Mackor, A.; Cooper, R. Electronic processes in semiconductor material studied by nanosecond time-resolved microwave conductivity. III. Al$_2$O$_3$, MgO and TiO$_2$ powders, *Radiat. Phys. Chem.* 1991, *37*, 433.

92. Warman, J. M.; de Haas, M. P.; Pichat, P.; Serpone, N. Effect of isopropyl alcohol on the surface localization and recombination of conduction-band electrons in Degussa

P25 TiO_2. A pulse-radiolysis time-resolved microwave conductivity study, *J. Phys. Chem.* 1991, *95*, 8858.

93. Fox, M. A.; Dulay, M. T. Heterogeneous photocatalysis, *Chem. Rev.* 1993, *93*, 341.

94. Li, Y.; Wang, L. Semiconductor-mediated photocatalysis for organic synthesis. In *Nanocrystalline Semiconductor Materials*, P. V. Kamat, D. Meisel, eds., Elsevier Science: Amsterdam, 1997, p. 391.

95. Kanemoto, M.; Ishihara, K.; Wada, Y.; Sakata, T.; Mori, H.; Yanagida, S. Semiconductor photocatalysis. Part 13. Visible-light induced effective photoreduction of CO2 to CO catalyzed by colloidal CdS microcrystallites, *Chem. Lett.* 1992, *6*.

96. Shiragami, T.; Fukami, S.; Pac, C.; Yanagida, S. Semiconductor photocatalysis: Quantised CdS-catalysed photoformation of 1-benzyl-1,4-dihydronicotinamide (BNAH) from 1-benzylnicotinamide (BNA+), *J. Chem. Soc., Faraday Trans.* 1993, *89*, 1857.

97. Murakoshi, K.; Kano, G.; Wada, Y.; Yanagida, S.; Miyazaki, H.; Matsumoto, M.; Murasawa, S. Importance of binding states between photosensitizing molecules and the TiO_2 surface for efficiency in a dye-sensitized solar cell, *J. Electroanal. Chem.* 1995, *396*, 27.

98. Kamat, P. V.; Fox, M. A. Primary photophysical and photochemical processes of dyes in polymer solutions and films. In *Lasers in Polymer Science and Technology: Applications*, vol. II, J. P. Fouassier, J. F. Rabek, eds., CRC Press: Boca Raton, FL, 1990, p. 185.

99. Nasr, C.; Vinodgopal, K.; Hotchandani, S.; Chattopadhyay, A. K.; Kamat, P. V. Photocatalytic reduction of azo dyes naphthol blue black and disperse blue 79, *Res. Chem. Intermed.* 1997, *23*, 219.

100. Henglein, A. Physicochemical properties of small metal particles in solution: "Microelectrode" reactions, chemisorption, composite metal particles, and the atom-to-metal transition, *J. Phys. Chem.* 1993, *97*, 5457.

101. Graetzel, M. Dynamics of interfacial electron transfer reactions in colloidal semiconductor systems and water cleavage by visible light, *Nato Asi Ser., Ser. C* 1986, *174*, 91.

102. Colombo, D. P. J.; Bowman, R. M. Does interfacial charge transfer compete with charge carrier recombination? Femtosecond diffuse reflectance investigation of TiO_2 nanoparticles, *J. Phys. Chem.* 1996, *100*, 18445.

103. Moser, J.; Graetzel, M. Photoelectrochemistry with colloidal semiconductors: Laser studies of halide oxidation in colloidal dispersions of TiO_2 and α-Fe_2O_3, *Helv. Chim. Acta* 1982, *65*, 1436.

104. Kamat, P. V.; Patrick, B. Photochemistry and photophysics of ZnO colloids, in *Symp. Electron. Ionic Prop. Silver Halides*, vol. 44, B. Levy, J. Deaton, P. V. Kamat, I. Leubner, A. Muenter, L. Slifkin, T. Tani, eds., The Society for Imaging Science and Technology: Springfield, VA, 1991, p. 293.

105. Kamat, P. V. Photoelectrochemistry in particulate systems. 3. Phototransformations in the colloidal TiO_2-thiocyanate system, *Langmuir* 1985, *1*, 608.

106. Micic, O. I.; Nenadovic, M. T.; Rajh, T.; Dimitrijevic, N. M.; Nozik, A. J. Electron transfer reactions on extremely small semiconductor colloids studied by pulse radiolysis, *Nato Asi Ser., Ser. C26*, 1986.

107. Schiek, R. C. Pigments. In *Kirk-Othmer Encyclopedia of Chemical Technology*, vol. 17, R. E. Kirk, D. F. Othmer, M. Grayson, D. Eckroth, eds., Wiley: New York, 1982, p. 788.

108. Kamat, P. V.; Dimitrijevic, N. M. Colloidal semiconductors as photocatalysts for solar energy conversion, *Sol. Energy* 1990, *44*, 83.

109. Graetzel, M. *Energy Resources Through Photochemistry and Catalysis*, Academic: New York, 1983.

110. Frank, S. N.; Bard, A. J. Heterogeneous photocatalytic oxidation of cyanide ion in aqueous solutions at TiO_2 powder, *J. Am. Chem. Soc.* 1977, *99*, 303.

111. Gerischer, H.; Heller, A. The role of oxygen in photooxidation of organic molecules on semiconductor particles, *J. Phys. Chem.* 1991, *95*, 5261.

112. Wang, C. M.; Heller, A.; Gerischer, H. Palladium catalysis of O_2 reduction by electrons accumulated on TiO_2 particles during photoassisted oxidation of organic compounds, *J. Am. Chem. Soc.* 1992, *114*, 5230.

113. Peterson, M. W.; Turner, J. A.; Nozik, A. J. Mechanistic studies of the photocatalytic behavior of TiO_2. Particles in a photoelectrochemical slurry cell and the relevance to photodetoxification reactions, *J. Phys. Chem.* 1991, *95*, 221.

114. Serpone, N.; Al-Ekabi, H.; Patterson, B.; Pelizzetti, B.; Minero, C.; Pramauro, E.; Fox, M. A.; Draper, R. B. Kinetic studies in heterogeneous photocatalysis. II. The TiO_2 mediated degradation of 4-chlorophenol alone and in three component mixture of chlorophenol, 2,4-dichlorophenol, and 2,4,5-trichlorophenol, *Langmuir* 1989, *5*, 250.

115. Al., S. G.; D'Oliveira, J. C.; Pichat, P. Semiconductor-sensitized photodegradation of 4-chlorophenol in water, *J. Photochem. Photobiol.* 1991, *A58*, 99.

116. Vinodgopal, K.; Hotchandani, S.; Kamat, P. V. Electrochemically assisted photocatalysis. TiO_2 particulate film electrodes for photocatalytic degradation of 4-chlorophenol, *J. Phys. Chem.* 1993, *97*, 9040.

117. Vinodgopal, K.; Stafford, U.; Gray, K. A.; Kamat, P. V. Electrochemically assisted photocatalysis. II. The role of oxygen and reaction intermediates in the degradation of 4-chlorophenol on immobilized TiO_2 particles, *J. Phys. Chem.* 1994, *98*, 6797.

118. Tang, W. Z.; Huang, C. P. Photocatalyzed oxidation pathways of 2,4-dichlorophenol by CdS in basic and acidic aqueous solutions, *Water Research* 1995, *29*, 745.

119. Cunningham, J.; Sedlak, P. Interrelationships between pollutant concentration, extent of adsorption, TiO_2-sensitized removal, photon flux and levels of electron or hole trapping additives. I. Aqueous monochlorophenol-TiO_2(P25) suspensions, *J. Photochem. Photobiol.* 1994, *A77*, 255.

120. Stafford, U.; Gray, K. A.; Kamat, P. V. Photocatalytic oxidation of 4-chlorophenol on TiO_2: A comparison with γ-radiolysis. In *Chemical Oxidation: Technologies for the 90s*, vol. 4, J. Roth and A. Bowers, eds., Technomic Publishing Co.: Lancaster, PA, 1996, p. 193.

121. Mills, A.; Morris, S.; Davies, R. Photomineralisation of 4-chlorophenol sensitised by titanium dioxide: A study of the intermediates, *J. Photochem. Photobiol.* 1993, *A70*, 183.

122. Suri, R. P. S.; L. J.; Hand, D. W.; Crittenden, J. C.; Perram, D. L., Mullins, M. E. Heterogeneous photocatalytic oxidation of hazardous organic contaminants in water, *Water Environ. Res.* 1993, *65*, 665.

123. Al-Ekabi, H.; Serpone, N. Kinetic studies in heterogeneous photocatalysis. I. Photocatalytic degradation of chlorinated phenols in aerated aqueous solutions over TiO_2 supported on a glass matrix, *J. Phys. Chem.* 1988, *92*, 5726.

124. D'Oliveira, J. C.; Minero, C.; Pelizzetti, E.; Pichat, P. Photodegradation of dichlorophenols and trichlorophenols in TiO_2 aqueous suspensions: Kinetic effects of the positions of the Cl atoms and identification of the intermediates, *J. Photochem. Photobiol.* 1993, *A72*, 261.

125. Mills, G.; Hoffmann, M. R. Photocatalytic degradation of pentachlorophenol on TiO_2 particles: Identification of intermediates and mechanism of reaction, *Environ. Sci. Technol.* 1993, *27*, 1681.

126. O'Shea, K. E.; Cardona, C. Hammett study on the TiO_2-catalyzed photooxidation of parasubstituted phenols. A kinetic and mechanistic analysis, *J. Org. Chem.* 1994, *59*, 5005.

127. Sehili, T.; Boule, P.; Lemaire, J. Photocatalyzed transformation of chloroaromatic derivatives on ZnO III. Chlorophenols, *J. Photochem. Photobiol. A: Chem.* 1989, *50*, 117.

128. Prairie, M. R.; Evans, L. R.; Strange, B. M.; Martinez, S. L. An investigation of TiO_2 photocatalysis for the treatment of water contaminated with metals and organic chemicals, *Environ. Sci. Technol.* 1993, *27*, 1776.

129. Pelizzetti, E.; Minero, C.; Borgarello, E.; Tinucci, L.; Serpone, N. Photocatalytic activity and selectivity of titania colloids and particles prepared by the sol-gel technique: Photooxidation of phenol and atrazine, *Langmuir* 1993, *9*, 2995.

130. Peral, J.; Casado, J.; Domenech, J. Light induced oxidation of phenol over ZnO power, *J. Photochem. Photobiol. A. Chem.* 1988, *44*, 209.

131. Wei, T. Y.; Wan, C. Kinetics of photocatalytic oxidation of phenol on TiO_2 surface, *J. Photochem. Photobiol.* 1992, *A69*, 241.

132. Augugliaro, V.; Davi, E.; Palmisano, L.; Schiavello, M.; Sclafani, A. Influence of hydrogen peroxide on the kinetics of phenol photodegradation in aqueous titanium dioxide dispersion, *Appl. Catal.* 1990, *65*, 101.

133. Matthews, R. W.; McEvoy, S. R. Photocatalytic degradation of phenol in the presence of near-UV illuminated titanium dioxide, *J. Photochem. Photobiol.* 1992, *A64*, 231.

134. Grabner, G.; Li, G.; Quint, R. M.; Quint, R.; Getoff, N. Pulsed laser-induced oxidation of phenol in acid aqueous TiO_2 sols, *J. Chem. Soc., Faraday Trans.* 1991, *87*, 1097.

135. Minero, C.; Aliberti, C.; Pelizzetti, E.; Terzian, R.; Serpone, N. Kinetic studies in heterogeneous photocatalysis. 6. AM1 simulated sunlight photodegradation over titania in aqueous media: A first case of fluorinated aromatics and identification of intermediates, *Langmuir* 1991, *7*, 928.

136. Tunesi, S.; Anderson, M. Influence of chemisorption on the photodecomposition of salicylic acid and related compounds using suspended TiO_2 ceramic membranes, *J. Phys. Chem.* 1991, *95*, 3399.

137. Cunningham, J.; Al, S. G. Factors influencing efficiencies of TiO_2-sensitised photodegradation. Part 1. Substituted benzoic acids: Discrepancies with dark-adsorption parameters, *J. Chem. Soc., Faraday Trans.* 1990, *86*, 3935.

138. Cunningham, J.; Srijaranai, S. Sensitized photo-oxidations of dissolved alcohols in

homogeneous and heterogeneous systems. Part 2. TiO_2-sensitized photodehydrogenations of benzyl alcohol, *J. Photochem. Photobiol.* 1991, *A58*, 361.

139. Terzian, R.; Serpone, N.; Minero, C.; Pelizzetti, E.; Hidaka, H. Kinetic studies in heterogeneous photocatalysis. 4. The photomineralization of a hydroquinone and a catechol, *J. Photochem. Photobiol.* 1990, *A55*, 243.

140. Terzian, R.; Serpone, N.; Minero, C.; Pelizzetti, E. Kinetic studies in heterogeneous photocatalysis. Part 5. Photocatalyzed mineralization of cresols in aqueous media with irradiated titania, *J. Catal.* 1991, *128*, 352.

141. Dieckmann, M.S.; Gray, K. A.; Kamat, P. V. Photocatalyzed degradation of adsorbed nitrophenolic compounds on semiconductor surfaces, *Wat. Sci. Tech.* 1992, *25*, 277.

142. Menassa, P. E.; Mak, M. K. S.; Langford, C. H. A study of the photodecomposition of different polychlorinated biphenyls by surface modified titanium (IV) oxide particles, *Environ. Technol. Lett.* 1988, *9*, 825.

143. Vinodgopal, K.; Kamat, P. V. Photochemistry on surfaces. Photodegradation of 1,3-diphenylisobenzofuran over metal oxide particles, *J. Phys. Chem.* 1992, *96*, 5053.

144. Kormann, C.; Bahnemann, D. W.; Hoffmann, M. R. Photolysis of chloroform and other organic molecules in suspensions, *Environ. Sci. Technol.* 1991, *25*, 494.

145. Hsiao, C. Y.; Lee, C. L.; Ollis, D. F. Heterogeneous photocatalysis: Degradation of dilute solutions of dichloromethane (CH_2Cl_2), chloroform ($CHCl_3$), and carbon tetrachloride (CCl_4) with illuminated TiO_2 photocatalyst, *J. Catal.* 1983, *82*, 418.

146. Dibble, L. A.; Raupp, G. B. Kinetics of the gas-solid heterogeneous photocatalytic oxidation of trichloroethylene by near UV illuminated TiO_2, *Catal. Let* 1990, *4*, 345.

147. Dibble, L. A.; Raupp, G. B. Fluidized-bed photocatalytic oxidation of trichloroethylene in contaminated airstreams, *Environ. Sci. Tech.* 1992, *26*, 492.

148. Mao, Y.; Schoeneich, C.; Asmus, K. D. Identification of organic acids and other intermediates in oxidative degradation of chlorinated ethanes on TiO_2 surfaces en route to mineralization. A combined photocatalytic and radiation chemical study, *J. Phys. Chem.* 1991, *95*, 10080.

149. Kim, D. H.; Anderson, M. A. Photoelectrocatalytic degradation of formic acid using a porous thin film electrode, *Environ. Sci. Technol.* 1994, *28*, 479.

150. Hidaka, H.; Asai, Y.; Zhao, J.; Nohara, K.; Pelizzetti, E.; Serpone, N. Photoelectrochemical decomposition of surfactants on a TiO_2/TCO particulate film electrode assembly, *J. Phys. Chem.* 1995, *99*, 8244.

151. Anpo, M.; Chiba, K.; Tomonari, M.; Coluccia, S.; Che, M.; Fox, M. A. Photocatalysis on native and platinum-loaded TiO_2 and ZnO catalysts. Origin of different reactivities on wet and dry metal oxides, *Bull. Chem. Soc. Jpn.* 1991, *64*, 543.

152. Pelizzetti, E.; Carlin, V.; Minero, C.; Graetzel, M. Enhancement of the rate of photocatalytic degradation on TiO_2 of 2-chlorophenol, 2,7-dichlorodibenzodioxin and atrazine by inorganic oxidizing species, *New J. Chem.* 1991, *15*, 351.

153. Harada, K.; Hisanaga, T.; Tanaka, K. Photocatalytic degradation of organophosphorus compounds in semiconductor suspension, *New J. Chem.* 1987, *11*, 597.

154. Muneer, M.; Das, S.; Manilal, V. B.; Haridas, A. Photocatalytic degradation of waste-water pollutants: Titanium dioxide-mediated oxidation of methyl vinyl ketone, *J. Photochem. Photobiol.* 1992, *A63*, 107.

155. Vinodgopal, K.; Bedja, I.; Hotchandani, S.; Kamat, P. V. A photocatalytic approach

for the reductive decolorization of textile azo dyes in colloidal semiconductor suspensions, *Langmuir* 1994, *10*, 1767.

156. Hustert, K.; Zepp, R. G. Photocatalytic degradation of selected azo dyes, *Chemosphere* 1992, *24*, 335.

157. Wang, Z.; Kutal, C. Photocatalytic mineralization of 2,4,6-trinitrotoluene in aqueous suspensions of titanium dioxide, *Chemosphere* 1995, *30*, 1125.

158. Schmelling, D.; Gray, K. A. Photocatalytic transformation of 2,4,6-trinitrotoluene, *Water Research* 1995, *29*, 2651.

159. Schmelling, D. C.; Gray, K. A.; Kamat, P. V. The influence of solution matrix on the photocatalytic degradation of TNT in TiO$_2$ slurries, *Water Research* 1997, *31*, 1439.

160. Augugliaro, V.; Palmisano, L.; Schiavello, M.; Sclafani, A.; Marchese, L.; Martra, G.; Miano, F. Photocatalytic degradation of nitrophenols in aqueous titanium dioxide dispersion, *Appl. Catal.* 1991, *69*, 323.

161. Gray, K. A.; Stafford, U.; Dieckmann, M. S.; Kamat, P. V. Mechanistic studies in TiO$_2$ systems: Photocatalytic degradation of chloro- and nitrphenols. In *Photocatalytic Purification and Treatment of Water and Air*, D. F. Ollis, H. Al-Ekabi, eds., Elsevier Science: Amsterdam, 1993, p. 8.

162. Piccinini, P.; Minero, C.; Vincenti, M.; Pelizzetti, E. Photocatalytic interconversion of nitrogen-containing benzene derivatives, *J. Chem. Soc., Faraday Trans.* 1997, *93*, 1997.

163. Gray, K. A.; Stafford, U. Probing photocatalytic reactions in semiconductor systems: Study of the chemical intermediates in 4-chlorophenol degradation by a variety of methods, *Res. Chem. Intermed.* 1994, *20*, 835.

164. Stafford, U.; Gray, K. A.; Kamat, P. V. Photocatalytic degradation of 4-chlorophenol 1. The effects of varying TiO$_2$ concentration and light wavelength., *J. Catal.* 1997, *167*, 25.

165. Stafford, U.; Gray, K. A.; Kamat, P. V. Photocatalytic degradation of 4-chlorophenol. 2. A. Model, *Res. Chem. Intermed.* 1997, *23*, 355.

166. Meier, H. Photosensitization of inorganic solids, *J. Photochem. Photobiol.* 1972, *16*, 219.

167. Gerischer, H.; Willig, F. Reaction of excited dye molecules at electrodes, *Top. Curr. Chem.* 1976, *61*, 31.

168. Spitler, M.; Parkinson, B. A. Efficient infrared dye sensitization of van der Waals surfaces of semiconductor electrodes, *Langmuir* 1986, *2*, 549.

169. Ryan, M. A.; Spitler, M. T. Photoelectrochemistry and photochemistry of dyes adsorbed at semiconductor surfaces, *J. Imaging Sci.* 1989, *33*, 46.

170. O'Regan, B.; Graetzel, M. A low-cost, high-efficiency solar cell based on dye-sensitized colloidal TiO$_2$ films, *Nature (London)* 1991, *353*, 737.

171. Graetzel, M. The artificial leaf, molecular photovoltaics achieve efficient generation of electricity from sunlight, *Coord. Chem. Rev.* 1991, *111*, 167.

172. Kamat, P. V.; Fox, M. A. Photosensitization of TiO$_2$ colloids by erythrosin B in acetonitrile, *Chem. Phys. Lett.* 1983, *102*, 379.

173. Moser, J.; Graetzel, M. Photosensitized electron injection in colloidal semiconductors, *J. Am. Chem. Soc.* 1984, *106*, 6557.

174. Rossetti, R.; Brus, L. E. Time-resolved Raman scattering study of adsorbed, semi-oxidized eosin Y formed by excited-state electron transfer into colloidal TiO_2 particles, *J. Am. Chem. Soc.* 1984, *106*, 4336.

175. Ryan, M. A.; Fitzgerald, E. C.; Spitler, M. T. Internal reflection flash photolysis study of the photochemistry of eosin at TiO_2 semiconductor electrodes, *J. Phys. Chem.* 1989, *93*, 6150.

176. Heleg, V.; Willner, I. Photocatalyzed CO_2-fixation to formate and H_2 evolution by eosin-modified Pd-TiO_2 powders, *J. Chem. Soc., Chem. Commun.* 1994, 2113.

177. Umapathy, S.; Cartner, A. M.; Parker, A. W.; Hester, R. E. Time-resolved resonance Raman spectroscopic studies of the photosensitization of colloidal titanium dioxide, *J. Phys. Chem.* 1990, *94*, 8880.

178. Fessenden, R. W.; Kamat, P. V. Photosensitized charge injection into TiO_2 particles as studied by microwave absorption, *Chem. Phys. Lett.* 1986, *123*, 233.

179. Gopidas, K. R.; Kamat, P. V. Photochemistry on surfaces. 4. Influence of support material on the photochemistry of an adsorbed dye, *J. Phys. Chem.* 1989, *93*, 6428.

180. Takizawa, T.; Watanabe, T.; Honda, K. Photocatalysis through excitation of adsorbates. 2. A comparative study of rhodamine B and methylene blue on CdS, *J. Phys. Chem.* 1978, *82*, 1391.

181. Hashimoto, K.; Hiramoto, M.; Sakata, T. Photo-induced electron transfer from adsorbed rhodamine B to oxide semiconductor substrates in vacuo: Semiconductor dependence, *Chem. Phys. Lett.* 1988, *148*, 215.

182. Hashimoto, K.; Hiramoto, M.; Sakata, T. Temperature-independent electron transfer: Rhodamine B/oxide semiconductor dye-sensitization system, *J. Phys. Chem.* 1988, *92*, 4272.

183. Liang, Y.; Moy, P. F.; P. J. A.; Ponte Goncalves, A. M. Fluorescence of rhodamine B on semiconductor and insulator surfaces: Dependence of quantum yield on surface coverage, *J. Phys. Chem.* 1984, *88*, 2451.

184. Bitterling, K.; Willig, F. Charge carrier dynamics in the picosecond time domain in photoelectrochemical cells, *J. Electroanal. Chem. Interfacial Electrochem.* 1986, *204*, 211.

185. Nasr, C.; Liu, D.; Hotchandani, S.; Kamat, P. V. Dye capped semiconductor colloids. Excited state and photosensitization aspects of Rhodamine 6G-H aggregates electrostatically bound to SiO2 and SnO2 colloids, *J. Phys. Chem.* 1996, *100*, 11054.

186. Crackel, R. L.; Struve, W. S. Non-radiative excitation decay of cresyl violet on TiO_2: Variation with dye-surface separation, *Chem. Phys. Lett.* 1985, *120*, 473.

187. Willig, F.; Eichberger, R.; Sundaresan, N. S.; Parkinson, B. A. Experimental time scale of Gerischer's distribution curves for electron-transfer reactions at semiconductor electrodes, *J. Am. Chem. Soc.* 1990, *112*, 2702.

188. Eichberger, R.; Willig, F. Ultrafast electron injection from excited dye molecules into semiconductor electrodes, *Chem. Phys. Lett.* 1990, *141*, 159.

189. Liu, D.; Kamat, P. V. Electrochemically active nanocrystalline SnO_2 films. Surface modifications with thiazine and oxazine dye aggregates, *J. Electrochem. Soc.* 1995, *142*, 835.

190. Liu, D.; Kamat, P. V. Picosecond dynamics of cresyl violet H-aggregates adsorbed on SiO_2 and SnO_2 nanocrystallites, *J. Phys. Chem.* 1996, *105*, 965.

191. Liu, D.; Fessenden, R. W.; Hug, G. L.; Kamat, P. V. Dye capped semiconductor

nanoclusters. Role of back electron transfer in the photosensitization of SnO_2 nanocrystallites with cresyl violet aggregates, *J. Phys. Chem.* 1997, *B101*, 2583.

192. Patrick, B.; Kamat, P. V. Photoelectrochemistry in semiconductor particulate systems. Part 17. Photosensitization of large-bandgap semiconductors. Charge injection from triplet excited thionine into ZnO colloids, *J. Phys. Chem.* 1992, *96*, 1423.

193. Hotchandani, S.; Kamat, P. V. Modification of electrode surface with semiconductor colloids and its sensitization with chlorophyll a, *Chem. Phys. Lett.* 1992, *191*, 320.

194. Bedja, I.; Kamat, P. V.; Hotchandani, S. Fluorescence and photoelectrochemical behavior of chlorophyll *a* adsorbed on a nanocrystalline SnO_2 film, *J. Appl. Phys.* 1996, *80*, 4637.

195. Yang, Y.; Zhou, R.; Han, Y.; Jiang, Y. Photoelectrochemical investigation of chlorophyll *a* adsorbed on SnO_2, *J. Photochem. Photobiol.* 1993, *A76*, 111.

196. Kay, A.; Graetzel, M. Artificial photosynthesis. 1. Photosensitization of TiO_2 solar cells with chlorophyll derivatives and related natural porphyrins, *J. Phys. Chem.* 1993, *97*, 6272.

197. Kay, A.; Humphry-Baker, R.; Graetzel, M. Artificial photosynthesis. 2. Investigations on the mechanism of photosensitization of nanocrystalline TiO_2 solar cells by chlorophyll derivatives, *J. Phys. Chem.* 1994, *98*, 952.

198. Bedja, I.; Hotchandani, S.; Carpentier, R.; Fessenden, R. W.; Kamat, P. V. Chlorophyll *b* modified nanocrystalline SnO_2 semiconductor thin film as a photosensitive electrode, *J. Appl. Phy.* 1994, *75*, 5444.

199. Kamat, P. V.; Chauvet, J. P.; Fessenden, R. W. Photoelectrochemistry in particulate systems. 4. Photosensitization of a TiO_2 semiconductor with a chlorophyll analogue, *J. Phys. Chem.* 1986, *90*, 1389.

200. Kamat, P. V. Photoelectrochemistry in particulate systems. 9. Photosensitized reduction in a colloidal TiO_2 system using anthracene-9-carboxylic acid as the sensitizer, *J. Phys. Chem.* 1989, *93*, 859.

201. Martini, I.; Hartland, G.; Kamat, P. V. Ultrafast study of interfacial electron transfer between 9-anthracene-carboxylate and TiO_2 semiconductor particles, *J. Chem. Phys.* 1997, *107*, 8064.

202. Burfeindt, B.; Hannappel, T.; Storck, W.; Willig, F. Measurement of temperature-independent femtosecond interfacial electron transfer from an anchored molecular electron donor to a semiconductor as acceptor, *J. Phys. Chem.* 1996, *100*, 16463.

203. Ferrere, S.; Zaban, A.; Gregg, B. A. Dye sensitization of nanocrystalline tin oxide by perylene derivative, *J. Phys. Chem.* 1997, *B101*, 4490.

204. Houlding, V. H.; Graetzel, M. Photochemical H_2 generation by visible light. Sensitization of TiO_2 particles by surface complexation with 8-hydroxyquinoline, *J. Am. Chem. Soc.* 1983, *105*, 5695.

205. Kalyanasundaram, K.; Vlachopoulos, N.; Krishnan, V.; Monnier, A.; Graetzel, M. Sensitization of TiO_2 in the visible light region using zinc porphyrins, *J. Phys. Chem.* 1987, *91*, 2342.

206. Fan, F. R. F.; Bard, A. J. Spectral sensitization of the heterogeneous photocatalytic oxidation of hydroquinone in aqueous solutions at phthalocyanine-coated TiO_2 powders, *J. Am. Chem. Soc.* 1979, *101*, 6139.

207. Arbour, C.; Sharma, D. K.; Langford, C. H. Picosecond flash spectroscopy of TiO_2 colloids with adsorbed dyes, *J. Phys. Chem.* 1990, *94*, 331.

208. Vrachnou, E.; Vlachopoulos, N.; Graetzel, M. Efficient visible light sensitization of TiO_2 by surface complexation with $Fe(CN)_6^{4-}$, *J. Chem. Soc., Chem. Commun.* 1987, 868.

209. Vrachnou, E.; Graetzel, M.; McEvoy, A. J. Efficient visible light photoresponse following surface complexation of titanium dioxide with transition metal cyanides, *J. Electroanal. Chem. Interfacial Electrochem.* 1989, *258*, 193.

210. Ghosh, P. K.; Spiro, T. G. Photoelectrochemistry of tris(bipyridyl)ruthenium(II) covalently attached to n-type SnO2, *J. Am. Chem. Soc.* 1980, *102*, 5543.

211. Gulino, D. A.; Drickamer, H. G. High-pressure studies of the dye-sensitized photocurrent spectrum of titanium dioxide, *J. Phys. Chem.* 1984, *88*, 1173.

212. Dabestani, R.; Bard, A. J.; Campion, A.; Fox, M. A.; Mallouk, T. E.; Webber, S. E.; White, J. M. Sensitization of titanium dioxide and strontium titanate electrodes by ruthenium(II) tris(2,2'-bipyridine-4,4'-dicarboxylic acid) and zinc tetrakis(4-carboxyphenyl)porphyrin: An evaluation of sensitization efficiency for component photoelectrodes in a multipanel device, *J. Phys. Chem.* 1988, *92*, 1872.

213. Nazeeruddin, M. K.; Liska, P.; Moser, J.; Vlachopoulos, N.; Graetzel, M. Conversion of light into electricity with trinuclear ruthenium complexes adsorbed on textured TiO_2 films, *Helv. Chim. Acta* 1990, *73*, 1788.

214. O'Regan, B.; Moser, J.; Anderson, M.; Graetzel, M. Vectorial electron injection into transparent semiconductor membranes and electric field effects on the dynamics of light-induced charge separation, *J. Phys. Chem.* 1990, *94*, 8720.

215. Taqui Khan, M. M.; Chatterjee, D.; Hussain, A.; Moiz, M. A. Synthesis and characteristics of mixed ligand Ru(III) complexes with EDTA-polypyridyl, and $Pt/TiO_2/RuO_2$ semiconductor particulate system modified by the complexes, *J. Photochem. Photobiol.* 1993, *A76*, 97.

216. Kim, Y. I.; Atherton, S. J.; Brigham, E. S.; Mallouk, T. E. Sensitized layered metal oxide semiconductor particles for photochemical hydrogen evolution from nonsacrificial electron donors, *J. Phys. Chem.* 1993, *97*, 11802.

217. Willner, I.; Eichen, Y.; Frank, A. J.; Fox, M. A. Photoinduced electron-transfer processes using organized redox-functionalized bipyridinium-polyethylenimine-TiO_2 colloids and particulate assemblies, *J. Phys. Chem.* 1993, *97*, 7264.

218. Ford, W. E.; Rodgers, M. A. J. Interfacial electron transfer in colloidal SnO_2 hydrosols photosensitized by electrostatically and covalently attached Ru(II) polypyridine complexes, *J. Phys. Chem.* 1994, *98*, 3822.

219. Iwasaki, T.; Oda, S.; Kamada, H.; Honda, K. Study of photochemical reaction of sensitizing dyes adsorbed on semiconductor powder by means of photoacoustic spectroscopy, *J. Phys. Chem.* 1980, *84*, 1060.

220. Hada, H.; Yonezawa, Y.; Inaba, H. Spectral sensitization of ZnO film electrodes by the J-aggregate of cyanine dye, *Ber. Bunseges. Phys. Chem.* 1981, *85*, 425.

221. Kavassalls, C.; Spitler, M. T. Photooxidation of thiacyanine dyes at ZnO single-crystal electrodes, *J. Phys. Chem.* 1983, *87*, 3166.

222. Natoli, L. M.; Ryan, M. A.; Spitler, M. T. J aggregate sensitization of ZnO electrodes as studied by internal reflection spectroscopy, *J. Phys. Chem.* 1985, *89*, 1448.

223. Hayashi, Y.; Ogawa, S.; Sanada, M.; Hirohashi, R. Spectral sensitization of thermally processed silver films by cyanine dyes, *J. Imagin. Sci.* 1989, *33*, 124.

224. Tani, T.; Suzumoto, T.; Ohzeki, K. Energy gap dependence of efficiency of photo-

induced electron transfer from cyanine dyes to silver bromide microcrystals in spectral sensitization, *J. Phys. Chem.* 1990, *94*, 1298.

225. Kietzmann, R.; Ehret, A.; Spitler, M.; Willig, F. Temperature-dependent electron-transfer quenching of dye monomer fluorescence on octahedral AgBr grains, *J. Am. Chem. Soc.* 1993, *115*, 1930.

226. Nasr, C.; Hotchandani, S.; Kamat, P. V.; Das, S.; George Thomas, K.; George, M. V. Electrochemical and photoelectrochemical properties of monoaza-15-Crown ether linked cyanine dyes: Photosensitization of nanocrystalline SnO_2 films, *Langmuir* 1995, *11*, 1777.

227. Hotchandani, S.; Das, S.; Thomas, K. G.; George, M. V.; Kamat, P. V. Interaction of semiconductor colloids with J-aggregates of squaraine dye and its role in sensitizing nanocrystalline semiconductor films, *Res. Chem. Intermed.* 1994, *20*, 927.

228. Kim, Y.-S.; Liang, K.; Law, K.-Y.; Whitten, D. G. An investigation of photocurrent generation by squaraine aggregates in monolayer-modified SnO2 electrodes, *J. Phys. Chem.* 1994, *98*, 984.

229. George Thomas, K.; Thomas, K. J.; Das, S.; George, M. V.; Liu, D.; Kamat, P. V. Photochemistry of squaraine dyes. 10. Excited state properties and photosensitization behavior of an IR sensitive cationic squaraine dye, *Faraday Trans.* 1996, *92*, 4913.

230. Liu, D.; Kamat, P. V.; George Thomas, K.; Thomas, K. J.; Das, S.; George, M. V. Picosecond dynamics of an IR sensitive squaraine dye. Role of singlet and triplet excited states in the photosensitization of TiO_2 nanoclusters, *J. Phys. Chem.* 1997, *106*, 6404.

231. Frei, H.; Fitzmaurice, D. J.; Graetzel, M. Surface chelation of semiconductors and interfacial electron transfer, *Langmuir* 1990, *6*, 198.

232. Nakahira, T.; Graetzel, M. Visible light sensitization of platinized TiO_2 photocatalyst by surface-adsorbed poly(4-vinylpyridine) derivatized with ruthenium trisbipyridyl complex, *Makromol. Chem., Rapid Commun.* 1985, *6*, 341.

233. Nakahira, T.; Inoue, Y.; Iwasaki, K.; Tanigawa, H.; Kouda, Y.; Iwabuchi, S.; Kojima, K.; Graetzel, M. Visible light sensitization of platinized TiO_2 photocatalyst by surface-coated polymers derivatized with ruthenium tris(bipyridyl), *Makromol. Chem., Rapid Commun.* 1988, *9*, 13.

234. Gerischer, H. Electrochemical techniques for the study of photosensitization, *Photochem. Photobiol.* 1972, *16*, 243.

235. Sonntag, L. P.; Spitler, M. T. Examination of the energetic threshold for dye-sensitized photocurrent at SrTiO3 electrodes, *J. Phys. Chem.* 1985, *89*, 1453.

236. Itoh, K.; Chiyokawa, Y.; Nakao, M.; Honda, K. Fluorescence quenching processes of rhodamine B on oxide semiconductors and light harvesting action of its dimers, *J. Am. Chem. Soc.* 1984, *106*, 1620.

237. Hashimoto, K.; Hiramoto, M.; Kajiwara, T.; Sakata, T. Luminescence decays and spectra of $Ru(bpy)_3^+$ adsorbed on TiO_2 in vacuo and in the presence of water vapor, *J. Phys. Chem.* 1988, *92*, 4636.

238. Hashimoto, K.; Hiramoto, M.; Lever, A. B. P.; Sakata, T. Luminescence decay of ruthenium(II) complexes adsorbed on metal oxide powders in vacuo: Energy gap dependence of the electron-transfer rate, *J. Phys. Chem.* 1988, *92*, 1016.

239. Kemnitz, K.; Nakashima, N.; Yoshihara, K.; Matsunami, H. Temperature depen-

dence of fluorescence decays of isolated rhodamine B molecules adsorbed on semiconductor single crystals, *J. Phys. Chem.* 1989, *93*, 6704.

240. Kamat, P. V. Photoelectrochemistry in semiconductor particulate systems. 14. Picosecond charge-transfer events in the photosensitization of colloidal TiO$_2$, *Langmuir* 1990, *6*, 512.

241. Kamat, P. V.; Das, S.; Thomas, K. G.; George, M. V. Ultrafast photochemical events associated with the photosensitization properties of a squaraine dye, *Chem. Phys. Lett.* 1991, *178*, 75.

242. Rehm, J. M.; McLendon, G. L.; Nagasawa, Y.; Yoshihara, K.; Moser, J.; Graetzel, M. Femtosecond electron-transfer dynamics at a sensitizing dye-semiconductor (TiO$_2$) interface, *J. Phys. Chem.* 1996, *100*, 9577.

243. Argazzi, R.; Bignozzi, C. A.; Heimer, T. A.; Castellano, F. N.; Meyer, G. J. Enhanced spectral sensitivity from ruthenium(II) polypyridyl based photovoltaic devices, *Inorg. Chem.* 1994, *33*, 5741.

244. Vinodgopal, K.; Hua, X.; Dahlgren, R. L.; Lappin, A. G.; Patterson, L. K.; Kamat, P. V. Photochemistry of Ru(bpy)$_2$(dcbpy)$^{2+}$ on Al$_2$O$_3$ and TiO$_2$ surfaces. An insight into the mechanism of photosensitization, *J. Phys. Chem.* 1995, *99*, 10883.

245. Fessenden, R. W.; Kamat, P. V. Rate constants for charge injection from excited sensitizer into SnO$_2$, ZnO, and TiO$_2$ semiconductor nanocrystallites, *J. Phys. Chem.* 1995, *99*, 12902.

246. Zepp, R. G.; Braun, A. M.; Leenheer, J. A. Photoproduction of hydrated electron from natural organic solutes in aquatic environments, *Environ. Sci. Technol.* 1987, *21*, 485.

247. Vinodgopal, K.; Kamat, P. V. Environmental photochemistry on surfaces. Charge injection from excited fulvic acid into semiconductor colloids, *Environ. Sci. Technol.* 1992, *26*, 1963.

248. Vinodgopal, K. Environmental photochemistry: Electron transfer from excited humic acid to TiO$_2$ colloids and semiconductor mediated reduction of oxazine dyes by humic acid, *Res. Chem. Intermed.* 1994, *20*, 825.

249. Vinodgopal, K.; Kamat, P. V. Photosensitization of semiconductor colloids by humic substances. In *Aquatic and Surface Photochemistry*, G. R. Helz, R. G. Zepp, D. G. Crosby, eds., CRC Press: Boca Raton, FL, 1994, p. 437.

250. Selli, E.; de Giorgi, A.; Bidoglio, G. Humic acid-sensitized photoreduction of Cr(VI) on ZnO particles, *Environ. Sci. Technol.* 1996, *30*, 598.

251. Gopidas, K. R.; Kamat, P. V.; George, M. V. Photochemical processes on oxide surfaces. A diffuse reflectance laser flash photolysis study, *Mol. Cryst. Liq. Cryst.* 1990, *183*, 403.

252. Gopidas, K. R.; Kamat, P. V. Photophysics and photochemistry of phenosafranin dye in aqueous and acetonitrile solutions, *J. Photochem. Photobiol.* 1989, *A48*, 291.

253. Gevaert, M.; Kamat, P. V. Visible laser-induced oxidation of C$_{70}$ on titanium dioxide particles, *J. Chem. Soc., Chem. Commun.* 1992, 1470.

254. Kamat, P. V.; Gevaert, M.; Vinodgopal, K. Photochemistry on semiconductor surfaces. Photochemical oxidation of C$_{60}$ on TiO$_2$ nanoparticles, *J. Phys. Chem.* 1997, *B101*, 4422.

255. Gray, K. A.; Kamat, P. V.; Stafford, U.; Dieckmann, M. Mechanistic studies of chloro- and nitrophenolic degradation on semiconductor surfaces. In *Aquatic and*

Surface Photochemistry, G. R. Helz, R. G. Zepp, D. G. Crosby, eds., CRC Press: Boca Raton, FL, 1994, p. 399.

256. Stafford, U.; Gray, K. A.; Kamat, P. V.; Varma, A. An in situ diffuse reflectance FTIR investigation of photocatalytic degradation of 4-chlorophenol on a TiO$_2$ powder surface, Chem. Phys. Lett. 1993, 205, 55.

257. Nasr, C. Vinodgopal, K.; Hotchandani, S.; Chattopadhyaya, A.; Kamat, P. V. Environmental photochemistry on semiconductor surfaces. Visible light induced degradation of a textile diazo dye, naphthol blue black on TiO$_2$ particles, J. Phys. Chem. 1996, 100, 8436.

258. Vinodgopal, K.; Wynkoop, D.; Kamat, P. V. Environmental photochemistry on semiconductor surfaces: A photosensitization approach for the degradation of a textile azo dye, Acid Orange 7, Environ. Sci. Technol. 1996, 30, 1660.

259. Dieckmann, M. S.; Gray, K. A.; Zepp, R. G. The sensitized photocatalysis of azo dyes in a solid system: a feasibility study, Chemosphere 1994, 28, 1021.

260. Matsui, K.; Shikata, K.; Takase, Y. Ozonolysis of 1-phenyl azo-2-naphthol, Dyes and Pigments 1984, 5, 325.

261. Matsui, M.; Kimura, T.; Namku, T.; Shibata, K.; Takase, Y. Reaction of water soluble dyes with ozone, J. Soc. Dyes Color 1984, 100, 125.

262. Sotiriou, C.; Lee, W.; Giese, R. W. Superoxide oxidation: A novel route to aromatic 1,2-dicarboxylic acids, J. Org. Chem. 1990, 55, 2159.

263. Fujishima, A.; Honda, K. Electrochemical photolysis at a semiconductor electrode, Nature 1972, 238.

264. Vinodgopal, K.; Kamat, P. V. Enhanced rates of photocatalytic degradation of an azo dye using SnO$_2$/TiO$_2$ coupled semiconductor thin films, Environ. Sci. Technol. 1995, 29, 841.

265. Vinodgopal, K.; Bedja, I.; Kamat, P. V. Nanostructured semiconductor films for photocatalysis. Photoelectrochemical behavior of SnO$_2$/TiO$_2$ coupled systems and its role in photocatalytic degradation of a textile azo dye, Chem. Mater. 1996, 8, 2180.

266. Gerischer, H.; Luebke, M. A particle size effect in the sensitization of TiO$_2$ electrodes by a CdS deposit, J. Electroanal. Chem. Interfacial Electrochem. 1986, 204, 225.

267. Hotchandani, S.; Kamat, P. V. Charge-transfer processes in coupled semiconductor systems. Photochemistry and photoelectrochemistry of the colloidal CdS-ZnO system, J. Phys. Chem. 1992, 96, 6834.

268. Liu, D.; Kamat, P. V. Photoelectrochemical behavior of thin CdSe and coupled TiO$_2$/CdSe semiconductor films, J. Phys. Chem. 1993, 97, 10769.

269. Vogel, R.; Hoyer, P.; Weller H. Quantum-sized PbS, CdS, Ag$_2$S, Sb$_2$S$_3$ and Bi$_2$S$_3$ particles as sensitizers for various nanoporous wide-bandgap semiconductors, J. Phys. Chem. 1994, 98, 3183.

270. Bedja, I.; Kamat, P. V. Capped semiconductor colloids. Synthesis and photoelectrochemical properties of TiO$_2$ capped SnO$_2$ surfaces, J. Phys. Chem. 1995, 99, 9182.

271. Fitzmaurice, D.; Frei, H.; Rabani, J. Time-resolved optical study on the charge carrier dynamics in a TiO$_2$/Agl sandwich colloid, J. Phys. Chem. 1995, 99, 9176.

272. Nasr, C.; Kamat, P. V.; Hotchandani, S. Photoelectrochemical behavior of coupled

SnO2/CdSe nanocrystalline semiconductor films, *J. Electroanal. Chem.* 1997, *420*, 201.

273. Nasr, C.; Hotchandani, S.; Kamat, P. V. Photoelectrochemical behavior of composite semiconductor thin films and their sensitization with ruthenium polypyridyl complex. In *Photoelectrochemistry*, K. Rajeshwar, ed., The Electrochemical Society: Pennington, NJ, 1997, in press.
274. Uekawa, N.; Suzuki, T.; Ozeki, S.; Kaneko, K. Rectification effect by a p-n junctioned oxide film, *Langmuir* 1992, *8*, 1.
275. Haesselbarth, A.; Eychmueller, A.; Eichberger, R.; Giersig, M.; Mews, A.; Weller, H. Chemistry and photophysics of mixed CdS/HgS colloids, *J. Phys. Chem.* 1993, *97*, 5333.
276. Eychmueller, A.; Vobmeyer, T.; Mews, A.; Weller, H. Transient photobleaching in the quantum dot quantum well CdS/HgS/CdS, *J. Lumin.* 1994, *58*, 223.
277. Kortan, A. R.; Hull, R.; Opila, R. L.; Bawendi, M. G.; Steigerwald, M. L.; Carroll, P. J.; Brus, L. E. Nucleation and growth of CdSe on ZnS quantum crystallite seeds and vice versa, in inverse micelle media, *J. Am. Chem. Soc.* 1990, *112*, 1327.
278. Haus, J. W.; Zhou, H. S.; Honma, I.; Komiyama, H. Quantum confinement in semiconductor heterostructure nanometer-size particles, *Phys. Rev.* 1993, *47*, 1359.
279. Eychmueller, A.; Haesselbarth, A.; Weller, H. Quantum-sized HgS in contact with quantum-sized CdS colloids, *J. Lumin.* 1992, *53*, 113.
280. Zhou, H. S.; Sasahara, H.; Honma, I.; Komiyama, H.; Haus, J. W. Coated semiconductor nanoparticles: The CdS/PbS system's photoluminescence properties, *Chem. Mater.* 1994, *6*, 1534.
281. Kamalov, V. F.; Little, R.; Logunov, S. L.; El-Sayed, M. A. Picosecond electronic relaxation in CdS/HgS/CdS quantum dot quantum well semiconductor nanoparticles, *J. Phys. Chem.* 1996, *100*, 6381.
282. Shanghavi, B.; Kamat, P. V. Interparticle electron transfer in metal/semiconductor composites. Picosecond dynamics of CdS capped gold nanoclusters, *J. Phys. Chem. B* 1997, *101*, 7675.
283. Serpone, N.; Borgarello, E.; Graetzel, M. Visible light induced generation of hydrogen from H₂S in mixed semiconductor dispersions: Improved efficiency through inter-particle transfer, *J. Chem. Soc., Chem. Commun.* 1984, 342.
284. Serpone, N.; Maruthamuthu, P.; Pichat, P.; Pelizzetti, E.; Hidaka, H. Exploiting the interparticle electron transfer process in the photocatalyzed oxidation of phenol, 4-chlorophenol and pentachlorophenol: Chemical evidence for electron and hole transfer between coupled semiconductors, *J. Photochem. Photobiol., A: Chem.* 1995, *85*, 247.
285. Stevenson, M.; Bullock, K.; Lin, W.-Y.; Rajeshwar, K. Sonolytic enhancement of the bacterial activity of irradiated titanium dioxide suspensions in water, *Res. Chem. Intermed.* 1997, *23*, 311.

Index

T
| Month